BIVARIATE DISCRETE DISTRIBUTIONS

STATISTICS: Textbooks and Monographs

A Series Edited by

D. B. Owen, Founding Editor, 1972–1991

W. R. Schucany, Coordinating Editor

Department of Statistics
Southern Methodist University
Dallas, Texas

R. G. Cornell, Associate Editor
for Biostatistics
University of Michigan

A. M. Kshirsagar, Associate Editor
for Multivariate Analysis and
Experimental Design
University of Michigan

W. J. Kennedy, Associate Editor
for Statistical Computing
Iowa State University

E. G. Schilling, Associate Editor
for Statistical Quality Control
Rochester Institute of Technology

1. The Generalized Jackknife Statistic, *H. L. Gray and W. R. Schucany*
2. Multivariate Analysis, *Anant M. Kshirsagar*
3. Statistics and Society, *Walter T. Federer*
4. Multivariate Analysis: A Selected and Abstracted Bibliography, 1957–1972, *Kocherlakota Subrahmaniam and Kathleen Subrahmaniam*
5. Design of Experiments: A Realistic Approach, *Virgil L. Anderson and Robert A. McLean*
6. Statistical and Mathematical Aspects of Pollution Problems, *John W. Pratt*
7. Introduction to Probability and Statistics (in two parts), Part I: Probability; Part II: Statistics, *Narayan C. Giri*
8. Statistical Theory of the Analysis of Experimental Designs, *J. Ogawa*
9. Statistical Techniques in Simulation (in two parts), *Jack P. C. Kleijnen*
10. Data Quality Control and Editing, *Joseph I. Naus*
11. Cost of Living Index Numbers: Practice, Precision, and Theory, *Kali S. Banerjee*
12. Weighing Designs: For Chemistry, Medicine, Economics, Operations Research, Statistics, *Kali S. Banerjee*
13. The Search for Oil: Some Statistical Methods and Techniques, *edited by D. B. Owen*
14. Sample Size Choice: Charts for Experiments with Linear Models, *Robert E. Odeh and Martin Fox*
15. Statistical Methods for Engineers and Scientists, *Robert M. Bethea, Ben- jamin S. Duran, and Thomas L. Boullion*
16. Statistical Quality Control Methods, *Irving W. Burr*

17. On the History of Statistics and Probability, *edited by D. B. Owen*
18. Econometrics, *Peter Schmidt*
19. Sufficient Statistics: Selected Contributions, *Vasant S. Huzurbazar (edited by Anant M. Kshirsagar)*
20. Handbook of Statistical Distributions, *Jagdish K. Patel, C. H. Kapadia, and D. B. Owen*
21. Case Studies in Sample Design, *A. C. Rosander*
22. Pocket Book of Statistical Tables, *compiled by R. E. Odeh, D. B. Owen, Z. W. Birnbaum, and L. Fisher*
23. The Information in Contingency Tables, *D. V. Gokhale and Solomon Kullback*
24. Statistical Analysis of Reliability and Life-Testing Models: Theory and Methods, *Lee J. Bain*
25. Elementary Statistical Quality Control, *Irving W. Burr*
26. An Introduction to Probability and Statistics Using BASIC, *Richard A. Groeneveld*
27. Basic Applied Statistics, *B. L. Raktoe and J. J. Hubert*
28. A Primer in Probability, *Kathleen Subrahmaniam*
29. Random Processes: A First Look, *R. Syski*
30. Regression Methods: A Tool for Data Analysis, *Rudolf J. Freund and Paul D. Minton*
31. Randomization Tests, *Eugene S. Edgington*
32. Tables for Normal Tolerance Limits, Sampling Plans and Screening, *Robert E. Odeh and D. B. Owen*
33. Statistical Computing, *William J. Kennedy, Jr., and James E. Gentle*
34. Regression Analysis and Its Application: A Data-Oriented Approach, *Richard F. Gunst and Robert L. Mason*
35. Scientific Strategies to Save Your Life, *I. D. J. Bross*
36. Statistics in the Pharmaceutical Industry, *edited by C. Ralph Buncher and Jia-Yeong Tsay*
37. Sampling from a Finite Population, *J. Hajek*
38. Statistical Modeling Techniques, *S. S. Shapiro*
39. Statistical Theory and Inference in Research, *T. A. Bancroft and C.-P. Han*
40. Handbook of the Normal Distribution, *Jagdish K. Patel and Campbell B. Read*
41. Recent Advances in Regression Methods, *Hrishikesh D. Vinod and Aman Ullah*
42. Acceptance Sampling in Quality Control, *Edward G. Schilling*
43. The Randomized Clinical Trial and Therapeutic Decisions, *edited by Niels Tygstrup, John M Lachin, and Erik Juhl*
44. Regression Analysis of Survival Data in Cancer Chemotherapy, *Walter H. Carter, Jr., Galen L. Wampler, and Donald M. Stablein*
45. A Course in Linear Models, *Anant M. Kshirsagar*
46. Clinical Trials: Issues and Approaches, *edited by Stanley H. Shapiro and Thomas H. Louis*
47. Statistical Analysis of DNA Sequence Data, *edited by B. S. Weir*
48. Nonlinear Regression Modeling: A Unified Practical Approach, *David A. Ratkowsky*
49. Attribute Sampling Plans, Tables of Tests and Confidence Limits for Proportions, *Robert E. Odeh and D. B. Owen*
50. Experimental Design, Statistical Models, and Genetic Statistics, *edited by Klaus Hinkelmann*
51. Statistical Methods for Cancer Studies, *edited by Richard G. Cornell*
52. Practical Statistical Sampling for Auditors, *Arthur J. Wilburn*

53. Statistical Methods for Cancer Studies, *edited by Edward J. Wegman and James G. Smith*
54. Self-Organizing Methods in Modeling: GMDH Type Algorithms, *edited by Stanley J. Farlow*
55. Applied Factorial and Fractional Designs, *Robert A. McLean and Virgil L. Anderson*
56. Design of Experiments: Ranking and Selection, *edited by Thomas J. Santner and Ajit C. Tamhane*
57. Statistical Methods for Engineers and Scientists: Second Edition, Revised and Expanded, *Robert M. Bethea, Benjamin S. Duran, and Thomas L. Boullion*
58. Ensemble Modeling: Inference from Small-Scale Properties to Large-Scale Systems, *Alan E. Gelfand and Crayton C. Walker*
59. Computer Modeling for Business and Industry, *Bruce L. Bowerman and Richard T. O'Connell*
60. Bayesian Analysis of Linear Models, *Lyle D. Broemeling*
61. Methodological Issues for Health Care Surveys, *Brenda Cox and Steven Cohen*
62. Applied Regression Analysis and Experimental Design, *Richard J. Brook and Gregory C. Arnold*
63. Statpal: A Statistical Package for Microcomputers—PC-DOS Version for the IBM PC and Compatibles, *Bruce J. Chalmer and David G. Whitmore*
64. Statpal: A Statistical Package for Microcomputers—Apple Version for the II, II+, and IIe, *David G. Whitmore and Bruce J. Chalmer*
65. Nonparametric Statistical Inference: Second Edition, Revised and Ex- panded, *Jean Dickinson Gibbons*
66. Design and Analysis of Experiments, *Roger G. Petersen*
67. Statistical Methods for Pharmaceutical Research Planning, *Sten W. Bergman and John C. Gittins*
68. Goodness-of-Fit Techniques, *edited by Ralph B. D'Agostino and Michael A. Stephens*
69. Statistical Methods in Discrimination Litigation, *edited by D. H. Kaye and Mikel Aickin*
70. Truncated and Censored Samples from Normal Populations, *Helmut Schneider*
71. Robust Inference, *M. L. Tiku, W. Y. Tan, and N. Balakrishnan*
72. Statistical Image Processing and Graphics, *edited by Edward J. Wegman and Douglas J. DePriest*
73. Assignment Methods in Combinatorial Data Analysis, *Lawrence J. Hubert*
74. Econometrics and Structural Change, *Lyle D. Broemeling and Hiroki Tsurumi*
75. Multivariate Interpretation of Clinical Laboratory Data, *Adelin Albert and Eugene K. Harris*
76. Statistical Tools for Simulation Practitioners, *Jack P. C. Kleijnen*
77. Randomization Tests: Second Edition, *Eugene S. Edgington*
78. A Folio of Distributions: A Collection of Theoretical Quantile-Quantile Plots, *Edward B. Fowlkes*
79. Applied Categorical Data Analysis, *Daniel H. Freeman, Jr.*
80. Seemingly Unrelated Regression Equations Models: Estimation and Inference, *Virendra K. Srivastava and David E. A. Giles*
81. Response Surfaces: Designs and Analyses, *Andre I. Khuri and John A. Cornell*
82. Nonlinear Parameter Estimation: An Integrated System in BASIC, *John C. Nash and Mary Walker-Smith*
83. Cancer Modeling, *edited by James R. Thompson and Barry W. Brown*

84. Mixture Models: Inference and Applications to Clustering, *Geoffrey J. McLachlan and Kaye E. Basford*
85. Randomized Response: Theory and Techniques, *Arijit Chaudhuri and Rahul Mukerjee*
86. Biopharmaceutical Statistics for Drug Development, *edited by Karl E. Peace*
87. Parts per Million Values for Estimating Quality Levels, *Robert E. Odeh and D. B. Owen*
88. Lognormal Distributions: Theory and Applications, *edited by Edwin L. Crow and Kunio Shimizu*
89. Properties of Estimators for the Gamma Distribution, *K. O. Bowman and L. R. Shenton*
90. Spline Smoothing and Nonparametric Regression, *Randall L. Eubank*
91. Linear Least Squares Computations, *R. W. Farebrother*
92. Exploring Statistics, *Damaraju Raghavarao*
93. Applied Time Series Analysis for Business and Economic Forecasting, *Sufi M. Nazem*
94. Bayesian Analysis of Time Series and Dynamic Models, *edited by James C. Spall*
95. The Inverse Gaussian Distribution: Theory, Methodology, and Applications, *Raj S. Chhikara and J. Leroy Folks*
96. Parameter Estimation in Reliability and Life Span Models, *A. Clifford Cohen and Betty Jones Whitten*
97. Pooled Cross-Sectional and Time Series Data Analysis, *Terry E. Dielman*
98. Random Processes: A First Look, Second Edition, Revised and Expanded, *R. Syski*
99. Generalized Poisson Distributions: Properties and Applications, *P. C. Consul*
100. Nonlinear L_p-Norm Estimation, *Rene Gonin and Arthur H. Money*
101. Model Discrimination for Nonlinear Regression Models, *Dale S. Borowiak*
102. Applied Regression Analysis in Econometrics, *Howard E. Doran*
103. Continued Fractions in Statistical Applications, *K. O. Bowman and L. R. Shenton*
104. Statistical Methodology in the Pharmaceutical Sciences, *Donald A. Berry*
105. Experimental Design in Biotechnology, *Perry D. Haaland*
106. Statistical Issues in Drug Research and Development, *edited by Karl E. Peace*
107. Handbook of Nonlinear Regression Models, *David A. Ratkowsky*
108. Robust Regression: Analysis and Applications, *edited by Kenneth D. Lawrence and Jeffrey L. Arthur*
109. Statistical Design and Analysis of Industrial Experiments, *edited by Subir Ghosh*
110. *U*-Statistics: Theory and Practice, *A. J. Lee*
111. A Primer in Probability: Second Edition, Revised and Expanded, *Kathleen Subrahmaniam*
112. Data Quality Control: Theory and Pragmatics, *edited by Gunar E. Liepins and V. R. R. Uppuluri*
113. Engineering Quality by Design: Interpreting the Taguchi Approach, *Thomas B. Barker*
114. Survivorship Analysis for Clinical Studies, *Eugene K. Harris and Adelin Albert*
115. Statistical Analysis of Reliability and Life-Testing Models: Second Edition, *Lee J. Bain and Max Engelhardt*
116. Stochastic Models of Carcinogenesis, *Wai-Yuan Tan*
117. Statistics and Society: Data Collection and Interpretation: Second Edition, Revised and Expanded, *Walter T. Federer*

118. Handbook of Sequential Analysis, *B. K. Ghosh and P. K. Sen*
119. Truncated and Censored Samples: Theory and Applications, *A. Clifford Cohen*
120. Survey Sampling Principles, *E. K. Foreman*
121. Applied Engineering Statistics, *Robert M. Bethea and R. Russell Rhinehart*
122. Sample Size Choice: Charts for Experiments with Linear Models: Second Edition, *Robert E. Odeh and Martin Fox*
123. Handbook of the Logistic Distribution, *edited by N. Balakrishnan*
124. Fundamentals of Biostatistical Inference, *Chap T. Le*
125. Correspondence Analysis Handbook, *J.-P. Benzécri*
126. Quadratic Forms in Random Variables: Theory and Applications, *A. M. Mathai and Serge B. Provost*
127. Confidence Intervals on Variance Components, *Richard K. Burdick and Franklin A. Graybill*
128. Biopharmaceutical Sequential Statistical Applications, *edited by Karl E. Peace*
129. Item Response Theory: Parameter Estimation Techniques, *Frank B. Baker*
130. Survey Sampling: Theory and Methods, *Arijit Chaudhuri and Horst Stenger*
131. Nonparametric Statistical Inference: Third Edition, Revised and Expanded, *Jean Dickinson Gibbons and Subhabrata Chakraborti*
132. Bivariate Discrete Distribution, *Subrahmaniam Kocherlakota and Kathleen Kocherlakota*

Additional Volumes in Preparation

Design and Analysis of Bioavailability and Bioequivalence Studies, *Shein-Chung Chow and Jen-pei Liu*

BIVARIATE DISCRETE DISTRIBUTIONS

SUBRAHMANIAM KOCHERLAKOTA
KATHLEEN KOCHERLAKOTA
Department of Statistics
University of Manitoba
Winnipeg, Manitoba, Canada

Marcel Dekker, Inc. New York • Basel • Hong Kong

Library of Congress Cataloging-in-Publication Data

Kocherlakota, Subrahmaniam.
Bivariate discrete distributions / Subrahmaniam Kocherlakota, Kathleen Kocherlakota.
p. cm. -- (Statistics, textbooks and monographs)
Includes bibliographical references and indexes.
ISBN 0-8247-8702-1 (acid-free paper)
1. Distribution (Probability theory) I. Kocherlakota, K. II. Title. III. Series.
QA273.6.K63 1992
519.2'4--dc20 92-8012
CIP

This book is printed on acid-free paper.

MARCEL DEKKER, INC.
270 Madison Avenue, New York, New York 10016

Current printing (last digit):
10 9 8 7 6 5 4 3 2 1

PRINTED IN THE UNITED STATES OF AMERICA

Dedicated to the late

Kocherlakota Satya Narayana Murthy

PREFACE

The first attempt at bringing together the statistical community deeply involved in the study and applications of discrete distributions was the International Symposium held at Montreal in 1963. Since then, the subject has received much attention on a global scale. The intervening period of nearly three decades has proved to be an explosive era, leading to the extensions of old results. More important it has been an exciting period of discovery of new distributions with applications in a variety of areas. New techniques relating to the estimation of the parameters and testing of hypotheses have been suggested. Characterizing properties of the distributions have also been presented. The computer age has made its imprint by providing the technology needed to generate random observations from a variety of populations, leading to the simulation of diverse types of populations.

In 1986 the present authors made an attempt at bringing together, as was done in 1963, the international community working in the various areas of discrete distributions. This was achieved by putting out a special issue of the *Communications in Statistics, Theory and Methods* with about 450 pages of contributions from around the world. However, we felt that a comprehensive documentation of the vast amount of literature that has accumulated over the past several years in the area of discrete distributions needed to be undertaken. Of particular interest to us was the development of the multidimensional discrete distributions. Unfortunately, we discovered that the general multivariate analytical tools, that have become commonplace in the normal case, are not available for the discrete distributions. What has been examined in this set-up are the structural properties of the multidimensional distributions. Also, in most multidimensional generalizations the distributions are of the "homogeneous" type in the sense of A. W. Kemp. Most of the non-homogeneous type of probability generating functions have to be restricted to the bivariate situations to facilitate mathematical tractability. This, in addition to the literature extant in inference and practical applications, led us to the decision to restrict the coverage in the text to the bivariate discrete distributions.

It should be made amply clear at the outset that this decision to limit the study to the bivariate case does not imply that we ignore or propose the ignoring of the multivariate distributions. To this end, we have widened the bibliography by incorporating relevant results in the multivariate situations. Hopefully, this will make the restricted coverage more palatable to the prospective user.

A general introduction to the structural properties of discrete distributions is presented in Chapter 1. These include the various generating functions and their relationships to the probability generating function; the moment relationships; polynomial representations of the bivariate probability functions; basic ideas of compounding and generalizing; and the computer simulations of random variables.

In Chapter 2 a general discussion of the problems of statistical inference is given with emphasis placed on specific techniques pertinent to the discrete case.

The book is laid out so that various chapters beginning with the third chapter are independent of each other. Although there is a common thread stretching across these topics, their treatment is very much individualized. The arrangement of the material exploits their interrelationships and thereby facilitates an easy understanding of the development of the distributions.

As in the univariate case, the Bernoulli trials form the basic building blocks of the distributions arising in the bivariate case. This is the lead-off topic of the text in Chapter 3, giving a natural access to the bivariate binomial distributions.

In Chapter 4 the bivariate Poisson distribution is introduced and studied both as a limit of the bivariate binomial and in its own right as a stochastic process.

Inverse sampling and waiting times in Bernoulli trials are considered in Chapter 5 under the general heading of the bivariate negative binomial distribution. The genesis of the distribution from the point of view of compounding is also discussed, although the general topic of generalized and compound distributions is deferred to a later chapter.

Sampling from a finite population without replacement is the thrust of Chapter 6 giving rise to the bivariate hypergeometric distributions of different forms. Inverse sampling from such populations is also examined in some detail in this chapter.

The bivariate logarithmic series distribution is derived as the Fisher-limit to the bivariate negative binomial distribution in Chapter 7. Various other models leading to this distribution and its modified version are also presented in this chapter.

The study of distributions when the parameter is subject to variability has been of interest in a variety of situations. Of particular interest for this purpose is the Poisson distribution; one reason for this seems to be that the parameter of the distribution has a natural tendency to be variable over the population. This has led to a rich class of distributions of practical value arising out of such compounding. These problems relating to compounding with the bivariate Poisson distribution are considered in the most general setting in Chapter 8.

The final chapter of the book deals with the distributions of more recent origin developed in connection with accident theory. These are the bivariate Waring and 'short' distributions. In addition, the bivariate generalized power series distribution is also discussed briefly in this chapter.

In most of the cases, the latest techniques for computer simulation of the distributions considered are discussed in some detail. While all attempts were made to include practical data in the relevant sections, in some instances we have had to compensate for its lack by using simulated data.

While we make no apology for the topics discussed in the book, it should be mentioned that we do feel that but for constraints of space we would have included several more contributions. To this extent we regret not having incorporated the material produced by a number of authors known to us personally. To them our sincere regrets. To others not known to us too we extend our personal regrets. The value of the manuscript has no doubt been enhanced by the inclusion of the vast amount of literature in the bibliography. This is made more useful by the provision of the keywords and phrases. A detailed index for the extended bibliography is appended to further enhance its usefulness.

A final word of thanks to all the authors who gave us hours of pleasure derived from going over their publications. The University of Manitoba is to be thanked for providing the necessary amenities as well as the study/research leaves to both of us. Without these, the work would have been much delayed.

Subrahmaniam Kocherlakota
Kathleen Kocherlakota

CONTENTS

NOTATION

$B(\nu, \theta)$ binomial distribution on ν trials with probability of success θ

$b(x; \nu, \theta)$ binomial probability function at x of $B(\nu, \theta)$

BVB Type I Type I bivariate binomial distribution defined in (3.3.2)

BVB Type II Type II bivariate binomial distribution defined in (3.4.2)

$P(\lambda)$ Poisson distribution with parameter λ

$p(x; \lambda)$ probability function at x of $P(\lambda)$

$BVP(\lambda_1, \lambda_2, \lambda_3)$ bivariate Poisson distribution defined in (4.3.4)

$NB(r, \theta)$ negative binomial distribution defined in (5.1.1)

$NTA(\lambda, \theta)$ Neyman Type A distribution obtained by $P(\theta\phi) \underset{\phi}{\wedge} P(\lambda)$ or

$P(\lambda) \vee P(\theta)$

$IG(\mu, \lambda)$ univariate inverse Gaussian distribution defined in (8.5.3)

$UVWD(\alpha, \kappa; \beta)$ univariate Waring distribution defined in (9.2.8)

$BVWD(\alpha; \tau, \kappa; \beta)$ bivariate Waring distribution defined in (9.2.7)

$UVSD(\Phi, \lambda, \theta)$ univariate 'short' distribution defined in (9.3.2)

BVSD bivariate 'short' distributions defined in (9.3.6), (9.3.7), (9.3.8), (9.3.9)

GPSD univariate generalized power series distribution defined in (9.4.1)

BVGPSD bivariate generalized power series distribution defined in (9.4.3)

$L(\underline{\theta}; \underline{x})$ likelihood function of $\underline{\theta}$ given the observation $\underline{x}$

log x natural logarithm of x

Γ information matrix: $\left\{- E\left[\frac{\partial^2 \log L}{\partial\theta_i \, \partial\theta_j}\right]\right\}$ or $\left\{E\left[\frac{\partial \log L}{\partial\theta_i} \frac{\partial \log L}{\partial\theta_j}\right]\right\}$

Γ^{-1} asymptotic variance matrix of $\underline{\hat{\theta}}$

~ distributed as

≈ asymptotically distributed as

pf probability function

pdf probability density function

pgf probability generating function

mgf moment generating function

cgf cumulant generating function

1

PRELIMINARIES

In this chapter we shall be considering the basic properties of bivariate discrete distributions. As is well known, central to the study of probability distributions is the generating function. In the case of discrete distributions, the probability generating function plays a vital role as, in most of the situations, it is much simpler to handle than is the moment generating function. However, in addition to the probability generating function we will define a variety of generating functions of which the importance and usefulness will be made clear in the context in which they appear. The relationships between the different types of moments will also be developed. The structure of bivariate discrete distributions has been studied by several authors by a canonical representation. These ideas will be introduced in this chapter. Specifics will be referred to a later section appropriate to the distribution under consideration. Finally, the computer age has been instrumental in making the abstraction of theory a reality. To this end, some general ideas concerning the computer generation of random samples from bivariate discrete distributions will also be presented in this chapter with, once again, the specifics being left to be given in the appropriate sections.

In the following, unless otherwise stated, we will be considering the joint distribution of the random variables X and Y. They will be assumed to have the probability mass function f(x, y) at the point (x, y) with $(x, y) \in T$, a subset of the Cartesian product of the set of nonnegative integers on the real line. In this case the pair (X, Y) will be said to have bivariate discrete distribution over T with the probability function f(x, y).

1.1 Generating functions

In the study of random variables a variety of generating functions have been introduced to facilitate the summarization of the properties in a compact but manageable functional form. There are a number of such properties that are of interest in general. Thus correspondingly, there are a number of generating functions defined, each giving rise to at least one specific property of the random variable. In this section we will define several such functions useful for developing the relationships between the various moment features of the random variables.

Definition 1.1.1 Probability generating function

The probability generating function (pgf) of the pair of random variables (X, Y) with probability function f(x, y) is the $E[t_1^X t_2^Y]$.

We shall denote the pgf by $\Pi(t_1, t_2)$. From the definition it is readily seen that

$$\Pi(t_1, t_2) = \sum_{(x,y)\in T} t_1^x t_2^y f(x, y) . \tag{1.1.1}$$

From equation (1.1.1) it is obvious that the series is absolutely convergent in the unit rectangle $|t_1| \le 1$, $|t_2| \le 1$. As such the pgf can be differentiated with respect to these variables at (0, 0) any number of times. It can also be shown that the pgf is unique in the sense of a one-to-one relationship to the probability function (pf). For determining the pf, given the pgf, we can either

(i) expand $\Pi(t_1, t_2)$ in powers of t_1 and t_2. Then the coefficient of $t_1^x t_2^y$ will be the probability function f(x, y) at (x, y),

or

(ii) use the fact that the pgf can be differentiated any number of times with respect to t_1 and t_2 and evaluated at (0, 0) yielding

$$f(x, y) = \frac{1}{x!}\frac{1}{y!}\frac{\partial^{x+y}}{\partial t_1^x \partial t_2^y} \Pi(t_1, t_2)\bigg|_{t_1=0, t_2=0} . \tag{1.1.2}$$

As its name implies, the probability generating function gives rise to the probability function of the random variable. However, our interest is often in the joint moments $\mu'_{r,s} = E(X^r Y^s)$ of the random variables X and Y. For this purpose we are led to the definition of the moment generating function (mgf). Unfortunately, the existence of the mgf is dependent on the existence of the moments. To ensure that it does exist, we use the restricted definition given by Hogg and Craig (1978, p. 77).

Definition 1.1.2 Moment generating function

The moment generating function (mgf) of the pair of random variables X and Y is the $E[\exp(t_1X + t_2Y)]$ provided the expectation exists for (t_1, t_2) lying in the rectangle $-h_1 < t_1 < h_1$, $-h_2 < t_2 < h_2$ where h_1 and h_2 are positive real numbers.

The moment generating function is represented by $M(t_1, t_2)$. From the definition,

$$M(t_1, t_2) = \sum_{x,y} \exp[t_1x + t_2y]\, f(x, y). \qquad (1.1.3)$$

As pointed out, this definition of the mgf assumes that all the moments do exist. If, however, the mgf does not exist, it is necessary to deal with the characteristic function defined over the complex plane. For details on the difficulties in this connection and other related problems we refer to Stuart and Ord (1987).

Expanding the exponential functions in (1.1.3), it is possible to write

$$M(t_1, t_2) = \sum_{r,s} \frac{t_1^r}{r!}\frac{t_2^s}{s!}\mu'_{r,s}\,. \qquad (1.1.4)$$

From this expansion and the fact that the order of differentiation and summation can be interchanged, it is possible to establish the following rules for the determination of the moments, given the mgf:

(i) $\mu'_{r,s}$ is the coefficient of $\frac{t_1^r}{r!}\frac{t_2^s}{s!}$ in the expansion of the mgf in powers of t_1 and t_2.

(ii) $\mu'_{r,s}$ is given by the mixed partial derivative

$$\frac{\partial^{r+s}}{\partial t_1^r \partial t_2^s} M(t_1, t_2)\Big|_{t_1=0, t_2=0}.$$

Comparing (1.1.1) and (1.1.3) we have the relationship

$$M(t_1, t_2) = \Pi[\exp t_1, \exp t_2].$$

Cumulants generating function

As in the univariate case the cumulants generating function, and consequently, the cumulants are defined by the relations

$$K(t_1, t_2) = \log M(t_1, t_2) \tag{1.1.5}$$

$$= \sum_{r} \sum_{s} \frac{t_1^r}{r!} \frac{t_2^s}{s!} \kappa_{r,s}. \quad (r,s) \neq (0,0) \tag{1.1.6}$$

In (1.1.6) the quantity $\kappa_{r,s}$ is called the cumulant of order (r, s).

Factorial moments and cumulants

In discrete distributions the factorial moments and factorial cumulants play an important role in describing the distribution as well as in the procedures for estimation.

The factorial moment generating function of (X, Y) is defined by the equation

$$G(t_1, t_2) = \Pi(t_1+1, t_2+1). \tag{1.1.7}$$

Substituting for the right hand side of (1.1.7), expanding the binomials and summing with respect to the variables x and y, we have

$$G(t_1, t_2) = \sum_{r,s} \frac{t_1^r}{r!} \frac{t_2^s}{s!} \mu_{[r,s]}, \tag{1.1.8}$$

where $\mu_{[r,s]}$ stands for $E[X^{[r]}Y^{[s]}]$ and $x^{[r]} = x(x-1)(x-2) \ldots (x-r+1)$. This expected value is called the factorial moment of order (r, s).

Thus it is readily seen that the factorial moment $\mu_{[r,s]}$ is generated as the coefficient of $\frac{t_1^r}{r!} \frac{t_2^s}{s!}$ in the power series expansion of the function $G(t_1, t_2)$. Another representation of the factorial moment is

$$\mu_{[r,s]} = \frac{\partial^{r+s}}{\partial t_1^r t_2^s} \Pi(t_1+1, t_2+1)\Big|_{t_1=0,t_2=0} . \qquad (1.1.9)$$

The factorial cumulants generating function is defined by the relationship

$$H(t_1, t_2) = \log G(t_1, t_2)$$
$$= \sum_{r} \sum_{s} \frac{t_1^r}{r!} \frac{t_2^s}{s!} \kappa_{[r,s]} . \qquad (1.1.10)$$
$$(r,s) \neq (0,0)$$

The quantity $\kappa_{[r,s]}$ is called the factorial cumulant of order (r, s).

Relationships among moments and cumulants

The question that arises is whether and, if so, how these moments and cumulants are interrelated. Before we answer this question, let us introduce the central moment of order r in the univariate case as $E[(X-\mu_x)^r]$, where μ_x is the expected value of X. Similarly the (r, s)th central moment in the bivariate case is $\mu_{r,s} = E[(X-\mu_x)^r(Y-\mu_y)^s]$. The following procedure is suggested by David and Barton (1962) to determine the relationship between the central moments and those defined earlier around zero . Consider the formal representation

$$\mu(r^4) = \mu'(r^4) - 4\,\mu'(r^3)\,\mu'(r) + 6\,\mu'(r^2)\,[\mu'(r)]^2 - 3\,[\mu'(r)]^4.$$

Operate on both sides of this equation by $s\frac{\partial}{\partial r}$ and cancel out the coefficient 4 from both sides of the resulting equation. Then changing the order of differentiation and summation, we have

$$\mu_{3,1} = \mu'_{3,1} - 3\mu'_{2,1}\mu'_{1,0} - \mu'_{3,0}\mu'_{0,1} + 3\mu'_{1,1}\mu'^2_{1,0}$$
$$+3\mu'_{2,0}\mu'_{1,0}\mu'_{0,1} - 3\mu'^3_{1,0}\mu'_{0,1}. \qquad (1.1.11)$$

This procedure can be applied repeatedly to obtain further relationships. Thus operating on $\mu[r^3s]$ by $s\frac{\partial}{\partial r}$ and cancelling out the factor 3, we have

$$\mu_{2,2} = \mu'_{2,2} - 2[\mu'_{2,1}\mu'_{0,1} + \mu'_{1,2}\mu'_{1,0}] + \mu'^2_{1,0}\mu'_{0,2} + \mu'^2_{0,1}\mu'_{2,0}$$

$$+ 4\mu'_{1,1}\mu'_{1,0}\mu'_{0,1} - 3\mu'^2_{1,0}\mu'^2_{0,1}. \tag{1.1.12}$$

This method can also be applied to set up the relationships between the moments and the cumulants. For example, recalling the expression for the fourth raw moment in terms of the cumulants

$$\mu'_4 = \kappa_4 + 4\kappa_3\kappa_1 + 3\kappa_2^2 + 6\kappa_2\kappa_1^2 + \kappa_1^4,$$

we can write formally

$$\mu'(r^4) = \kappa(r^4) + 4\kappa(r^3)\kappa(r) + 3[\kappa(r^2)]^2 + 6\kappa(r^2)[\kappa(r)]^2 + [\kappa(r)]^4. \tag{1.1.13}$$

Treating this as a function of r and operating on it by $s\frac{\partial}{\partial r}$ and then replacing the mixed products by their corresponding moments, we have

$$\mu'_{3,1} = \kappa_{3,1} + 3\kappa_{2,1}\kappa_{1,0} + \kappa_{3,0}\kappa_{0,1} + 3\kappa_{1,1}\kappa_{2,0} + 3\kappa_{1,1}\kappa_{1,0}^2$$
$$+ 3\kappa_{2,0}\kappa_{1,0}\kappa_{0,1} + \kappa_{1,0}^3\kappa_{0,1}.$$

If the means are zero, then this equation simplifies to $\mu_{3,1} = \kappa_{3,1} + 3\kappa_{1,1}\kappa_{2,0}$. Using the same technique and treating the means equal to zero, we have the reverse relation $\kappa_{3,1} = \mu_{3,1} - 3\mu_{1,1}\mu_{2,0}$.

Similarly, the operation with $(s\frac{\partial}{\partial r})^2$ on (1.1.13) gives, upon setting the means equal to zero, $\mu_{2,2} = \kappa_{2,2} + 2\kappa_{1,1}^2 + \kappa_{2,0}\kappa_{0,2}$. The reverse relation can also be shown to be $\kappa_{2,2} = \mu_{2,2} - 2\mu_{1,1}^2 - \mu_{2,0}\mu_{0,2}$. Reference may be made to Stuart and Ord (1987, p. 104) for further examples in the computation of the relationships of this nature.

Employing expansion techniques, Kocherlakota (1991) has studied the relationship between the factorial cumulants and the ordinary cumulants. It is shown that

$$\kappa_{[1,1]} = \mu_{1,1} = \kappa_{1,1}$$
$$\kappa_{[2,1]} = \mu_{2,1} - \mu_{1,1} = \kappa_{2,1} - \kappa_{1,1}$$

$$\kappa_{[3,1]} = \mu_{3,1} - 3\mu_{2,1} + 2\mu_{1,1} - 3\mu_{2,0}\mu_{1,1}$$
$$= \kappa_{3,1} - 3\kappa_{2,1} + 2\kappa_{1,1}$$
$$\kappa_{[2,2]} = \mu_{2,2} - \mu_{2,1} - \mu_{1,2} + \mu_{1,1} - 2\mu_{1,1}^2 - \mu_{2,0}\mu_{0,2}$$
$$= \kappa_{2,2} - \kappa_{2,1} - \kappa_{1,2} + \kappa_{1,1}$$
$$\kappa_{[4,1]} = \mu_{4,1} - 6\mu_{3,1} + 11\mu_{2,1} - 6\mu_{1,1} - 4\mu_{1,1}\mu_{3,0} + 18\mu_{2,0}\mu_{1,1} - 6\mu_{2,0}\mu_{2,1}$$
$$= \kappa_{4,1} - 6\kappa_{3,1} + 11\kappa_{2,1} - 6\kappa_{1,1}$$
$$\kappa_{[3,2]} = \mu_{3,2} - 3\mu_{2,2} - \mu_{3,1} + 3\mu_{2,1} + 2\mu_{1,2} - 2\mu_{1,1} - \mu_{0,2}(\mu_{3,0} - 3\mu_{2,0}) - 3\mu_{2,0}(\mu_{1,2} - \mu_{1,1}) - 6\mu_{1,1}(\mu_{2,1} - \mu_{1,1})$$
$$= \kappa_{3,2} - 3\kappa_{2,2} - \kappa_{3,1} + 3\kappa_{2,1} + 2\kappa_{1,2} - 2\kappa_{1,1}$$
$$\kappa_{[5,1]} = \kappa_{5,1} - 10\kappa_{4,1} + 35\kappa_{3,1} - 50\kappa_{2,1} + 24\kappa_{1,1}$$
$$\kappa_{[4,2]} = \kappa_{4,2} - 6\kappa_{3,2} - \kappa_{4,1} + 6\kappa_{3,1} + 11\kappa_{2,2} - 6\kappa_{1,2} - 11\kappa_{2,1} + 6\kappa_{1,1}$$
$$\kappa_{[3,3]} = \kappa_{3,3} - 3\kappa_{3,2} - 3\kappa_{2,3} + 2\kappa_{3,1} + 9\kappa_{2,2} + 2\kappa_{1,3} - 6\kappa_{2,1} - 6\kappa_{1,2} + 4\kappa_{1,1}.$$

The inverse relations are correspondingly given by

$$\kappa_{1,1} = \kappa_{[1,1]}$$
$$\kappa_{2,1} = \kappa_{[2,1]} + \kappa_{[1,1]}$$
$$\kappa_{3,1} = \kappa_{[3,1]} + 3\kappa_{[2,1]} + \kappa_{[1,1]}$$
$$\kappa_{2,2} = \kappa_{[2,2]} + \kappa_{[2,1]} + \kappa_{[1,2]} + \kappa_{[1,1]}$$
$$\kappa_{4,1} = \kappa_{[4,1]} + 6\kappa_{[3,1]} + 7\kappa_{[2,1]} + \kappa_{[1,1]}$$
$$\kappa_{3,2} = \kappa_{[3,2]} + \kappa_{[3,1]} + 3\kappa_{[2,2]} + 3\kappa_{[2,1]} + \kappa_{[1,2]} + \kappa_{[1,1]}$$
$$\kappa_{5,1} = \kappa_{[5,1]} + 10\kappa_{[4,1]} + 25\kappa_{[3,1]} + 15\kappa_{[2,1]} + \kappa_{[1,1]}$$
$$\kappa_{4,2} = \kappa_{[4,2]} + 6\kappa_{[3,2]} + \kappa_{[4,1]} + 6\kappa_{[3,1]} + 7\kappa_{[2,2]} + 7\kappa_{[2,1]} + \kappa_{[1,2]} + \kappa_{[1,1]}$$
$$\kappa_{3,3} = \kappa_{[3,3]} + 3(\kappa_{[3,2]} + \kappa_{[2,3]}) + 9\kappa_{[2,2]} + \kappa_{[1,3]} + \kappa_{[3,1]} + 3(\kappa_{[1,2]} + \kappa_{[2,1]}) + \kappa_{[1,1]}.$$

David and Barton (1962, p. 51) show that the factorial and ordinary moments have the relationship

$$\kappa_{[r]} = \sum_{i=0}^{r} s(r, i)\, \kappa_r, \qquad \kappa_r = \sum_{i=0}^{r} S(r, i)\, \kappa_{[r]}, \tag{1.1.14}$$

where s(r, i) and S(r, i) are Stirling numbers of the first and second kind, respectively. [See, for example, Berg (1988), p. 776.] An examination of the above equations shows that the expression for $\kappa_{[r,u]}$ in terms of $\kappa_{t,v}$ for $t \le r$, $v \le u$ has the coefficient s(r, t) s(u, v). On the other hand the expression for $\kappa_{r,u}$ in terms of $\kappa_{[t,v]}$ for $t \le r$, $v \le u$ has the coefficient S(r, t) S(u, v).

1.2 Convolutions

Sometimes bivariate distributions can be generated by convolutions of random variables. Thus if we consider $X = X_1+X_3$ and $Y = X_2+X_3$ with X_1, X_2 and X_3 being independently distributed, then the random variables X and Y are jointly distributed. The method, termed the trivariate reduction method, is discussed in Mardia (1970).

Let the pgf's of the random variables under consideration be $\Pi_i(t)$, i=1, 2, 3. Then the joint pgf of (X, Y) is

$$\Pi(t_1, t_2) = \Pi_1(t_1)\, \Pi_2(t_2)\, \Pi_3(t_1 t_2). \tag{1.2.1}$$

Similarly, if the mgf of X_i, i = 1, 2, 3 is $M_i(t)$ then the mgf of (X, Y) is seen to be

$$M(t_1, t_2) = M_1(t_1)\, M_2(t_2)\, M_3(t_1+t_2). \tag{1.2.2}$$

Probability function

From the joint pgf (1.2.1), by successive differentiation with respect to the arguments, we have

$$\Pi^{(r,s)}(t_1, t_2) = \sum_{i=0}^{r} \sum_{h=0}^{s} \sum_{k=0}^{\min(r-i,s-h)} \frac{r!s!}{i!h!k!(r-i-k)!(s-h-k)!}\, t_1^{s-h-k} t_2^{r-i-k} \cdot$$
$$\Pi_1^{(i)}(t_1)\, \Pi_2^{(h)}(t_2)\, \Pi_3^{[r+s-i-h-k]}(t_1 t_2). \tag{1.2.3}$$

Dividing (1.2.3) by r! s! and setting $t_1 = t_2 = 0$, we get the joint pf of

(X, Y) as

$$P\{X=r,\ Y=s\} = \sum_{k=0}^{\min(r,s)} f_1(r-k)f_2(s-k)f_3(k), \tag{1.2.4}$$

where $f_i(w)$ is the pf of the random variable X_i. This equation can also be obtained directly from the definition of the random variables X and Y.

Moments

Setting $t_1 = t_2 = 1$ in (1.2.3) yields

$$\mu_{[r,s]} = \sum_{i=0}^{r} \sum_{h=0}^{s} \sum_{k=0}^{\min(r-i,s-h)} \frac{r!s!}{i!h!k!(r-i-k)!(s-h-k)!} \mu_{[r-i-k]}^{(1)} \mu_{[s-h-k]}^{(2)} \mu_{[i+h+k]}^{(3)}. \tag{1.2.5}$$

The moment generating function (1.2.2) gives the (raw) moments

$$\mu'_{r,s} = \sum_{i=0}^{r} \sum_{j=0}^{s} \binom{r}{i}\binom{s}{j} \mu_i^{(1)'} \mu_j^{(2)'} \mu_{r+s-i-j}^{(3)'}. \tag{1.2.6}$$

Since

$$E\left[e^{t_1(X-\mu_X) + t_2(Y-\mu_Y)}\right] = E\left[e^{t_1(X_1-\mu_1)} e^{t_2(X_2-\mu_2)} e^{(t_1+t_2)(X_3-\mu_3)}\right],$$

the equation for the central moments is

$$\mu_{r,s} = \sum_{i=0}^{r} \sum_{j=0}^{s} \binom{r}{i}\binom{s}{j} \mu_i^{(1)} \mu_j^{(2)} \mu_{r+s-i-j}^{(3)}. \tag{1.2.7}$$

Factorial Cumulants

The factorial cumulant generating function is in this case

$$H(t_1,t_2) = \log \pi_1(1+t_1) + \log \pi_2(1+t_2) + \log \pi_3\{(1+t_1)(1+t_2)\}. \tag{1.2.8}$$

Now

$$\log \pi_1(1+t_1) = \sum_{r=1}^{\infty} \frac{t_1^r}{r!} \kappa_{[r]}^{(1)}, \qquad \log \pi_2(1+t_2) = \sum_{s=1}^{\infty} \frac{t_2^s}{s!} \kappa_{[s]}^{(2)},$$

and

$$\log \pi_3\{(1+t_1)(1+t_2)\} = \sum_{u=1}^{\infty} \frac{(t_1+t_2+t_1t_2)^u}{u!} \kappa_{[u]}^{(3)}$$

$$= \sum_{u=1}^{\infty} \frac{\kappa_{[u]}^{(3)}}{u!} \sum_{i=0}^{u} \sum_{\substack{j=0 \\ i+j\le u}}^{u} \frac{u!\, t_1^{u-j} t_2^{u-i}}{i!j!(u-i-j)!} .$$

Hence

$$H(t_1,t_2) = \sum_{r=1}^{\infty} \frac{t_1^r}{r!} \kappa_{[r]}^{(1)} + \sum_{s=1}^{\infty} \frac{t_2^s}{s!} \kappa_{[s]}^{(2)} + \sum_{u=1}^{\infty} \frac{\kappa_{[u]}^{(3)}}{u!} \sum_{i=0}^{u} \sum_{\substack{j=0 \\ i+j\le u}}^{u} \frac{u!\, t_1^{u-j} t_2^{u-i}}{i!j!(u-i-j)!} . \tag{1.2.9}$$

The marginal factorial cumulants are readily seen to be

$$\kappa_{[r,0]} = \kappa_{[r]}^{(1)} + \kappa_{[r]}^{(3)}, \qquad \kappa_{[0,s]} = \kappa_{[s]}^{(1)} + \kappa_{[s]}^{(3)}.$$

Also the last term of (1.2.9) can be written as

$$\sum_{r=1}^{\infty} \frac{t_1^r}{r!} \sum_{s=1}^{\infty} \frac{t_2^s}{s!} \sum_{u=\max(r,s)}^{r+s} \frac{r!s!}{(u-r)!(u-s)!(r+s-u)!} \kappa_{[u]}^{(3)}. \tag{1.2.10}$$

Hence, for $r \ne 0$, $s \ne 0$,

$$\kappa_{[r,s]} = \sum_{u=\max(r,s)}^{r+s} \frac{r!s!}{(u-r)!(u-s)!(r+s-u)!} \kappa_{[u]}^{(3)}$$

$$= \kappa_{[s,r]} . \tag{1.2.11}$$

Cumulants

Using (1.2.2), the cumulant generating function is seen to be

$$K(t_1,t_2) = \log M_1(t_1) + \log M_2(t_2) + \log M_3(t_1 + t_2)$$

$$= \sum_{r=1}^{\infty} \frac{t_1^r}{r!} \kappa_r^{(1)} + \sum_{s=1}^{\infty} \frac{t_2^s}{s!} \kappa_s^{(2)} + \sum_{u=1}^{\infty} \frac{(t_1+t_2)^u}{u!} \kappa_u^{(3)}. \tag{1.2.12}$$

Therefore, the cumulants are seen to be

$$\kappa_{r,0} = \kappa_r^{(1)} + \kappa_r^{(3)}, \quad \kappa_{0,s} = \kappa_s^{(2)} + \kappa_s^{(3)}, \quad \kappa_{r,s} = \kappa_{r+s}^{(3)} = \kappa_{s,r}. \qquad (1.2.13)$$

Factorial cumulants in terms of cumulants

Recalling that $\kappa_{[u]} = \sum_{j=0}^{u} s(u, j)\, \kappa_j$ and $\kappa_u = \sum_{j=0}^{u} S(u, j)\, \kappa_{[j]}$, we can also write

$$\left.\begin{aligned} \kappa_{[r,t]} &= \sum_{u=\max(r,t)}^{r+t} \frac{r!t!}{(u-r)!(u-t)!(r+t-u)!} \sum_{j=0}^{u} s(u, j)\kappa_j^{(3)} \\ \kappa_{r,t} &= \sum_{j=0}^{r+t} S(r+t, j)\kappa_{[j]}^{(3)} \end{aligned}\right\}. \qquad (1.2.14)$$

Example 1.2.1: From the relations between ordinary and factorial cumulants, we have

$$\kappa_{[1,1]} = \kappa_2^{(3)} = \kappa_{[2]}^{(3)}$$

$$\kappa_{[2,1]} = \kappa_3^{(3)} - \kappa_2^{(3)} = \kappa_{[3]}^{(3)} + 2\kappa_{[2]}^{(3)}$$

$$\kappa_{[3,1]} = \kappa_4^{(3)} - 3\,\kappa_3^{(3)} + 2\kappa_2^{(3)} = \kappa_{[4]}^{(3)} + 3\kappa_{[3]}^{(3)}$$

$$\kappa_{[2,2]} = \kappa_4^{(3)} - 2\,\kappa_3^{(3)} + \kappa_2^{(3)} = \kappa_{[4]}^{(3)} + 4\,\kappa_{[3]}^{(3)}$$

$$\kappa_{[4,1]} = \kappa_5^{(3)} - 6\,\kappa_4^{(3)} + 11\,\kappa_3^{(3)} - 6\,\kappa_2^{(3)} = \kappa_{[5]}^{(3)} + 4\,\kappa_{[4]}^{(3)}$$

$$\kappa_{[3,2]} = \kappa_5^{(3)} - 4\,\kappa_4^{(3)} + 5\,\kappa_3^{(3)} - 2\,\kappa_2^{(3)} = \kappa_{[5]}^{(3)} + 6\,\kappa_{[4]}^{(3)} + 6\,\kappa_{[3]}^{(3)}$$

$$\kappa_{[5,1]} = \kappa_6^{(3)} - 10\,\kappa_5^{(3)} + 35\,\kappa_4^{(3)} - 50\,\kappa_3^{(3)} + 24\,\kappa_2^{(3)} = \kappa_{[6]}^{(3)} + 5\,\kappa_{[5]}^{(3)}$$

$$\kappa_{[4,2]} = \kappa_6^{(3)} - 7\,\kappa_5^{(3)} + 17\,\kappa_4^{(3)} - 17\,\kappa_3^{(3)} + 6\,\kappa_2^{(3)} = \kappa_{[6]}^{(3)} + 8\,\kappa_{[5]}^{(3)} + 12\,\kappa_{[4]}^{(3)}$$

$$\begin{aligned} \kappa_{[3,3]} &= \kappa_6^{(3)} - 6\,\kappa_5^{(3)} + 13\,\kappa_4^{(3)} - 12\,\kappa_3^{(3)} + 4\,\kappa_2^{(3)} \\ &= \kappa_{[6]}^{(3)} + 9\,\kappa_{[5]}^{(3)} + 18\,\kappa_{[4]}^{(3)} + 6\,\kappa_{[3]}^{(3)} \end{aligned}$$

1.3 Marginal and conditional distributions

It is of interest to consider the marginal and conditional distributions. The former describes the individual behavior of each of the random variables. On the other hand, the interdependence of the random variables is brought out by the conditional distributions. Thus a determination of the conditional distribution in a form that can be given simple interpretations is useful. These will be shown in a later section to be helpful in the computer simulation of the bivariate distribution. One aspect of the interrelationships that is useful in prediction is the regression function.

It will be shown here that it is possible to relate the pgf's of the conditional distributions with the pgf of the joint distribution. Further, It will be seen that in most situations the resulting representation leads to an interesting interpretation of the conditional distribution. Although the usual method of determining the conditional pf directly from the joint pf and the marginal pf's can be used, this will entail the determination of the joint probability function. Unfortunately, in the case of the bivariate discrete distributions this is not always a trivial problem. Also the resulting conditional pf does not lend itself to useful interpretation, being quite intractable in most of the cases that arise in practice. On the other hand, it will be seen that the use of the pgf in determining the conditional distribution without having to find the bivariate pf is much more general. It is also helpful in giving a greater insight into the nature of the conditional distributions. The resulting pgf is seen in most situations to yield general, but simple, techniques for the computer simulation of the bivariate distribution.

Marginal distributions

Let the joint pf of (X, Y) be f(x, y). Then the marginal probability functions are $g(x) = \sum_{y} f(x, y)$ and $h(y) = \sum_{x} f(x, y)$. The generating functions of the marginal distributions can be determined from the joint generating functions. Thus the marginal pgf's are

$$\left.\begin{aligned} \Pi_x(t) &= \sum_x g(x)t^x = \sum_x t^x \sum_y f(x, y) = \sum_x \sum_y f(x, y)t^x \\ &= \Pi(t, 1) \\ \Pi_y(t) &= \sum_y h(y)t^y = \sum_y t^y \sum_x f(x, y) = \sum_x \sum_y f(x, y)t^y \\ &= \Pi(1, t) \end{aligned}\right\}. \tag{1.3.1}$$

Similarly from the definition of mgf it is possible to see that the mgf's of the marginal distributions are $M_x(t) = M(t,0)$ and $M_y(t) = M(0,t)$.

Conditional distributions

By definition, the conditional pf of Y, given X=x, is

$$f(y|x) = \frac{f(x, y)}{g(x)}. \tag{1.3.2}$$

The following result due to Subrahmaniam (1966) gives the pgf of this conditional distribution in terms of the joint pgf of (X, Y).

Theorem 1.3.1

Let $\Pi(t_1, t_2)$ be the joint pgf of (X, Y). Then the pgf of the conditional distribution of Y given X=x is

$$\Pi_y(t|x) = \frac{\Pi^{(x,0)}(0, t)}{\Pi^{(x,0)}(0, 1)}, \tag{1.3.3}$$

where

$$\Pi^{(x,y)}(u, v) = \frac{\partial^{x+y}}{\partial t_1^x \partial t_2^y} \Pi(t_1, t_2)\Big|_{t_1=u, t_2=v}.$$

Proof:

By definition the conditional pgf is

$$\Pi_y(t|x) = \sum_y f(y|x)t^y = \frac{\sum_y f(x, y)t^y}{\sum_y f(x, y)}. \tag{1.3.4}$$

From the definition of $\Pi(t_1, t_2)$ it can be readily seen that

$$\Pi^{(x,0)}(t_1, t_2) = \sum_{r,s} \frac{r!}{(r-x)!} t_1^{r-x} t_2^{s} f(r, s);$$

therefore,

$$\Pi^{(x,0)}(0, t) = x! \sum_{s} f(x, s)\, t^s, \qquad \Pi^{(x,0)}(0, 1) = x! \sum_{s} f(x, s),$$

which yield the result in (1.3.3). □

The theorem can be used to determine the regression of Y on X as

Corollary 1.3.1

Regression of Y on X is

$$E[Y|X=x] = \frac{\Pi^{(x,1)}(0, 1)}{\Pi^{(x,0)}(0, 1)} . \qquad (1.3.5)$$

Proof.

By definition

$$E[Y|X=x] = \frac{\partial}{\partial t} \Pi(t|x)\Big|_{t=1}$$

$$= \frac{\partial}{\partial t} [\Pi^{(x,0)}(0, t)/\Pi^{(x,0)}(0, 1)]\Big|_{t=1},$$

which is the result in (1.3.5). □

Definition 1.3.1 Independence

Random variables X and Y are said to be independent if and only if $f(x, y) = g(x)h(y)$ for all $(x, y) \in T$.

A property of the pgf that characterizes independence is given by the following theorem.

Theorem 1.3.2

Let the joint pgf of the random variables X and Y be $\Pi(t_1, t_2)$ with marginal pgf's $\Pi_x(t)$ and $\Pi_y(t)$, respectively. Then X and Y are independent if and only if $\Pi(t_1, t_2) = \Pi_x(t_1)\, \Pi_y(t_2)$.

Proof:

If X and Y are independent, then

$$\Pi(t_1, t_2) = \sum_r \sum_s t_1^r t_2^s f(r, s) = \sum_r \sum_s t_1^r t_2^s g(r) h(s) = \sum_r t_1^r g(r) \sum_s t_2^s h(s)$$

$$= \Pi_x(t) \Pi_y(t). \tag{1.3.6}$$

On the other hand, if $\Pi(t_1, t_2) = \Pi_x(t_1) \Pi_y(t_2)$, then substituting for the appropriate expressions in terms of the summations, we have

$$\sum_r \sum_s t_1^r t_2^s f(r, s) = \sum_r t_1^r g(r) \sum_s t_2^s h(s) = \sum_r \sum_s t_1^r t_2^s g(r) h(s). \tag{1.3.7}$$

Equating like powers of t_1 and t_2 on the two sides of the equation (1.3.7), we have $f(r, s) = g(r)h(s)$ for all $(r, s) \in T$. □

1.4 Sum and difference of the random variables

If the joint pgf of (X, Y) is $\Pi(t_1, t_2)$ then the distribution of the sum Z = X + Y has the pgf

$$\Pi_z(t) = E[t^Z] = E[t^{X+Y}] = \Pi(t, t). \tag{1.4.1}$$

Similarly, it is possible to see that the pgf of the difference W = X–Y has the pgf

$$\Pi_w(t) = E[t^W] = E[t^{X-Y}] = \Pi(t, -t). \tag{1.4.2}$$

Kemp (1981) has considered the conditional distribution of the random variable X given the sum and difference. We will examine these problems in the following sections.

1.4.1 Conditional distribution of X given the sum

The joint pgf of X and Z is

$$\Pi_{x,z}(t_1, t_2) = E[t_1^X t_2^Z] = E[t_1^X t_2^{X+Y}] = \Pi(t_1 t_2, t_2). \tag{1.4.3}$$

Similarly, the joint pgf of Y and Z is

$$\Pi_{y,z}(t_1, t_2) = E[t_1^Y t_2^Z] = E[t_1^Y t_2^{X+Y}] = \Pi(t_2, t_1 t_2). \qquad (1.4.4)$$

To find the conditional pgf of X given the sum Z = z there are two possible techniques available:

(i) From Theorem 1.3.1 we know that

$$\Pi_x(t|z) = \frac{\left.\frac{\partial^z}{\partial t_2^z} \Pi(t_1 t_2, t_2)\right|_{t_1=t, t_2=0}}{\left.\frac{\partial^z}{\partial t_2^z} \Pi(t_1 t_2, t_2)\right|_{t_1=1, t_2=0}}. \qquad (1.4.5)$$

(ii) The following approach is simpler in some instances. Writing f(r, s) for the probability function of (X, Z), we have

$$\Pi_{x,z}(t_1, t_2) = \sum_{r,s} t_1^r t_2^s f(r, s) = \sum_s t_2^s \sum_r t_1^r f(r, s). \qquad (1.4.6)$$

In this equation, the coefficient of t_2^s is $\sum_r t_1^r f(r, s)$. The marginal probability of {Z = z} can be found as the coefficient of t_2^s in $\Pi_{x,z}(1, t_2)$. Therefore the conditional pgf of X given Z = z is

$$\Pi_x(t|z) = \sum_r \frac{f(r, z)}{f(., z)} t^r = \frac{\sum_r f(r, z) t^r}{f(., z)}.$$

Or

$$\Pi_x(t|z) = \frac{\text{coeff. of } t_2^z \text{ in } \sum_{r,s} t_2^s t^r f(r, s)}{\text{coeff. of } t_2^z \text{ in } \sum_{r,s} t_2^s t^r f(r, s) \text{ at } t=1}, \qquad (1.4.7)$$

where f(., s) denotes the marginal pf of Z. Hence, the conditional pgf is the ratio

$$\Pi_X(t|z) = \frac{\text{coeff. of } t_2^z \text{ in } \Pi(t_2 t,\ t_2)}{\text{coeff. of } t_2^z \text{ in } \Pi(t_2,\ t_2)}. \qquad (1.4.8)$$

1.4.2 Conditional distribution of X given the difference

There are two differences that one can consider in this case, namely, W=X−Y and V=Y−X. The joint distributions in each case are found by using the techniques similar to those developed for the sum in 1.4.1. It can be seen that the joint pgf's are:

$$\Pi_{X,W}(t_1,\ t_2) = \Pi(t_1 t_2,\ 1/t_2), \qquad \Pi_{X,V}(t_1,\ t_2) = \Pi(t_1/t_2,\ t_2).$$

The conditional pgf can be found by applying the result in Theorem 1.3.1. Thus from the joint pgf's given above for the two distributions, we have

$$\Pi_X(t|w) = \frac{\left.\dfrac{\partial^w}{\partial t_2^w} \Pi(t_1 t_2,\ 1/t_2)\right|_{t_1=t, t_2=0}}{\left.\dfrac{\partial^w}{\partial t_2^w} \Pi(t_1 t_2,\ 1/t_2)\right|_{t_1=1, t_2=0}} \qquad (1.4.9)$$

and

$$\Pi_X(t|v) = \frac{\left.\dfrac{\partial^v}{\partial t_2^v} \Pi(t_1/t_2,\ t_2)\right|_{t_1=t, t_2=0}}{\left.\dfrac{\partial^v}{\partial t_2^v} \Pi(t_1/t_2,\ t_2)\right|_{t_1=1, t_2=0}}. \qquad (1.4.10)$$

Alternatively, it is possible to use the expansion technique suggested by Kemp (1981). Expanding $\Pi(t_1 t_2,\ 1/t_2)$ in powers of t_1 and t_2, it is possible to see that the numerator in (1.4.9) is proportional to the coefficient of t_2^w in $\Pi(t_2 t,\ 1/t_2)$. By the same argument the denominator is proportional to the coefficient of t_2^w in the expansion of $\Pi(t_2,\ 1/t_2)$. Therefore,

$$\Pi_x(t|w) = \frac{\text{coefficient of } t_2^w \text{ in } \Pi(t_2 t, 1/t_2)}{\text{coefficient of } t_2^w \text{ in } \Pi(t_2, 1/t_2)}. \qquad (1.4.11)$$

Similarly,

$$\Pi_x(t|v) = \frac{\text{coefficient of } t_2^v \text{ in } \Pi(t/t_2, t_2)}{\text{coefficient of } t_2^v \text{ in } \Pi(1/t_2, t_2)}. \qquad (1.4.12)$$

If $w = v = 0$, then the corresponding conditional pgf is given by

$$\Pi_x(t|0) = \frac{\text{constant term in } \Pi(t_2 t, 1/t_2)}{\text{constant term in } \Pi(t_2, 1/t_2)}. \qquad (1.4.13)$$

1.4.3   Homogeneous probability generating function

Kemp (1981) has introduced the idea of homogeneous pgf's.

Definition 1.4.1   Homogeneous pgf
Let X and Y be jointly distributed with the pgf $\Pi(t_1, t_2)$. The pgf is said to be of the homogeneous type if

$$\Pi(t_1, t_2) = H(at_1 + bt_2), \qquad (1.4.14)$$

with $H(a + b) = 1$.

Several of the bivariate pgf's belong to this class. The following theorem characterizes the homogeneous pgf's. It relates the conditional distribution of the random variable X given the sum of the two random variables to the binomial distribution. In this respect the binomial distribution seems to play a unique role in the bivariate discrete distributions. Subrahmaniam (1967) noted a similar role of the binomial distribution in the case on non-homogeneous bivariate distributions. This will be discussed in the next section.

Theorem 1.4.1
The pgf $\Pi(t_1, t_2)$ is of the homogeneous type if and only if the conditional distribution of X given $X + Y = z$ is binomial with index parameter z and probability $p = a/(a+b)$.

Proof:

    To establish that the conditional distribution is of the form stated in the theorem, recall that the joint pgf of X and Z = X + Y is

$$\Pi_{x,z}(t_1, t_2) = \Pi(t_1\, t_2, t_2) = H(at_1\, t_2 + bt_2). \qquad (1.4.15)$$

From Theorem 1.3.1, the conditional pgf of X given Z = z is

$$\Pi_x(t|z) = \frac{\left.\dfrac{\partial^z}{\partial t_2^z}\, \Pi_{x,z}(t_1, t_2)\right|_{t_1=t,t_2=0}}{\left.\dfrac{\partial^z}{\partial t_2^z}\, \Pi_{x,z}(t_1, t_2)\right|_{t_1=1,t_2=0}}. \qquad (1.4.16)$$

However, from (1.4.15)

$$\frac{\partial^z}{\partial t_2^z} H[(at_1+b)t_2] = (at_1+b)^z\, H^{(z)}[(at_1+b)t_2].$$

Hence

$$\Pi_x(t|z) = \frac{(at+b)^z}{(a+b)^z}\,,$$

which is the pgf of B[z, a/(a+b)], proving the result.

    To prove the converse proposition, assume that

$$\Pi_x(t|z) = (\theta_2+\theta_1 t)^z \text{ with } \theta_1+\theta_2 = 1.$$

Also for any pair of random variables

$$\Pi(t_1, t_2) = E[t_2^Y \Pi_x(t_1|y)]. \qquad (1.4.17)$$

In the present instance this yields

$$\Pi_{x,z}(t_1, t_2) = E[t_2^Z \Pi_x(t_1|z)] = E[t_2^z(\theta_2+\theta_1 t_1)^z]\,, \qquad \theta_1+\theta_2 = 1$$

Hence the joint pgf of X and Z is

$$\Pi_{x,z}(t_1, t_2) = E[(t_2\theta_2+\theta_1 t_1 t_2)^z] = \Pi_z(t_2\theta_2+\theta_1 t_1 t_2),$$

which is the pgf of Z. This can be represented as $H(t_2\theta_2+\theta_1 t_1 t_2)$ with $\theta_1+\theta_2 = 1$. But the joint pgf of X and Z is $\Pi(t_1 t_2, t_2)$; therefore,

$$\Pi(t_1 t_2, t_2) = H(t_2\theta_2+\theta_1 t_1 t_2)$$

with $\theta_1+\theta_2 = 1$. This shows that the pgf Π is of the homogeneous type. □

1.4.4 Non-homogeneous probability generating functions

The question that arises at this point is "what can one say about the non-homogeneous pgf's?" Unfortunately no general answer seems to be available for the question if the conditional distribution being considered is that of X (or, of Y) given the sum Z = z. Subrahmaniam (1967) has provided a partial response to this problem by studying the role of the binomial distribution in some of the special bivariate pgf's that arise in practice that are not of the homogeneous type. He has shown that the conditional distribution of Y given X = x is, in all the cases examined by him, a convolution. One component of the convolution is a binomial and the other component is specific to the bivariate pgf under consideration. The following theorem is of importance in this connection.

Theorem 1.4.2

Let X and Y have a bivariate distribution with the pgf $\Pi(t_1, t_2)$. Also, let the conditional distribution of Y for given X be the convolution of the random variable χ_1 with the random variable χ_2. Let the conditional pgf of χ_i be represented by $\Pi_i(t|x)$, i = 1, 2. Then

$$\Pi(t_1, t_2) = \sum_x \Pi_1(t_2|x)\Pi_2(t_2|x))t_1^x f(x) \qquad (1.4.18)$$

Proof:

Consider the representation for the joint pgf

$$\Pi(t_1, t_2) = E[t_1^X \Pi_y(t_2|x)]. \qquad (1.4.19)$$

Under the assumption regarding the conditional distribution of Y given X, the conditional pgf appearing in the summation is

$$\Pi_y(t_2|x) = \Pi_1(t_2|x)\Pi_2(t_2|x).$$

The result in (1.4.18) follows upon substituting this in (1.4.19). □

The following theorem illustrates the role that the binomial distribution plays in the bivariate discrete distributions. It appears that the role is more general than is apparent from these results.

Theorem 1.4.3

Let X and Y be jointly distributed with the pgf $\Pi(t_1, t_2)$. Let the marginal pgf of X be $\Pi_x(t)$ and the conditional distribution of Y given X be the convolution of χ_i, i= 1, 2 with the respective pgf's $\Pi_i(t|x)$, i= 1, 2.

(i) If

$$\Pi_x(t) = \exp[\lambda(t-1)], \qquad \Pi_2(t|x) = \exp[\theta(t-1)], \tag{1.4.20}$$

then the joint pgf

$$\Pi(t_1, t_2) = \exp[\lambda_1(t_1-1)+\lambda_2(t_2-1)+\lambda_3(t_1t_2-1)] \tag{1.4.21}$$

if and only if χ_1 has the binomial distribution.

Proof:

From the preceding theorem we see that

$$\Pi(t_1, t_2) = \sum_x \Pi_1(t_2|x)\Pi_2(t_2|x))t_1^x f(x).$$

Taking $\Pi_1(t_2|x) = (q+pt_2)^x$, we have

$$\Pi(t_1, t_2) = \sum_{x=0}^{\infty} \exp[\theta(t_2-1)](q+pt_2)^x \frac{t_1^x}{x!} \lambda^x \exp(-\lambda)$$

which yields (1.4.21) with an appropriate choice of the parameters $\lambda_1 = \lambda q$, $\lambda_2 = \theta$ and $\lambda_3 = \lambda p$.

Conversely, if the joint pgf of X and Y is of the form in (1.4.21) and

$$\Pi_x(t) = \exp[-(\lambda_1+\lambda_3)(t-1)] , \qquad \Pi_2(t|x) = \exp[-\lambda_2(t-1)],$$

then the pgf $\Pi_1(t|x)$ satisfies the condition (with $t = t_2$)

$$\exp[\lambda_1(t_1-1)+\lambda_2(t_2-1)+\lambda_3(t_1t_2-1)]$$
$$=\sum_{x=0}^{\infty} \exp[\lambda_2(t_2-1)]\Pi_1(t_2|x)\frac{t_1^x}{x!}(\lambda_1+\lambda_3)^x \exp[-(\lambda_1+\lambda_3)] \cdot$$

Equating the coefficients of t_1^x on the two sides, we have

$$\Pi_1(t_2|x) = \left[\frac{\lambda_1+\lambda_3 t_2}{\lambda_1+\lambda_3}\right]^x,$$

which is a binomial pgf. □

(ii) If

$$\Pi_x(t) = [q+pt]^n, \qquad \Pi_2(t|x) = [q_2+p_2t]^{n-x}, \tag{1.4.22}$$

then the joint pgf

$$\Pi(t_1, t_2) = [a+bt_1+ct_2+dt_1t_2]^n, \tag{1.4.23}$$

with a+b+c+d = 1, if and only if χ_1 has the binomial distribution.

Proof:

Let us assume that $\Pi_1(t|x) = [q_1+p_1t]^x$. Then substituting appropriately in the equation

$$\Pi(t_1, t_2) = \sum_x \Pi_1(t_2|x)\Pi_2(t_2|x))t_1^x f(x),$$

the joint pgf of X and Y is seen to be

$$\Pi(t_1, t_2) = \sum_x [q_1+p_1t_2]^x [q_2+p_2t_2]^{n-x} \binom{n}{x} p^x q^{n-x} t_1^x$$

$$= \left\{p[q_1+p_1t_2]t_1+q[q_2+p_2t_2]\right\}^n,$$

which is of the form stated in the theorem with $a = qq_2$, $b = pq_1$, $c = qp_2$ and $d = pp_1$.

Conversely, if

$$\Pi(t_1, t_2) = [a+bt_1+ct_2+dt_1t_2]^n, \quad \Pi_x(t) = [q+pt]^n, \quad \Pi_2(t|x) = [q_2+p_2t]^{n-x}$$

where $p = b+d$, $p_2 = c/(a+c)$ and $a+b+c+d = 1$, then $\Pi_1(t|x)$ satisfies the equation (for $t = t_2$)

$$[a+bt_1+ct_2+dt_1t_2]^n = \sum_{x=0}^{n} [q_2+p_2t_2]^{n-x} \Pi_2(t_2|x) \binom{n}{x} p^x q^{n-x} t_1^x .$$

Equating the coefficients of t_1^x on the two sides and replacing the parameters in terms of a, b, c and d, we have

$$\Pi_2(t_2|x) = \left\{ \frac{dt_2+b}{d+b} \right\}^x,$$

as was to be shown. □

Subrahmaniam (1967) has considered the problem in greater detail and has given examples to establish the role of the binomial distribution in the bivariate discrete distributions.

1.5 Compounding and generalizing

In the univariate case a wide class of distributions has been constructed by a process known as *compounding* or *mixing*. Gurland (1957) defines this operation as:

Definition 1.5.1 Compounding

Let the random variable X_1 have the distribution function $F_1(x_1|\,\theta)$ for a given value of the parameter θ. Let θ be regarded as a random variable X_2 with the distribution function $F_2(x_2)$. The random variable with the distribution function

$$G(x_1) = \int_{-\infty}^{\infty} F_1(x_1|\, cx_2)\, dF_2(x_2) \qquad (1.5.1)$$

(where c is a constant which is arbitrary or restricted in some prescribed sense) is said to be a compound X_1 variable with respect to the compounder X_2.

This process is denoted by writing $X_1 \wedge X_2$ and is referred to as a compound of F_1 with F_2.

An example of the compounding and the compound distribution is the negative binomial distribution that has already been studied in Chapter 5. As discussed by Gurland there is a closely related process which also generates a variety of distributions. This is called *generalization* :

Definition 1.5.2 Generalizing

Let the random variables X and Y have the respective probability generating functions $\Pi_X(t)$ and $\Pi_Y(t)$. The random variable having the pgf $\Pi_X[\Pi_Y(t)]$ is called the generalized X variable with respect to the generalizer Y.

This process will be denoted by writing $X \vee Y$. In practice the process of generalizing arises when we consider random sum of identical random variables. Thus let $Y = \sum_{i=1}^{N} X_i$, with the summands being independent identical random variables having the pgf $\Pi_2(t)$ while N is itself a random variable with the pgf $\Pi_1(t)$. Then the pgf of Y is seen to be

$$\Pi_Y(t) = \sum_{n=1}^{\infty} \Pi_Y(t \mid N=n) \, P\{N=n\} = \sum_{n=1}^{\infty} [\Pi_2(t)]^n \, P\{N=n\}$$

$$= \Pi_1[\Pi_2(t)]. \tag{1.5.2}$$

In order to relate these two processes, we need to recognize the equivalence of random variables. Two random variables are said to be equivalent when their distributions are of the same *form* but with different sets of parameters. But it may be possible to relate the parameters so that the values of the distribution functions are identical at all points.

Definition 1.5.3 Equivalence

Consider the random variables X_1 and X_2 with the distribution functions $F_1(x|\alpha)$ and $F_2(x|\beta)$. If for each α there exists some β and for each β there exists some α such that the values of the distribution functions are equal whatever be x, then we say that the random variables are *equivalent*. This is written as $X_1 \sim X_2$.

Gurland (1957) has established that under certain conditions there is a close relationship between compounding and generalizing in that they give rise to equivalent random variables. This correspondence is given by the following theorem:

Theorem 1.5.1

Let X_1 be a random variable with the pgf $[\Pi_1(t)]^{\theta}$, where θ is a parameter. Suppose that θ is regarded as a random variable X_2 with the pgf $\Pi_2(t)$. Then $X_1 \wedge X_2$ is equivalent to $X_2 \vee X_1$.

Proof:

The pgf of $X_1 \wedge X_2$ is given by $\sum [\Pi_1(t)]^{\theta} f(\theta) = \Pi_2[\Pi_1(t)]$; but this is the pgf of $X_2 \vee X_1$. From the uniqueness of pgf's it follows that the random variables $X_1 \wedge X_2$ and $X_2 \vee X_1$ are equivalent. □

In later chapters these results will be extended to the bivariate case.

1.6 Structure of bivariate distributions: Polynomial expansions

The problem of constructing bivariate distributions of given marginals and the 'strength' of dependence as measured by the correlation coefficient has been of interest. An aspect of the answer to this question has been provided, among others, by Lancaster (1958,1963). For recent developments in this and related problems reference may be made to Griffiths (1985, p. 530). Some results and definitions relevant to the topics considered in later sections are presented in this section.

1.6.1 Canonical variables

The canonical variables or functions of them are two sets of orthonormal functions defined on the marginal distributions in a recursive manner such that the correlation between corresponding members of the two sets is maximal. Unity is taken to be a member of zero order of each of the set of variables. These conditions can be represented formally as follows:

$$\left.\begin{array}{ll}\xi^{(i)} = \xi^{(i)}(x), & \eta^{(i)} = \eta^{(i)}(y) \\ \left.\begin{array}{l}\int \xi^{(i)}\, dG(x) = \int \eta^{(i)}\, dH(y) = 0 \\ \int \xi^{(i)2}\, dG(x) = \int \eta^{(i)2}\, dH(y) = 1\end{array}\right\} & i = 1, 2, \ldots \\ \int \xi^{(i)}\xi^{(j)}\, dG(x) = \int \eta^{(i)}\eta^{(j)} dH(y) = 0 \quad i \neq j & \end{array}\right\} \quad (1.6.1)$$

The condition of maximal correlation is that

$$\rho_i = \mathrm{Corr}(\xi^{(i)}, \eta^{(i)}) = \int \int \xi^{(i)}, \eta^{(i)}\, dF(x, y) \qquad (1.6.2)$$

should be maximal for each i, given the preceding canonical variables. The ρ_i are called canonical correlations. By convention they are taken to be positive.

Let F(x, y) be the joint distribution function of X and Y, while G(x) and H(y) are the marginal distribution functions. Let

$$\Phi^2 = \int \int \frac{[dF(x, y)]^2}{dG(x)\, dH(y)} - 1$$

$$= \int \int \Omega^2(x, y)\, dG(x)\, dH(y) - 1, \qquad (1.6.3)$$

where $\Omega(x, y) = dF(x, y) / dG(x)\, dH(y)$ is the Radon-Nikodyn derivative of F(x, y) with respect to G(x) and H(y). It is taken to be zero if the point (x, y) does not correspond to points of increase of both G(x) and H(y).

It may be noted that Φ^2 can be regarded as the limit of the sum

$$\sum \sum \frac{f_{ij}^2}{f_{i+}\, f_{+j}} - 1,$$

where f_{ij} is the weight of the bivariate distribution corresponding to the marginal sets A_i, B_j and f_{i+}, f_{+j} are, respectively, the weights of the marginal distributions over the same sets.

The following theorem due to Lancaster (1958) establishes the relationship between the joint and marginal distributions.

Theorem 1.6.1 [Lancaster (1958; 1969, p. 95)]

If a bivariate distribution function F(x, y) is Φ^2-bounded with marginal distributions G(x) and H(y), then the complete sets of orthonormal functions can be defined on the marginal distributions such that each member of a set of canonical variables appears as a member of the complete set of orthonormal functions. The element of the frequency can be expressed in terms of the marginal distributions as

$$dF(x, y) = \left\{1 + \sum_{i=1}^{\infty} \rho_i \, x^{(i)} \, y^{(j)} \right\} dG(x) \, dH(y) \tag{1.6.4}$$

almost everywhere and

$$\Phi^2 = \sum_{i=1}^{\infty} \rho_i^2 . \tag{1.6.5}$$

Proof:

It should be noted that in this theorem $x^{(i)}$ and $y^{(j)}$ are orthonormal functions. For a detailed discussion of their construction and the proof of the theorem, we refer to the source paper. □

The representation given in (1.6.4) for the bivariate distribution in terms of the canonical variables is called its canonical representation. Any bivariate distribution admitting such a representation can be shown to have simple linear regressions

$$x^{(i)} = \rho_i \, y^{(i)}, \qquad y^{(i)} = \rho_i \, x^{(i)}, \tag{1.6.6}$$

while the regression of $x^{(i)}$ on $y^{(j)}$ and $y^{(i)}$ on $x^{(j)}$ are both zero if $i \neq j$. Lancaster (1969, p. 96) shows that the converse of theorem 1.6.1 is also true.

1.6.2 Polynomial representation

Eagleson (1964) has examined a class of bivariate distributions whose canonical variables are the orthogonal polynomials of the marginal distributions. This class consists, among others, of the bivariate Poisson, bivariate binomial, bivariate negative binomial and bivariate hypergeometric.

The special classes of distributions examined by Eagleson are generated from independent additive random variables; that is, they are closed under the addition of independent, identically distributed random variables. Thus

$$X = W_1 + W_2, \qquad Y = W_2 + W_3$$

with W_1, W_2, W_3 being independent additive random variables. The marginal distributions of X and Y belong to the same family of random variables as W_i for i = 1, 2, 3. They are correlated with

$$\rho = \frac{\mathrm{Var}(W_2)}{\{\mathrm{Var}(W_1 + W_2)\ \mathrm{Var}(W_2 + W_3)\}^{\frac{1}{2}}}.$$

This implies that $\rho = 0$ is a necessary and sufficient condition for X and Y to be independent. Also, $\rho = 1$ if and only if X and Y are linearly dependent.

Theorem 1.6.2 [Eagleson (1964)]

If, for a particular distribution, we have:

(i) the orthogonal polynomials are generated by a function of the form f(t) exp [x u(t)].

(ii) the distribution is additive.

(iii) the bivariate distribution is generated by using the addition property given above,

then the matrix of correlations of the pairs of orthonormal polynomials on the marginals is diagonal. Further the rth correlation ρ_r depends only on the normalizing factor of the rth orthogonal polynomial.

Proof:

For a proof of the theorem we refer to the source paper. □

In this case, if the distribution is Φ^2-bounded and the orthonormal polynomials generated by a function of the form f(t) exp [x u(t)] are represented by $\rho_r^*(x)$, then its probability density function (or, in the discrete case, the probability function) is almost everywhere equal to

$$dG(x)\, dH(y) \{1 + \sum_{r=1}^{\infty} \rho_r \; \rho_r^*(x) \; \rho_r^*(y)\} \tag{1.6.7}$$

and

$$\Phi^2 = \sum_{i=1}^{\infty} \rho_i^2 .$$

Also the regression of $\rho_i^*(x)$ on $\rho_i^*(y)$ is linear for all i.

1.7 Computer simulation

Computer generation of *univariate* discrete random variables is widely available using subroutines found in IMSL and standard statistical packages, such as SAS, MINITAB and S. The simulation of random variables, both discrete and continuous, uses uniform random numbers as building blocks; hence, the adequacy of the simulation depends, to a large extent, on the 'randomness' of the underlying uniform random variables.

For generating univariate discrete random variables, several procedures are commonly used:

(i) inverse method or a table look up technique

(ii) methods using the stochastic nature of a particular distribution

(iii) combining random variables through convolutions or compounding.

Until recently the literature concerning the generation of observations from a bivariate discrete distribution has been quite sparse. A summary of general techniques available in this case can be found in a series of papers by Kemp and Loukas (1978, 1981, 1986).

As in the univariate case, the inverse method can be used for generating observations from a bivariate discrete distribution. In this case a r+1 by s+1 matrix of the probability function f(x, y) with $x = 0, 1, \dots, r$; $y = 0, 1, \dots, s$ has to be formed. Then these probabilities are cumulated in some prescribed order. A realization u from a uniform distribution on (0,1) is then compared with the cumulated probabilities in the same order until the point is reached where the cumulated probability equals or exceeds u. This defines a bivariate observation (x, y) from f(x, y). This general method can be used for any probability function for which it is possible to construct the matrix of probabilities f(x, y). Kemp and Loukas (1978) have discussed this technique and modifications of it. As pointed out by them, even with modifications, this procedure is very time intensive.

In this monograph we have exploited the role of conditional distributions in constructing bivariate distributions. Recall that if we know the conditional distribution $f_1(x|y)$ and the marginal distribution $f_2(y)$, then the joint distribution of X and Y can be found as product of the conditional and marginal distributions. Computer simulation based on this procedure was introduced by Kemp and Loukas (1978).

We have, also, considered techniques which are based on the stochastic nature of particular bivariate discrete distributions: trivariate reduction method, convolutions, compounding and mixtures. Often these procedures are used in conjunction with the conditional distribution technique.

1.7.1 Conditional distribution technique

We will assume that we can simulate observations from both the marginal distribution $f_2(y)$ and the conditional distribution $f_1(x|y)$. In this technique, a realization y is generated from the distribution $f_2(y)$. The parameters in the conditional distribution of X given Y = y are usually functions of y. For this specified value of the parameters, a realization of x from $f_1(x|y)$ is generated; hence, the resulting pair (x, y) is an observation from the joint distribution f(x, y). This procedure is repeated n times to give a random sample of size n. The sample of n observations can be summarized in a matrix $\{n_{xy}\}$ with the entries representing the frequency of the outcome (X = x, Y = y).

In the simplest cases, both the distributions, $f_2(y)$ and $f_1(x|y)$ will be of a recognizable univariate form, such as a binomial, Poisson or negative binomial. Standard algorithms can be used to generate observations from them. If, however, these distributions are more complicated, techniques specific to the particular distribution have to be introduced.

1.7.2 Convolutions

Frequently the conditional distribution $f_1(x|y)$ can be expressed as a convolution (see section 1.2) of two or more random variables. Suppose that the random variable X given Y = y can be expressed as the convolution

$$X = W_1 + W_2 + W_3$$

with pf (or pdf) $g_i(w_i)$ for i = 1, 2, 3. The parameters in $g_i(w_i)$ are usually functions of the realized value of Y. Since the W_i's are independent, a realization w_i can be obtained from each $g_i(w_i)$; hence, a realization of X is given by

$$x = w_1 + w_2 + w_3.$$

1.7.3 Mixtures

It may be possible to express the conditional distribution $f_1(x|y)$ as a finite mixture of probability functions:

$$f_1(x|y) = \sum_{i=1}^{y} \omega_i(y)\, f_i(x|y)$$

with $\sum_{i=1}^{y} \omega_i(y) = 1$. Here the parameters in the ith component $f_i(x|y)$ are a function of i and y.

To determine a realization of X, $\omega_i(y)$ is evaluated for i = 1, 2, ..., y and a table of S_i, the cumulative partial sums of $\omega_i(y)$, is formed. A value u is generated from a uniform distribution on the interval 0 to 1. The ith component is selected such that u lies in the interval (S_{i-1}, S_i). The realized value of X is then generated from $f_i(x|y)$.

1.7.4 Trivariate reduction

Suppose that X can be expressed as $W_1 + W_3$ and Y as $W_2 + W_3$ with the W_i's as independent and usually identically distributed random variables. Realizations can be generated independently from each of $g_i(w_i)$ and then a pair of observations from f(x, y) is obtained by

$$x = w_1 + w_3, \qquad y = w_2 + w_3.$$

2

STATISTICAL INFERENCE

This chapter deals with the general problem of statistical inference with particular reference to discrete distributions. For a detailed exposition on problems of estimation theory reference may be made to Rao (1973). The interest here is mainly to present results that are used extensively in later chapters. Reference will be made often to the general formulations presented in this chapter. In this respect the material presented here is a brief review. There are two main sections -- estimation and testing of hypotheses.

2.1 Estimation

Under regularity conditions, the method of maximum likelihood has been shown to be superior to the other methods. Asymptotically it yields estimators which are normally distributed, unbiased and with minimum variance. This technique does, however, suffer from some drawbacks, not the least of which are the computational difficulties encountered. For many discrete distributions, iterative procedures are needed for solving the equations.

Several techniques have been suggested in the literature for estimation of the parameters of discrete distributions. Most of them are not nearly as efficient as the method of maximum likelihood. While the performance of all these alternative procedures is not equally 'good', their importance rests to a great extent on the ease of solution of the resulting equations. Kemp and Kemp (1988) describe techniques useful for univariate discrete distributions, which provide 'rapid' estimates. Although these estimators may not possess the desirable properties, such as unbiasedness and small variance, they may be useful for a quick look at the data or for initial trial values for iterative procedures.

The type of estimator to be used depends on the particular distribution under study and the nature of its parameters. In some situations, several procedures may have to be tried in order to obtain a solution. Papageorgiou and Kemp (1988) point out that, for moderate sample sizes of less than 500 observations, estimation procedures may often fail; that is, the estimated value of the parameter may not lie within the parameter space. In practice data sets of far less than 500 observations are typical; hence, several methods may have to be used to estimate the parameters.

The techniques which seem to be most useful for bivariate distributions are maximum likelihood and some type of method of moments. We will discuss these and their properties in the following sections.

2.1.1 Method of moments

The classical method of moments consists of equating the sample moments to their populations equivalents, expressed in terms of the parameters. The number of moments required depends on the number of parameters. The choice of particular moments is arbitrary, but lower order moments are usually preferred since they are usually simpler functions of the parameters. In addition, the higher order moments have larger variances.

A variation on this method has been suggested by Papageorgiou and Kemp (1988a, b). They use both marginal and conditional means.

In a bivariate discrete distribution, the data can be summarized in an array $\{n_{ij}\}$, the observed frequency corresponding to the observation $\{X = i, Y = j\}$ for $i = 0, 1, 2, \ldots; j = 0, 1, 2, \ldots$. Here

$$\sum_i \sum_j n_{ij} = n,$$

the total sample size.

The probability function

$$f(r, s) = P\{X = r, Y = s)$$

is assumed to be a function of the parameters $\theta_1, \theta_2, \ldots, \theta_q$.

The sample moment (around zero) of the order (r, s) will be denoted by

$$m'_{r,s} = \frac{1}{n}\sum_i \sum_j i^r j^s n_{ij}. \qquad (2.1.1)$$

In particular

$$m'_{1,0} = \bar{x}, \quad m'_{0,1} = \bar{y}.$$

The central moments are correspondingly represented by

$$m_{r,s} = \frac{1}{n}\sum_i \sum_j (i - \bar{x})^r (j - \bar{y})^s n_{ij}. \qquad (2.1.2)$$

It is obvious that, in general, the population moments are functions of the parameters $\theta_1, \theta_2, ..., \theta_q$ or a subset of them.

To be specific, let $q = 3$ with c_1, c_2, c_3 denoting the appropriate sample moments being used to estimate $\theta_1, \theta_2, \theta_3$. Then

$$\left.\begin{aligned} c_1 &= u_1(\tilde{\theta}_1,\tilde{\theta}_2,\tilde{\theta}_3) \\ c_2 &= u_2(\tilde{\theta}_1,\tilde{\theta}_2,\tilde{\theta}_3) \\ c_3 &= u_3(\tilde{\theta}_1,\tilde{\theta}_2,\tilde{\theta}_3) \end{aligned}\right\}, \qquad (2.1.3)$$

where $(\tilde{\theta}_1, \tilde{\theta}_2, \tilde{\theta}_3)$ are the *moment estimators* of $(\theta_1, \theta_2, \theta_3)$.

It should be noted that, while in the univariate case, there may not be an ambiguity in the choice of the moments c_1, c_2, c_3, in the bivariate case this choice is not unique. Thus in the univariate case c_1, c_2, c_3 are usually taken to be the first three sample moments. In the bivariate situation, c_1 and c_2 can be chosen to be the first two marginal moments and c_3 is equated to the mixed moment.

The solution $(\tilde{\theta}_1,\tilde{\theta}_2,\tilde{\theta}_3)$ to (2.1.3) can be obtained directly or iteratively. One such technique is the Newton-Raphson method presented here for illustration.

The right hand side of (2.1.3) is expanded around a 'trial' solution

$\underline{\theta}^0$ in a Taylor's series. Then

$$\underline{c} = \underline{u}(\underline{\theta}^0) + U\,\underline{\delta},$$

with

$$U = \left(\frac{\partial u_i}{\partial \theta_1}, \frac{\partial u_i}{\partial \theta_2}, \frac{\partial u_i}{\partial \theta_3}\right)\Bigg|_{\underline{\theta} = \underline{\theta}^0} \qquad i = 1, 2, 3$$

and

$$\underline{\delta} = \underline{\tilde{\theta}} - \underline{\theta}^0.$$

Solving for $\underline{\delta}$, we have

$$\underline{\delta} = U^{-1}\,[\underline{c} - \underline{u}(\underline{\theta}^0)]. \tag{2.1.4}$$

This yields $\underline{\tilde{\theta}} = \underline{\theta}^0 + \underline{\delta}$. The procedure is terminated when $\underline{\delta}$ is 'small'.

Moments of the moment estimators

It is possible to determine the moments of the estimators $\underline{\tilde{\theta}}$ by expanding (2.1.3) in a Taylor's series around the expected values of the estimators. Let $E(\underline{\tilde{\theta}}) = \underline{\mu}$, then

$$\underline{c} = \underline{u}(\underline{\mu}) + V(\underline{\tilde{\theta}} - \underline{\mu}), \tag{2.1.5}$$

where

$$V = \left\{\frac{\partial u_i}{\partial \theta_j}\right\}\Bigg|_{\underline{\theta} = \underline{\mu}}.$$

Since $\underline{\mu}$ is the expected value of $\underline{\tilde{\theta}}$, it is readily seen that the variance matrix of $\underline{\tilde{\theta}}$ is

$$\Sigma_{MM} = T\,\Sigma_{\underline{c}}\,T' \tag{2.1.6}$$

where $T = V^{-1}$ and $\Sigma_{\underline{c}}$ is the variance matrix of the moments $\underline{c}$.

Standard errors of the sample moments

The method of moments estimation procedure in a bivariate discrete distribution may involve the sample factorial moments $m_{[r,s]}$ in addition to the moments $m'_{r,s}$ and $m_{r,s}$. In order to find $\Sigma_{\underline{c}}$ in (2.1.6), it is necessary to determine the variances of the sample moments. Here the asymptotic results on the first two moments of $m'_{r,s}$ are derived. From these results the covariance matrix of $m_{r,s}$ is determined. For the factorial moments only the lower order moments are examined.

Let n_{ij} be the observed frequency in the (i, j) cell corresponding to the event $\{X = i, Y = j\}$ with probability p_{ij} for $i = 0, 1, 2, \ldots; j = 0, 1, 2, \ldots$. Then the marginal totals and corresponding probabilities are given by

$$n_{i+} = \sum_j n_{ij}, \quad n_{+j} = \sum_i n_{ij}, \quad p_{i+} = \sum_j p_{ij}, \quad p_{+j} = \sum_i p_{ij}. \tag{2.1.7}$$

It is known the $\{n_{ij}\}$ have a multinomial distribution with the parameters n and $\{p_{ij}\}$. Thus

$$E(n_{ij}) = n\, p_{ij}, \quad E(n_{ij}^2) = n\, p_{ij}\, [1 + (n-1)\, p_{ij}], \quad E(n_{ij} n_{kh}) = n\,(n-1)\, p_{ij}\, p_{kh}$$

and

$$Var(n_{ij}) = n\, p_{ij}\,(1 - p_{ij}), \quad Cov(n_{ij}, n_{kh}) = -\, n\, p_{ij}\, p_{kh}$$

for $i \neq j$, $k \neq h$.

Recalling the definition of $m'_{r,s}$, we have

$$E(m'_{r,s}) = \mu'_{r,s}, \tag{2.1.8}$$

where

$$\mu'_{r,s} = \sum_i \sum_j i^r j^s p_{ij}.$$

Also

$$E[(m'_{r,s})^2] = E\left[\frac{1}{n^2}\{\sum_i \sum_j i^r j^s n_{ij}\}^2\right]$$

$$= E\left[\frac{1}{n^2}\{\sum_i \sum_j i^{2r} j^{2s} n_{ij}^2 + \sum_{i_1} \sum_{j_1} \sum_{i_2} \sum_{j_2} i_1^r\, i_2^r\, j_1^s\, j_2^s\, n_{i_1 j_1}\, n_{i_2 j_2}\}\right]$$

$$= \frac{1}{n^2}\{\sum_i \sum_j i^{2r} j^{2s} [n\, p_{ij} + n\,(n-1)\, p_{ij}^2]$$

$$+ \sum_{i_1} \sum_{j_1} \sum_{i_2} \sum_{j_2} i_1^r \, i_2^r \, j_1^s \, j_2^s [n(n-1) p_{i_1 j_1} \, p_{i_2 j_2}]\}$$

Hence,

$$E[(m'_{r,s})^2] = \frac{1}{n} \mu'_{2r,2s} + \frac{n-1}{n} \mu'^2_{r,s}. \tag{2.1.9}$$

Similarly,

$$E[m'_{r_1,s_1} m'_{r_2,s_2}] = E\{\frac{1}{n^2} [\sum_i \sum_j i^{r_1} j^{s_1} n_{ij}] [\sum_i \sum_j i^{r_2} j^{s_2} n_{ij}]\}$$

$$= E\{\frac{1}{n^2} \sum_i \sum_j i^{r_1+r_2} j^{s_1+s_2} n_{ij}^2$$

$$+ \frac{1}{n^2} \sum_{i_1} \sum_{j_1} \sum_{i_2} \sum_{j_2} i_1^{r_1} j_1^{s_1} i_2^{r_2} j_2^{s_2} n_{i_1 j_1} n_{i_2 j_2}\},$$

which yields

$$E[m'_{r_1,s_1} m'_{r_2,s_2}] = \frac{1}{n} \sum_i \sum_j i^{r_1+r_2} j^{s_1+s_2} p_{ij} +$$

$$\frac{n-1}{n} [\sum_i \sum_j i^{r_1+r_2} j^{s_1+s_2} p_{ij}^2 + \sum_{i_1} \sum_{j_1} \sum_{i_2} \sum_{j_2} i_1^{r_1} j_1^{s_1} i_2^{r_2} j_2^{s_2} p_{i_1 j_1} p_{i_2 j_2}]$$

$$= \frac{1}{n} \mu'_{r_1+r_2,s_1+s_2} + \frac{n-1}{n} \mu'_{r_1,s_1} \mu'_{r_2,s_2}. \tag{2.1.10}$$

Combining (2.1.9) and (2.1.10), we have the variance matrix of $(m'_{r_1,s_1}, m'_{r_2,s_2})$ is given by

$$\frac{1}{n} \begin{bmatrix} \mu'_{2r_1,2s_1} - \mu'^2_{r_1,s_1} & \mu'_{r_1+r_2,s_1+s_2} - \mu'_{r_1,s_1} \mu'_{r_2,s_2} \\ \cdots & \mu'_{2r,2s} - \mu'^2_{r_2,s_2} \end{bmatrix}. \tag{2.1.11}$$

For $(m'_{r_1,s_1}, m'_{r_2,s_2}, m'_{r_3,s_3})$ the matrix is a 3 x 3 extension of (2.1.11).

For the special case of the sample sums of squares and cross products in the above expressions, assuming $\mu'_{1,0} = \mu'_{0,1} = 0$, we can see that

$$\left.\begin{aligned} \mathrm{Var}(m'_{2,0}) &= \frac{1}{n}[\mu_{4,0} - \mu^2_{2,0}] \\ \mathrm{Var}(m'_{0,2}) &= \frac{1}{n}[\mu_{0,4} - \mu^2_{0,2}] \\ \mathrm{Var}(m'_{1,1}) &= \frac{1}{n}[\mu_{2,2} - \mu^2_{1,1}] \end{aligned}\right\} \quad (2.1.12)$$

$$\left.\begin{aligned} \mathrm{Cov}(m'_{2,0}, m'_{0,2}) &= \frac{1}{n}[\mu_{2,2} - \mu_{2,0}\,\mu_{0,2}] \\ \mathrm{Cov}(m'_{2,0}, m'_{1,1}) &= \frac{1}{n}[\mu_{3,1} - \mu_{2,0}\,\mu_{1,1}] \\ \mathrm{Cov}(m'_{0,2}, m'_{1,1}) &= \frac{1}{n}[\mu_{1,3} - \mu_{0,2}\,\mu_{1,1}] \end{aligned}\right\}. \quad (2.1.13)$$

Instead of the raw moments, often factorial moments are used to obtain the method of moments estimates. In this case

$$\left.\begin{aligned} \mathrm{Var}(m_{[2,0]}) &= \frac{1}{n}[(\mu_{4,0} - \mu^2_{2,0}) + \mu_{2,0} - 2\mu_{3,0}] \\ \mathrm{Var}(m_{[0,2]}) &= \frac{1}{n}[(\mu_{0,4} - \mu^2_{0,2}) + \mu_{0,2} - 2\mu_{0,3}] \;] \\ \mathrm{Var}(m_{[1,1]}) &= \mathrm{Var}(m'_{1,1}) = \frac{1}{n}[\mu_{2,2} - \mu^2_{1,1} \end{aligned}\right\} \quad (2.1.14)$$

$$\left.\begin{aligned} \mathrm{Cov}(m_{[2,0]}, m_{[0,2]}) &= \frac{1}{n}[(\mu_{2,2} - \mu_{2,0}\mu_{0,2}) + \mu_{1,1} - \mu_{2,1} - \mu_{1,2}] \\ \mathrm{Cov}(m_{[2,0]}, m_{[1,1]}) &= \frac{1}{n}[\mu_{3,1} - \mu_{2,0}\mu_{1,1} - \mu_{2,1}] \\ \mathrm{Cov}(m_{[0,2]}, m_{[1,1]}) &= \frac{1}{n}[\mu_{1,3} - \mu_{0,2}\mu_{1,1} - \mu_{1,2}] \end{aligned}\right\}. \quad (2.1.15)$$

In some instances the central moments are required for determining the moment estimators. In this case Stuart and Ord (1988, p. 250) give the following variances and covariances to the first order in n^{-1}:

$$\mathrm{Var}(m_{r,s}) = \frac{1}{n}[\mu_{2r,2s} - \mu_{r,s}^2 + r^2\mu_{2,0}\mu_{r-1,s}^2 + s^2\mu_{0,2}\mu_{r,s-1}^2$$

$$+ 2rs\mu_{1,1}\mu_{r-1,s}\mu_{r,s-1} - 2r\mu_{r+1,s}\mu_{r-1,s} - 2s\mu_{r,s-1}\mu_{r,s+1}] \qquad (2.1.16)$$

$$\mathrm{Cov}(m_{r,s}, m_{u,v}) = \frac{1}{n}[\mu_{r+u,s+v} - \mu_{r,s}\mu_{u,v} + ru\mu_{2,0}\mu_{r-1,s}\mu_{u-1,v}$$

$$+ sv\mu_{0,2}\mu_{r,s-1}\mu_{u,v-1} + rv\mu_{1,1}\mu_{r-1,s}\mu_{u,v-1}$$

$$+ su\mu_{1,1}\mu_{r,s-1}\mu_{u-1,v} - u\mu_{r+1,s}\mu_{u-1,v} - v\mu_{r,s+1}\mu_{u,v-1}$$

$$- r\mu_{r-1,s}\mu_{u+1,v} - s\mu_{r,s-1}\mu_{u,v+1}]. \qquad (2.1.17)$$

In particular, to the order n^{-1},

$$\left.\begin{aligned} \mathrm{Var}(m_{2,0}) &= \frac{1}{n}[\mu_{4,0} - \mu_{2,0}^2] \\ \mathrm{Var}(m_{0,2}) &= \frac{1}{n}[\mu_{0,4} - \mu_{0,2}^2] \\ \mathrm{Var}(m_{1,1}) &= \frac{1}{n}[\mu_{2,2} - \mu_{1,1}^2] \end{aligned}\right\} \qquad (2.1.18)$$

$$\left.\begin{aligned} \mathrm{Cov}(m_{2,0}, m_{0,2}) &= \frac{1}{n}[\mu_{2,2} - \mu_{2,0}\mu_{0,2}] \\ \mathrm{Cov}(m_{2,0}, m_{1,1}) &= \frac{1}{n}[\mu_{3,1} - \mu_{2,0}\mu_{1,1}] \\ \mathrm{Cov}(m_{0,2}, m_{1,1}) &= \frac{1}{n}[\mu_{1,3} - \mu_{0,2}\mu_{1,1}] \end{aligned}\right\}. \qquad (2.1.19)$$

Expressions for other moment formulas, including factorial moments, can be found using the preceding results.

2.1.2 Method of even-points

As has been seen, in the method of moments for bivariate distributions, estimates for two of the parameters can be obtained by equating the marginal moments with their observed analogs. If there are more than two parameters to be estimated, we then have the difficulty of deciding which of the moments to use. In the method of moments another type of problem can arise. As in the univariate case, the moment estimators may not meet the constraints of the parameter space. For example, in the univariate negative binomial distribution, if s^2 is not greater than the mean, we will have a negative estimate for a positive parameter. This type of problem is more pronounced in bivariate distributions, which have more parameters, depicting not only their marginal means and variances, but also a correlation structure.

The even-points method of estimation provides a modification for the method of moments. In the univariate case this technique was first introduced by Patel (1976) for estimating the parameters in the univariate Hermite distribution. As in the method of moments, the first estimation equation is obtained by equating $\bar{x}$ to μ, which is a function of the unknown parameters. The second equation is obtained by equating the sum of the relative frequencies of the even values of the random variable X in the sample to the sum of the corresponding even probabilities:

$$\frac{1}{n}\sum_x n_{2x} = \sum_x f(2x; \underline{\theta}),$$

where $f(x; \underline{\theta})$ is the pf of the random variable X. If the pgf of X is denoted by $\Pi_X(t; \underline{\theta})$, then Kemp and Kemp (1988) show that

$$\begin{aligned}\Pi_X(0; \underline{\theta}) + \Pi_X(-1; \underline{\theta}) &= \sum_x f(x; \underline{\theta})\,[1 + (-1)^x] \\ &= 2\sum_x f(2x; \underline{\theta});\end{aligned}$$

that is, the even-points method may be considered as a special case of estimation using the empirical pgf.

In the bivariate case $\bar{x} = \mu_x$ and $\bar{y} = \mu_y$ form the first two of the estimating equations. For the third equation the joint pgf $\Pi_{X,Y}(t_1, t_2; \underline{\theta})$ evaluated at $(t_1, t_2) = (1, 1)$ and $(-1, -1)$ gives

$$\begin{aligned}&\Pi_{X,Y}(1, 1; \underline{\theta}) + \Pi_{X,Y}(-1, -1; \underline{\theta}) = \\ &\qquad 2\Big\{\sum_x\sum_y f(2x, 2y; \underline{\theta}) + \sum_x\sum_y f(2x+1, 2y+1; \underline{\theta})\Big\}.\end{aligned}$$

The probabilities on the right hand side are then replaced by the corresponding relative frequencies.

If further parameters are to be estimated, $\Pi_{X,Y}(-1, 1; \underline{\theta})$ and $\Pi_{X,Y}(1, -1; \underline{\theta})$ will also be needed.

Papageorgiou and Loukas (1987, 1988) have used a variation of this technique in which even point estimators are based on the conditional distribution of X given Y = 0.

2.1.3 Zero-zero cell frequency technique

The method of moments can also be modified by using the cell relative frequencies and equating them to the population probabilities. In a univariate distribution with two unknown parameters, the sample mean would again be used as the first equation. Then the second equation would be obtained by equating the relative frequency in the zero cell with the corresponding probability function.

A natural extension to this in the bivariate case is the use of the zero-zero cell frequency. Usually the marginal moments $m'_{1,0}$, $m'_{0,1}$ will be used in combination with the zero-zero cell frequency. Denoting the zero-zero cell frequency by n_{00} and the corresponding probability by f(0, 0), we have

$$E(n_{00}) = n\, f(0, 0).$$

In order to evaluate the performance of these estimators the following sample moments are of interest:

$$Var(n_{00}) = n\, f(0, 0)\, [1 - f(0, 0)]$$

$$Cov(n_{00}, m'_{r,s}) = Cov[n_{00}, \frac{1}{n} \Sigma\, \Sigma\, i^r j^s n_{ij}]$$

$$= \frac{1}{n} \Sigma\, \Sigma\, i^r j^s\, Cov(n_{00}, n_{ij})$$

$$= - f(0, 0)\, \Sigma\, \Sigma\, i^r j^s f(i, j)$$

or

$$Cov(n_{00}, m'_{r,s}) = - f(0, 0)\, \mu'_{r,s}. \qquad (2.1.20)$$

The result in (2.1.20) can be used to derive the covariances in special cases of interest:

$$\left.\begin{aligned} \mathrm{Cov}(n_{00}, m'_{1,0}) &= -f(0,0)\,\mu'_{1,0} \\ \mathrm{Cov}(n_{00}, m'_{2,0}) &= -f(0,0)\,\mu'_{2,0} \\ \mathrm{Cov}(n_{00}, m'_{1,1}) &= -f(0,0)\,\mu'_{1,1} \end{aligned}\right\}. \tag{2.1.21}$$

In the case of factorial moments, we have

$$\begin{aligned} \mathrm{Cov}(n_{00}, m_{[2,0]}) &= \mathrm{Cov}(n_{00}, m'_{2,0}) - \mathrm{Cov}(n_{00}, m'_{1,0}) \\ &= -f(0,0)\,\mu_{[2,0]}. \end{aligned} \tag{2.1.22}$$

2.1.4 Method of maximum likelihood

The likelihood function is denoted by $L(\underline{\theta}; \underline{x})$ with $\underline{\theta}$ representing the vector of unknown parameters and $\underline{x}$ standing for the observations. The Principle of Maximum Likelihood [Rao (1973, p. 353)] consists of choosing for $\underline{\theta}$ the estimates obtained by maximizing L with respect to these parameters. Thus

$$\frac{\partial \log L}{\partial \theta_i} = 0 \qquad i = 1, 2, \ldots, q \tag{2.1.23}$$

can be solved for $\theta_1, \theta_2, \ldots, \theta_q$ in terms of the observations $\underline{x}$ .The solutions are denoted by $\hat{\theta}_1, \hat{\theta}_2, \ldots, \hat{\theta}_q$. Unfortunately, it is not always possible to solve (2.1.23) for $\hat{\theta}_1, \hat{\theta}_2, \ldots, \hat{\theta}_q$ directly. There are two possible iterative methods for solving these equations:

A. Newton-Raphson Method: The equations in (2.1.23) can be expanded in a Taylor's series around a trial solution $\underline{\theta}^0$ as

$$\begin{aligned} \frac{\partial \log L}{\partial \theta_i} &= \left.\frac{\partial \log L}{\partial \theta_i}\right|_{\underline{\theta} = \underline{\theta}^0} \\ &+ \left\{(\theta_1 - \theta_1^0)\frac{\partial}{\partial \theta_1} + (\theta_2 - \theta_2^0)\frac{\partial}{\partial \theta_2} + \ldots + (\theta_q - \theta_q^0)\frac{\partial}{\partial \theta_q}\right\} \left.\frac{\partial \log L}{\partial \theta_i}\right|_{\underline{\theta} = \underline{\theta}^0}. \end{aligned} \tag{2.1.24}$$

If $\underline{\theta}$ is the solution to (2.1.24), then the left hand side of (2.1.23) will be zero; hence, at each of the iterations the new value $\underline{\theta}_{j+1}$ can be found from $\underline{\theta}_j$, by adding $\underline{\delta}$ to it. The vector $\underline{\delta}$ is given by

$$\underline{\delta} = - A^{-1} \underline{D} \tag{2.1.25}$$

with A^{-1} and $\underline{D}$ both evaluated at $\underline{\theta} = \underline{\theta}_j$ and where

$$A = \begin{bmatrix} \frac{\partial^2 \log L}{\partial \theta_1^2} & \frac{\partial^2 \log L}{\partial \theta_1 \partial \theta_2} & \cdots & \frac{\partial^2 \log L}{\partial \theta_1 \partial \theta_q} \\ \cdots & \frac{\partial^2 \log L}{\partial \theta_2^2} & \cdots & \frac{\partial^2 \log L}{\partial \theta_2 \partial \theta_q} \\ \cdots & \cdots & \cdots & \frac{\partial^2 \log L}{\partial \theta_q^2} \end{bmatrix}, \quad \underline{D} = \begin{bmatrix} \frac{\partial \log L}{\partial \theta_1} \\ \frac{\partial \log L}{\partial \theta_2} \\ \cdots \\ \frac{\partial \log L}{\partial \theta_q} \end{bmatrix}. \tag{2.1.26}$$

This process is continued until $\underline{\delta}$ becomes small. The convergence properties and other problems in connection with this procedure are discussed in Rao (*loc. cit.*).

B. Method of Scoring: Rao (1973, p. 366) defines the quantity $\frac{\partial \log L}{\partial \theta_i}$ as the *efficient score* for θ_i. Thus the maximum likelihood estimator is the value of θ_i for which the efficient score becomes zero. As above, we can expand around a trial solution $\underline{\theta}^0$ to obtain the value of $\underline{\delta}$ given in (2.1.25). Rao suggests, however, that in order to stabilize the value of $\underline{\delta}$, it is better to replace the matrix A by its expected value evaluated at $\underline{\theta}^0$. Clearly the expected value of $-A$ is the information matrix at $\underline{\theta} = \underline{\theta}^0$

$$\underline{\Gamma}(\underline{\theta}^0) = \left\{ E\left[-\frac{\partial^2 \log L}{\partial\theta_i \partial\theta_j} \right] \right\} \Bigg|_{\underline{\theta} = \underline{\theta}^0}. \tag{2.1.27}$$

Here

$$\underline{\delta} = \Gamma^{-1}(\underline{\theta}^0)\, \underline{D}.$$

In this technique, the new value $\underline{\theta}_{j+1}$ is again found by adding $\underline{\delta}$ to the value $\underline{\theta}_j$ obtained at the jth iteration.

For large samples the values of $\underline{\theta}$ obtained by the two techniques will be very close. The scoring method, however, does tend to converge more rapidly.

Properties of the maximum likelihood estimators

The exact distribution of the maximum likelihood estimators is not available in general. If, however, the sample size n is large, it can be shown that the estimator $\hat{\underline{\theta}}$ is consistent, asymptotically normal and efficient for $\underline{\theta}$. The asymptotic variance matrix of $\hat{\underline{\theta}}$ is given by $\Gamma^{-1}(\underline{\theta})$.

Grouped data

In the special case when the data are classified in a two way frequency table, as is usually the situation in a bivariate discrete distribution, the above discussion can be further simplified. Let n_{rs} be the observed frequency in the (r, s) cell for $r = 0, 1, 2, \ldots, k_1$; $s = 0, 1, 2, \ldots, k_2$ with

$$\sum_{r}^{k_1} \sum_{s}^{k_2} n_{rs} = n.$$

Then the likelihood can be written as

$$L \propto \prod_{r,s} f(r, s)^{n_{rs}}$$

or

$$\log L = C + \sum_{r}^{k_1} \sum_{s}^{k_2} n_{rs} \log f(r, s).$$

To obtain the maximum likelihood estimators, we need the first two derivatives of the log of the likelihood with respect to θ_i:

$$\frac{\partial \log L}{\partial \theta_i} = \sum_r^{k_1} \sum_s^{k_2} n_{rs} \frac{1}{f(r, s)} \frac{\partial f(r, s)}{\partial \theta_i}$$

$$\frac{\partial^2 \log L}{\partial \theta_i \, \partial \theta_j} = \sum_r^{k_1} \sum_s^{k_2} n_{rs} \left[\frac{1}{f(r, s)} \frac{\partial^2 f(r, s)}{\partial \theta_i \, \partial \theta_j} - \frac{1}{f^2(r, s)} \frac{\partial f(r, s)}{\partial \theta_i} \frac{\partial f(r, s)}{\partial \theta_j}\right].$$

Since the n_{rs} are random variables having a multinomial distribution with probabilities prescribed by the probability function f(r, s), $E(n_{rs}) = n\, f(r, s)$; hence,

$$E\left\{\frac{\partial^2 \log L}{\partial \theta_i \, \partial \theta_j}\right\} = n \sum_r^{k_1} \sum_s^{k_2} \left[\frac{\partial^2 f(r, s)}{\partial \theta_i \, \partial \theta_j} - \frac{1}{f(r, s)} \frac{\partial f(r, s)}{\partial \theta_i} \frac{\partial f(r, s)}{\partial \theta_j}\right]$$

$$= - n \sum_r^{k_1} \sum_s^{k_2} f(r, s) \frac{\partial \log f(r, s)}{\partial \theta_i} \frac{\partial \log f(r, s)}{\partial \theta_j}$$

or

$$\Gamma(\underline{\theta}^0) = \left\{ n \sum_r^{k_1} \sum_s^{k_2} f(r, s) \frac{\partial f(r, s)}{\partial \theta_i} \frac{\partial f(r, s)}{\partial \theta_j} \right\} \Bigg|_{\underline{\theta} = \underline{\theta}^0} . \tag{2.1.28}$$

Therefore, using the method of scoring,

$$\underline{\delta} = \Gamma^{-1} (\underline{\theta}^0) \, \underline{D}^*,$$

where

$$\underline{D}^* = \begin{bmatrix} \sum_r^{k_1} \sum_s^{k_2} \frac{1}{f(r, s)} \frac{\partial f(r, s)}{\partial \theta_1} \\ \sum_r^{k_1} \sum_s^{k_2} \frac{1}{f(r, s)} \frac{\partial f(r, s)}{\partial \theta_2} \\ \cdots \\ \sum_r^{k_1} \sum_s^{k_2} \frac{1}{f(r, s)} \frac{\partial f(r,s)}{\partial \theta_q} \end{bmatrix},$$

with $\underline{\theta} = \underline{\theta}^0$. The asymptotic variance matrix of the maximum likelihood estimator $\hat{\underline{\theta}}$ is $\Gamma^{-1}(\underline{\theta})$ and is estimated by evaluating Γ^{-1} at $\hat{\underline{\theta}}$.

2.2 Tests of hypotheses

In testing of hypotheses we are usually interested in several types of problems: (i) goodness-of-fit tests for model building and (ii) tests for composite hypotheses in terms of the parameters of the distribution. Each of these will be discussed in a general setting that is applicable to bivariate discrete distributions. Specific applications will be presented in later chapters.

2.2.1 Tests for goodness-of-fit

The data will be assumed to be presented in the form of a two way array with the observed frequency n_{rs} in the (r, s) cell and the corresponding probability f(r, s). What is of interest to us in a goodness-of-fit procedure is testing of the hypothesis

$$H_0: \; f(r, s) = f^0(r, s) \text{ (prescribed)}$$

with $r = 0, 1, 2, \ldots, k_1$; $s = 0, 1, 2, \ldots, k_2$ and $\sum_r^{k_1} \sum_s^{k_2} f(r, s) = 1$. In general the prescribed functional form may involve unknown parameters, say $\theta_1, \theta_2, \ldots, \theta_q$, which have to be estimated from the data. Using estimates of $\theta_1, \theta_2, \ldots, \theta_q$, we find the estimated probabilities $\hat{f}^0(r, s)$. Then the hypothesis is essentially one of testing if the data $\{n_{rs}\}$ conform to the prescribed distribution estimated by $\hat{f}^0(r, s)$. Several approaches are available for performing this type of test:

Pearson's χ^2 test

The statistic used for this purpose is

$$\underline{X}^2 = \sum_r^{k_1} \sum_s^{k_2} \frac{[n_{rs} - n\hat{f}^0(r, s)]^2}{n\hat{f}^0(r, s)}, \tag{2.2.1}$$

which has the χ^2 distribution for large n. The degrees of freedom are given by $(k_1 \times k_2) - 1 -$ (number of estimated parameters).

In order for this statistic to be asymptotically χ^2, the data have to be grouped to ensure that the expected frequency in any cell does not fall below a minimum number, which is usually taken to be 5. If the data set is large, some cells are often allowed to have expected frequency as small as 1. In the univariate case, the cells in the tail with smaller probabilities are usually combined. For bivariate data there is no such natural way of combining the cells. Loukas and Kemp (1986c) have studied three systematic grouping procedures: (i) row based, (ii) column based and (iii) reordering the classes so that they may be treated as a J-shaped univariate distributions. They assess these three procedures by examining the asymptotic power of the test. They advocate that the third procedure, which is less arbitrary than the first two should be used.

Likelihood ratio test:

The generalized likelihood ratio statistic is given by

$$G^2 = 2 \sum_{r}^{k_1} \sum_{s}^{k_2} n_{rs} \log \left\{ \frac{n_{rs}}{n\hat{f}^0(r, s)} \right\}, \tag{2.2.2}$$

which, like the statistic in (2.2.1) can be shown to be asymptotically distributed as a χ^2 with degrees of freedom are given by $k_1 \times k_2 - 1 -$ (number of estimated parameters). As in the previous test, the data have to be grouped to maintain a minimum expected frequency.

Probability generating function technique

An alternative technique for testing the goodness-of-fit has been suggested by Kocherlakota and Kocherlakota (1986). This method, based on the probability generating function is computationally simpler since it does not involve the pooling of the cells.

Defining the empirical pgf by

$$P(t_1, t_2) = \sum_{r}^{k_1} \sum_{s}^{k_2} n_{rs}\, t_1^r\, t_2^s, \tag{2.2.3}$$

the suggested test statistic is

$$Z = \frac{P(t_1, t_2) - \Pi(t_1, t_2; \hat{\underline{\theta}})}{\sigma}, \tag{2.2.4}$$

which is approximately N(0,1) under the null hypothesis H_0.

Here

$$\sigma^2 = \frac{1}{n}\,[\Pi(t_1^2, t_2^2; \underline{\theta}) - \Pi^2(t_1, t_2; \underline{\theta})] - \sum_i^q \sum_j^q \sigma_{ij} \frac{\partial \Pi(t_1, t_2; \underline{\theta})}{\partial \theta_i} \frac{\partial \Pi(t_1, t_2; \underline{\theta})}{\partial \theta_j}, \tag{2.2.5}$$

where $\{\sigma_{ij}\}$ is the inverse of the information matrix. To perform the test, σ^2 in (2.2.5) is estimated by replacing the parameters by their maximum likelihood estimates. Thus the level α rejection region is

$$Z > z_{\frac{\alpha}{2}} \tag{2.2.6}$$

with z_γ being the upper 100 γ percent point in the standard normal distribution.

2.2.2 Tests for composite hypotheses

The problem of constructing tests for composite hypotheses has been discussed in Rao (1973, pp. 415-420).

Let $X_1, X_2, \ldots, X_n$ be n independent observations from a distribution with pf (or pdf) $f(x; \underline{\theta})$. Without loss of generality we can assume that the distribution is one dimensional. The parameter vector $\underline{\theta}$ is q-dimensional. Then the logarithm of the corresponding likelihood function is

$$\log L(\underline{\theta}; \underline{x}) = \sum_{i=1}^{n} \log f(x_i; \underline{\theta}). \tag{2.2.7}$$

Let us write

$$u_i(\underline{\theta}) = \frac{1}{\sqrt{n}} \frac{\partial \log(\underline{\theta})}{\partial \theta_i}, \quad i = 1, 2, \ldots q,$$

the ith efficient score [Rao (1973, p. 415)].

Here the test of interest is

$$H_0:\ \theta_1 = \theta_1^0,\ \theta_2 = \theta_2^0,\ \ldots,\ \theta_k = \theta_k^0;\ \theta_{k+1},\ \ldots,\ \theta_q \text{ being unspecified.}$$

or

$$H_0:\ \underline{\theta}_1 = \underline{\theta}_1^0;\ \underline{\theta}_2 \ \text{unspecified}$$

against the alternative

$$H_a:\ \underline{\theta}\ \text{unspecified.}$$

A brief discussion of the various procedures available to test this hypothesis is presented below.

Generalized likelihood ratio test

Under the null hypothesis, let $\underline{\theta}$ satisfy the set of k restrictions $R_i(\underline{\theta}) = 0$, i = 1, 2, ..., k. Then the generalized likelihood ratio procedure is based on the statistic

$$\lambda = \frac{\sup\limits_{R_i(\underline{\theta})=0} L(\underline{\theta};\ \underline{x})}{\sup\limits_{\underline{\theta}} L(\underline{\theta};\ \underline{x})}.$$

The supremum in the numerator is evaluated under the set of restrictions imposed by H_0, while that in the denominator is unrestricted.

The test statistic here is given by

$$-2 \log \lambda = 2\,[\log L(\hat{\underline{\theta}}; \underline{x}) - \log L(\hat{\underline{\theta}}^*; \underline{x})], \tag{2.2.8}$$

where $\hat{\underline{\theta}}$ is the maximum likelihood estimate of $\underline{\theta}$ with no restrictions, while $\hat{\underline{\theta}}^*$ is the maximum likelihood estimate of the unspecified part of $\underline{\theta}$ under the null hypothesis H_0.

The statistic $-2 \log \lambda$ given in (2.2.8) is asymptotically distributed as a χ^2 with k degrees of freedom.

Efficient score test

Rao (1947) has suggested the statistic

$$S_c = \underline{\phi}^{*'} (\Gamma^*)^{-1} \underline{\phi}^*, \tag{2.2.9}$$

where

$$\underline{\phi}^{*'} = \{u_1(\underline{\theta}), \ldots, u_q(\underline{\theta})\} \Big|_{\underline{\theta}_1 = \underline{\theta}_1^0;\ \underline{\theta}_2 = \hat{\underline{\theta}}^*} \tag{2.2.10}$$

and Γ^* is the information matrix for the complete vector $\underline{\theta}$ evaluated under $\underline{\theta}_1^0$ and $\hat{\underline{\theta}}^*$. S_c is asymptotically distributed as a χ^2 on k degrees of freedom.

Neyman's C(α) test

Neyman (1959) suggested, independently, a test of the hypothesis H_0, which becomes identical to Rao's efficient score test under certain conditions.

Let

$$\underline{\phi}^{0'} = \{u_1(\underline{\theta}), \ldots, u_q(\underline{\theta})\} \Big|_{\underline{\theta}_1 = \underline{\theta}_1^0}. \tag{2.2.11}$$

It can be shown that the vector $\underline{\phi}^0$ has asymptotically, under H_0, the multivariate normal distribution with mean zero and variance matrix

$$\Sigma^0 = \left\{ E\left[\frac{\partial \log f(x;\underline{\theta})}{\partial \theta_i} \frac{\partial \log f(x;\underline{\theta})}{\partial \theta_j} \right] \right\} \Big|_{\underline{\theta}_1 = \underline{\theta}_1^0}. \tag{2.2.12}$$

Neyman's locally asymptotically most powerful test consists of constructing the residual vector

$$\underline{\eta} = \underline{\phi}_1^0 - B\underline{\phi}_2^0$$

with

$$B = \Sigma_{12}^0 \Sigma_{22}^{0\ -1}.$$

It can be seen that, when H_0 is true, $\underline{\eta}$ has the k-dimensional multivariate normal distribution with mean zero and variance matrix

$$\Sigma_{11.2}^0 = \Sigma_{11}^0 - \Sigma_{12}^0 \Sigma_{22}^{0\ -1} \Sigma_{21}^0 .$$

Hence, the quadratic form

$$N_c = \underline{\eta}' \Sigma_{11.2}^{0\;-1} \underline{\eta} \tag{2.2.13}$$

is asymptotically distributed as a χ^2 with k degrees of freedom. Since in (2.2.13) the parameter vector $\underline{\theta}_2$ is unspecified, Neyman suggests using any $\sqrt{n}$ consistent estimator.

An important feature of this test is that it does not require that maximum likelihood estimators be used for the nuisance parameters. If, however, maximum likelihood estimators are used, then the test in (2.2.13) is equivalent to Rao's efficient score test given in (2.2.9). [See (Kocherlakota and Kocherlakota (1991) for details of this result.]

3

SAMPLING WITH REPLACEMENT

Sampling from a finite population can be carried out *with replacement* in which the items drawn successively are returned to the population. Or it may be conducted *without replacement* in which the selected items are successively discarded. In the former scheme the nature of the population remains unaffected as a consequence of the sampling. In the latter scheme, the nature of the population is altered. In addition from a probabilistic point of view, the two schemes are vitally different. In sampling with replacement the successive trials (or draws) are *independent* while in a without replacement scheme the trials are no longer independent.

In the univariate situation where one character or attribute is under study, it can be shown that the indicator random variables associated with either of the sampling schemes are identically distributed; however, they are independently distributed only when the sampling is with replacement. In the literature the sequence of trials in sampling with replacement has been referred to as Bernoulli trials. Bivariate extensions of the Bernoulli scheme will be studied in this chapter. The distributions of random variables associated with two characteristics will also be presented.

3.1 Bivariate Bernoulli trials

In the univariate case, Bernoulli trials are characterized by the following properties:

(i) Each trial has two possible outcomes, say success (S) or failure (F).

(ii) Trials are independent.

(iii) The probability of success, say p, and consequently that of a failure (1 - p) remains constant over the sequence of trials.

Associated with each trial is a random variable I which takes on the values

$$I = 1 \quad \text{if the trial results in a success}$$
$$I = 0 \quad \text{otherwise.}$$

Thus, we have a sequence of independent random variables I_1, I_2, ... with each of the random variables assuming one of two values: zero or one. These random variables are independent as the trials are independent.

It is possible to generalize the preceding by considering several characteristics that may be studied simultaneously on each individual. For example, let us suppose that we are sampling with replacement from a population and studying two characteristics on each individual: hair color and eye color. Let

$$I_1 = 1 \quad \text{if the individual has dark hair}$$
$$I_1 = 0 \quad \text{otherwise}$$

and

$$I_2 = 1 \quad \text{if the individual has dark eyes}$$
$$I_2 = 0 \quad \text{otherwise.}$$

The outcome of each trial can be displayed in a 2 x 2 table as follows:

		Eye Color		
		Dark	Not Dark	
Hair Color	Dark	$I_1 = 1, I_2 = 1$	$I_1 = 1, I_2 = 0$	$I_1 = 1$
	Not Dark	$I_1 = 0, I_2 = 1$	$I_1 = 0, I_2 = 0$	$I_1 = 0$
		$I_2 = 1$	$I_2 = 0$	

Marginally, the table describes the behavior of an individual by each of the characteristics being studied. The table itself portrays the behavior of the random variables (I_1, I_2). We can then generalize the notion of Bernoulli trials to include the study of two characteristics as follows:

(i) Each trial has four possible outcomes.

(ii) Probabilities of the outcomes remain constant over the trials.

(iii) The trials are independent.

Writing

$$p_{rs} = P\{I_1 = r,\ I_2 = s\}, \qquad r = 0, 1; s = 0, 1$$

and

$$p_{r+} = \sum_s p_{rs} = p_{r0} + p_{r1} \qquad r = 0, 1$$

$$p_{+s} = \sum_r p_{rs} = p_{0s} + p_{1s} \qquad s = 0, 1,$$

we have

$$\sum_r \sum_s p_{rs} = \sum_r p_{r+} = \sum_s p_{+s} = 1.$$

Suppose that n such trials are performed. What is of interest is the number of times the outcome ($I_1 = r$, $I_2 = s$) occurs in the n trials. Let

$$n_{rs} = \text{number of times } (I_1 = r,\ I_2 = s), \qquad r = 0, 1; s = 0, 1$$

with

$$n_{r+} = \sum_s n_{rs} \qquad r = 0, 1$$

$$n_{+s} = \sum_r n_{rs} \qquad s = 0, 1$$

and

$$\sum_r \sum_s n_{rs} = \sum_r n_{r+} = \sum_s n_{+s} = n.$$

In this situation we can consider two types of distributions:

(i) The joint distribution of (n_{11}, n_{10}, n_{01}, n_{00}), which is quadrinomial with parameters n and (p_{11}, p_{10}, p_{01}, p_{00}).

(ii) The joint distribution of (n_{1+}, n_{+1}), which is the *bivariate binomial distribution.*

Unfortunately, some confusion has arisen in the literature by the use of the term *bivariate binomial* synonymously with the trinomial. [See, *e.g.* Mardia (1970, p. 82).] The sampling scheme by which the trinomial arises is quite distinct. In this case, each trial is assumed to give rise to one of three possible mutually exclusive outcomes:

$$A_1,\ A_2,\ \bar{A}_1\bar{A}_2 \text{ (neither } A_1 \text{ nor } A_2)$$

with probabilities

$$p_1, p_2, 1 - p_1 - p_2.$$

The random variables of interest are the number of occurrences of the events A_1, A_2, $\bar{A}_1 \bar{A}_2$; that is n_1, n_2, n_3, respectively. Since $n = n_1 + n_2 + n_3$, in n trials there are only two random variables which are functionally independent. The joint distribution of these two, say (n_1, n_2) is the trinomial distribution. This distribution can be generated from a 2 x 2 table by setting any cell probability, say p_{11}, equal to zero as in the table below.

	A_2	$\bar{A}_2$	
A_1	0	$p_{10} = p_1$	$p_{1+} = p_1$
$\bar{A}_1$	$p_{01} = p_2$	$p_{00} = 1 - p_1 - p_2$	$p_{0+} = 1 - p_1$
	$p_{+1} = p_2$	$p_{+0} = 1 - p_2$	1

Equivalently, we can collapse any two of the cells and examine the resulting three cells. Hamdan (1975) has considered this relationship between the bivariate binomial distribution and the trinomial.

3.2 Bivariate Bernoulli distribution

Let us consider the random variables I_1 and I_2 as defined on a single Bernoulli trial with the joint probabilities $P\{ I_1 = r, I_2 = s\} = p_{rs}$, $r = 0, 1;\ s = 0, 1$. Then the joint pgf of (I_1, I_2) is

$$\Pi(t_1, t_2) = E\,[t_1^{I_1}\, t_2^{I_2}]$$

$$= (t_1\, t_2\, p_{11} + t_1\, p_{10} + t_2\, p_{01} + p_{00}). \qquad (3.2.1)$$

Using the pgf given in (3.2.1), it is possible to determine the moments. In particular

$$E\,[\,I_1\, I_2] = p_{11}$$

and

$$\mathrm{Cov}[I_1, I_2] = p_{11} - p_{1+}p_{+1};$$

hence, the correlation coefficient is

$$\rho(I_1, I_2) = \frac{p_{11} - p_{1+}p_{+1}}{\{p_{1+}(1 - p_{1+})p_{+1}(1 - p_{+1})\}^{\frac{1}{2}}}, \tag{3.2.2}$$

which takes on values in the interval (-1, 1). A necessary and sufficient condition for independence is $p_{11} = p_{1+}p_{+1}$. This implies that in this case $\rho(I_1, I_2) = 0$ is a necessary and sufficient condition for independence.

An interesting property that emerges from the joint pgf (3.2.1) is that the marginal pgf's are:

$$\Pi_1(t) = \Pi(t, 1) = [p_{0+} + p_{1+}t]$$
$$\Pi_2(t) = \Pi(1, t) = [p_{+0} + p_{+1}t],$$

each of which is a Bernoulli distribution.

3.3 Type I bivariate binomial distribution

As in section (3.2) consider a sequence of n bivariate Bernoulli trials with a pair of characteristics being studied at each trial. Let

$$P\{I_1 = r, I_2 = s\} = p_{rs} \text{ for } r = 0, 1;\ s = 0, 1$$

and define

$$X = \sum_{i=1}^{n} I_{1i} \text{ and } Y = \sum_{i=1}^{n} I_{2i}.$$

Then the joint pgf of (I_1, I_2) is given by (3.2.1) and

$$\Pi_{X,Y}(t_1, t_2) = E[t_1^X t_2^Y] = \{E[t_1^{I_1} t_2^{I_2}]\}^n$$

$$= [t_1 t_2 p_{11} + t_1 p_{10} + t_2 p_{01} + p_{00}]^n.$$

$$= [1 + p_{1+}(t_1 - 1) + p_{+1}(t_2 - 1) + p_{11}((t_1 - 1)(t_2 - 1)]^n. \tag{3.3.1}$$

From (3.3.1) the pf of (X, Y) is found to be

$$g(x, y) = \sum_{n_{11}} \frac{n!\, p_{11}^{\;n_{11}} (p_{1+} - p_{11})^{x - n_{11}} (p_{+1} - p_{11})^{y - n_{11}}}{n_{11}!\,(x - n_{11})!\,(y - n_{11})!\,(n - x - y + n_{11})!} \cdot$$

$$(1 - p_{+1} - p_{1+} + p_{11})^{n - x - y + n_{11}} \qquad (3.3.2)$$

where the summation extends over max $(0, x + y - n) \le n_{11} \le$ min (x, y). This approach was used by Wicksell (1916), who termed this distribution the bivariate binomial distribution. [See Marshall and Olkin (1985).]

An alternative derivation was given by Aitken and Gonin (1935) in connection with a fourfold sampling scheme with replacement. Consider sampling with replacement from a population in which individuals are classified as either A or $\bar{A}$ and at the same time as either B or $\bar{B}$. This is precisely the model examined in section (3.1).

A random sample of size n selected with replacement from this population has the outcomes:

		Characteristic B		
		B	$\bar{B}$	
Characteristic A	A	n_{11}	n_{10}	n_{1+}
	$\bar{A}$	n_{01}	n_{00}	n_{0+}
		n_{+1}	n_{+0}	n

The joint distribution of $(n_{11}, n_{10}, n_{01}, n_{00})$ is quadrinomial with parameters n and $(p_{11}, p_{10}, p_{01}, p_{00})$:

$$f(n_{11}, n_{10}, n_{01}, n_{00}) = \frac{n!}{n_{11}!\, n_{10}!\, n_{01}!\, n_{00}!} p_{11}^{\;n_{11}} p_{10}^{\;n_{10}} p_{01}^{\;n_{01}} p_{00}^{\;n_{00}}. \qquad (3.3.3)$$

The random variables of interest to us are (n_{1+}, n_{+1}). For simplifying the notation, we will take $X = n_{1+}$ and $Y = n_{+1}$. The joint pf of these random variables can be obtained from (3.3.3) by recalling that, if we fix X and Y, then n_{11} is the only random component in (3.3.3). Hence the marginal pf of (X, Y) is obtained as

$$g(x, y) = \sum_{n_{11}} f(n_{11}, x - n_{11}, y - n_{11}, n - x - y + n_{11}), \tag{3.3.4}$$

which yields (3.3.2).

Using this approach the pgf of (X, Y) can also be found from that of the quadrinomial pgf corresponding to the pf given in (3.3.3). Recalling that $X = n_{11} + n_{10}$ and $Y = n_{01} + n_{11}$, the joint pgf can be written as

$$\Pi_{X,Y}(t_1, t_2) = E\,[(t_1 t_2)^{n_{11}} t_1^{n_{10}} t_2^{n_{01}}].$$

Noting that the pgf of the pf in (3.3.3) is

$$\Pi(T_1, T_2, T_3) = E\,[T_1^{n_{10}} T_2^{n_{01}} T_3^{n_{11}}]$$

$$= [T_3 p_{11} + T_1 p_{10} + T_2 p_{01} + p_{00}]^n;$$

hence,

$$\Pi_{X,Y}(t_1, t_2) = \Pi(t_1, t_2, t_1 t_2),$$

yielding the pgf as in equation (3.3.1).

Although in the literature the distribution of (X, Y) has been referred to as the bivariate binomial, we will be more specific and refer to it as the Type I form of the bivariate binomial (BVB Type I).

Marginal distributions

From (3.3.1) we can easily see that the marginal pgf of X and of Y are

$$\left.\begin{aligned} \Pi_X(t) &= \Pi_{X,Y}(t, 1) = [p_{0+} + p_{1+}\, t\,]^n \\ \Pi_Y(t) &= \Pi_{X,Y}(1, t) = [p_{+0} + p_{+1}\, t\,]^n \end{aligned}\right\} \tag{3.3.5}$$

each of which is a binomial distribution on the same index n, but with probability of success being p_{1+} for X and p_{+1} for Y.

Factorial moment and factorial cumulant generating function

The fmgf is obtained from (3.3.1) as

$$G(t_1, t_2) = \Pi(t_1 + 1, t_2 + 1)$$
$$= [1 + p_{1+}t_1 + p_{+1}t_2 + p_{11}t_1t_2]^n. \tag{3.3.6}$$

On the other hand the fcgf is

$$H(t_1, t_2) = \log G(t_1, t_2)$$
$$= n \log [1 + p_{1+}t_1 + p_{+1}t_2 + p_{11}t_1t_2]. \tag{3.3.7}$$

Factorial moments

The (r, s)th factorial moment $\mu_{[r,s]}$ is obtained by taking the coefficient of $t_1^r t_2^s / r!\, s!$ in the expansion of (3.3.6) in powers of t_1 and t_2. This can be readily seen to be

$$\mu_{[r,s]} = (r+s)!\left\{ \sum_{i=0}^{\min(r,s)} \frac{\binom{r}{i}\binom{s}{i}\binom{n}{r+s-i}}{\binom{r+s}{i}} \left(\frac{p_{11}}{p_{1+}\, p_{+1}}\right)^i \right\} p_{1+}^r\, p_{+1}^s. \tag{3.3.8}$$

The joint factorial moment $\mu_{[r,s]}$ can be written in terms of the marginal moments $\mu_{[r]}^{(x)}, \mu_{[s]}^{(y)}$ of X and Y, respectively. Here

$$\mu_{[r]}^{(x)} = n^{[r]}\, p_{1+}^r \text{ and } \mu_{[s]}^{(y)} = n^{[s]}\, p_{+1}^s;$$

hence, with $\tau = p_{11} / (p_{1+}\, p_{+1})$,

$$\mu_{[r,s]} = \left\{ \sum_i \frac{r!\, s!}{(r-i)!\,(s-i)!} \frac{n!\, \tau^i}{(n-r-s+i)!\, i!} \right\} p_{1+}^r\, p_{+1}^s$$

$$= \left\{ \sum_i \frac{r!}{(r-i)!\, i!} \frac{s!\, n!}{(s-i)!\,(n-r-s+i)!} \frac{\tau^i}{n^{[r]}\ n^{[s]}} \right\} \mu_{[r]}^{(x)}\, \mu_{[s]}^{(y)}.$$

In the above expression the quantity within the curly brackets can be rewritten as

$$\sum_i \frac{\binom{r}{i}\binom{n-r}{s-i}}{\binom{n}{s}} \tau^i ;$$

hence,

$$\mu_{[r,s]} = \mu_{[r]}^{(x)} \mu_{[s]}^{(y)} \sum_i \frac{\binom{r}{i}\binom{n-r}{s-i}}{\binom{n}{s}} \tau^i .$$

When X and Y are independent, satisfying the condition $p_{11} = p_{1+}\, p_{+1}$ or $\tau = 1$, and using van der Monde's identity

$$\mu_{[r,s]} = \mu_{[r]}^{(x)} \mu_{[s]}^{(y)} .$$

For the particular case of r = s = 1,

$$E(X\,Y) = \mu_{[1,1]} = 2\left[\binom{n}{2} p_{1+}\, p_{+1} + \frac{n}{2}\, p_{11}\right].$$

Since $E(X) = n\, p_{1+}$ and $E(Y) = n\, p_{+1}$,

$$\begin{aligned} \text{Cov}(X, Y) &= [n(n-1)\, p_{1+}\, p_{+1} + n\, p_{11}] - n^2\, p_{1+}\, p_{+1} \\ &= n\,[p_{11} - p_{1+}\, p_{+1}]. \end{aligned}$$

Hence,

$$\rho\,(X, Y) = \frac{p_{11} - p_{1+}\, p_{+1}}{\sqrt{p_{1+}\,(1 - p_{1+})\, p_{+1}\,(1 - p_{+1})}} \tag{3.3.9}$$

From (3.3.9) we see that $\rho\,(X, Y) = 0$ if and only if $p_{11} = p_{1+}\, p_{+1}$, which in turn implies that $\rho\,(X, Y) = 0$ is a necessary and sufficient condition for independence.

Conditional distributions and regression

From Theorem 1.3.1, the pgf of the conditional distribution of Y given X = x, is

$$\Pi_Y(t \mid x) = \Pi^{(x,0)}(0, t) \,/\, \Pi^{(x,0)}(0, 1).$$

Taking the partial derivative of $\Pi(t_1, t_2)$ with respect to t_1, x times, and then setting $t_1 = 0$, $t_2 = t$, we have

$$\Pi^{(x,0)}(0, t) = n^{[x]} [(1 - p_{1+} - p_{+1} + p_{11}) + (p_{+1} - p_{11})\, t\,]^{n-x} \cdot [(p_{1+} - p_{11}) + p_{11}\, t\,]^{x}.$$

Setting t = 1,

$$\Pi^{(x,0)}(0, 1) = n^{[x]} [1 - p_{1+}]^{n-x} [p_{1+}]^{x}.$$

Hence, the conditional distribution of Y given X = x has the pgf

$$\Pi_Y(t \mid x) = [Q + P\,t]^{\,n-x} [q + p\,t]^{x}, \tag{3.3.10}$$

where

$$Q = 1 - P, \quad P = \frac{p_{+1} - p_{11}}{1 - p_{1+}}$$

and

$$q = 1 - p, \quad p = \frac{p_{11}}{p_{1+}}.$$

From (3.3.10) we see that the conditional distribution of Y given X = x is the convolution of B (n - x, P) with B (x, p), where B (ν, θ) denotes a binomial distribution on ν trials with probability of success θ.

Using this conditional distribution, the regression of Y given X = x is

$$E[Y \mid x] = (n - x)\,P + x\,p$$
$$= n\,P + x\,(p - P), \tag{3.3.11}$$

which is linear in x with regression coefficient

$$p - P = \frac{p_{11} - p_{1+}p_{+1}}{p_{1+}(1 - p_{1+})}. \tag{3.3.12}$$

By the same argument the pgf of the conditional distribution of X given Y = y is

$$\Pi_X(t \mid y) = [Q^* + P^*\,t]^{\,n-y} [q^* + p^*\,t]^{y}, \tag{3.3.13}$$

where

$$Q^* = 1 - P^*, \quad P^* = \frac{p_{1+} - p_{11}}{1 - p_{+1}}$$

and

$$q^* = 1 - p^*, \quad p^* = \frac{p_{11}}{p_{+1}};$$

that is, the conditional distribution of X given Y = y is the convolution of B (n - y, P^*) with B (y, p^*). Here the regression of X on Y is

$$\begin{aligned} E[X|y] &= (n - y) P^* + y p^* \\ &= n P^* + y (p^* - P^*), \end{aligned} \tag{3.3.14}$$

which is linear in y with regression coefficient given by

$$p^* - P^* = \frac{p_{11} - p_{1+}p_{+1}}{p_{+1}(1 - p_{+1})}. \tag{3.3.15}$$

For both of these regressions the conditional variance can be determined directly. For example

$$\begin{aligned} \mathrm{Var}[Y|x] &= (n - x) P (1 - P) + x p (1 - p) \\ &= n P (1 - P) + x [p (1 - p) - P (1 - P)] \\ &= n \frac{p_{01}p_{00}}{p_{0+}^2} - x (p - P) (p - Q), \end{aligned} \tag{3.3.16}$$

which is linear in x.

From equations (3.3.11) and (3.3.14) and the linearity of the regression, it is seen that ρ [X, Y] is of the from given in equation (3.3.9). In the case when X and Y are independent, $p_{11} = p_{1+}\, p_{+1}$ from which it follows that $P = p_{+1}$ and $P^* = p_{1+}$.

3.4 Type II bivariate binomial distribution

In section 3.3 we considered the Type I bivariate binomial distribution. It can be seen from (3.3.5) that in this case each of the marginal pf's is a binomial distribution with the same index parameter, n. A more general version of the distribution was introduced by Hamdan (1972) and Hamdan and Jensen (1976). The generalization permits the index parameters as well as the probability parameters in the marginal pf's to be *unequal*. We will refer to this model as the Type II bivariate binomial distribution (BVB Type II).

Consider the following modified sampling scheme, which gives rise to a BVB Type II distribution. Three independent samples are drawn with replacement from a population in which each individual can be classified as in section 3.3 according to one of two characteristics A and

B. In the first sample k individuals are classified according to both characteristics. In this sample let W_1 count the number of times that A occurs and W_2 the number of times the B occurs. In a second sample of size $n_1 - k$, only characteristic A is studied. Let S be the number of times that A occurs in this sample. In a third sample of size $n_2 - k$, B alone is studied and it occurs T times. Then the total number of times A occurs is $W_1 + S$ while the total number of times B occurs is $Y = W_2 + T$. Here we are again interested in the joint pf of X and Y. Suppose that the proportions of A, B and AB in the population are taken to be p_{1+}, p_{+1} and p_{11}. The marginal distributions of S and T are

$$S \sim B(n_1 - k, p_{1+}), \quad T \sim B(n_2 - k, p_{+1})$$

while

$$(W_1, W_2) \sim \text{BVB Type I } (k; p_{00}, p_{10}, p_{01}, p_{11})$$

with S, T and (W_1, W_2) independent due to the sampling scheme. As will be seen below the random variables (X, Y) have the Type II bivariate binomial distribution.

The following development of the BVB Type II and its properties using pgf's is based on Kocherlakota (1989). Appealing to the laws for pgf's, the joint pgf of (X, Y) is seen to be

$$\Pi(t_1, t_2) = [p_{0+} + p_{1+}t_1]^{n_1 - k} [p_{+0} + p_{+1}t_2]^{n_2 - k}$$
$$[p_{00} + p_{10}t_1 + p_{01}t_2 + p_{11}t_1t_2]^k; \tag{3.4.1}$$

hence, the marginals of X and Y are B (n_1, p_{1+}) and B (n_2, p_{+1}), respectively. The joint pf of (X, Y) can be found by expanding the three terms and determining the coefficient of $t_1^r\, t_2^s$ in their product:

$$[p_{0+} + p_{1+}t_1]^{n_1 - k} = \sum_{j_1=0}^{n_1-k} a_{j_1} t_1^{j_1},$$

$$[p_{+0} + p_{+1}t_2]^{n_2 - k} = \sum_{j_2=0}^{n_2-k} b_{j_2} t_2^{j_2}$$

and

$$[p_{00} + p_{10}t_1 + p_{01}t_2 + p_{11}t_1t_2]^k = \sum_{j_3} \sum_{j_4} c_{j_3,j_4} t_1^{j_3} t_2^{j_4}$$

with

$$a_{j_1} = \binom{n_1 - k}{j_1} p_{1+}^{j_1} p_{0+}^{n_1 - k - j_1}, \quad b_{j_2} = \binom{n_2 - k}{j_2} p_{+1}^{j_2} p_{+0}^{n_2 - k - j_2}$$

and

$$c_{j_3,j_4} = \sum_{i=0} \frac{k!\, p_{00}^{k + i - j_3 - j_4}\, p_{10}^{j_3 - i}\, p_{01}^{j_4 - i}\, p_{11}^{i}}{(k + i - j_3 - j_4)!\,(j_3 - i)!\,(j_4 - i)!\, i!}.$$

Taking the product of the three terms and considering the term $t_1^r t_2^s$, we have for the $P\{X = r, Y = s\}$

$$f(r, s) = \sum_{j_1=0} \sum_{j_2=0} \binom{n_1 - k}{j_1} \binom{n_2 - k}{j_2} p_{1+}^{j_1} p_{0+}^{n_1 - k - j_1} \cdot$$

$$p_{+1}^{j_2} p_{+0}^{n_2 - k - j_2} \cdot$$

$$\left\{ \sum_{i=0} \frac{k!\, p_{00}^{k + i + j_1 + j_2 - r - s}\, p_{10}^{r - j_1 - i}\, p_{01}^{s - j_2 - i}\, p_{11}^{i}}{(k + i + j_1 + j_2 - r - s)!\,(r - j_1 - i)!\,(s - j_2 - i)!\, i!} \right\}.$$

(3.4.2)

The distribution defined by the pgf in (3.4.1) and the pf in (3.4.2) will be referred to as the Type II bivariate binomial (BVB Type II).

A special case of this distribution is considered by Hamdan (1972). He assumes that $p_{01} = p_{10} = 0$ or

$$p_{1+} = p_{11} = p_{+1} = p \text{ and } p_{0+} = p_{00} = p_{+0} = 1 - p = q.$$

Then the joint pgf of (X, Y) reduces to

$$\Pi(t_1, t_2) = [q + pt_1]^{n_1 - k}[q + pt_2]^{n_2 - k}[q + pt_1t_2]^k. \qquad (3.4.3)$$

In this special case the Type II bivariate binomial has binomial marginals with unequal index parameters, n_1 and n_2, but with *equal* probability of success p.

Conditional distributions and regression

The conditional distribution of X given Y = y can be found directly by using Theorem 1. 3.1. From the pgf given in (3.4.1) we see that

$$\Pi^{(0,y)}(t_1, t_2) = [p_{0+} + p_{1+}t_1]^{n_1 - k} \sum_{i=0}^{y} \left\{ \binom{y}{i} \frac{(n_2 - k)!}{(n_2 - k - i)!} p_{+1}^{i} \right.$$

$$[p_{+0} + p_{+1}t_2]^{n_2 - k - i} \frac{k!}{(k - y + i)!} [p_{01} + p_{11}t_1]^{y - i}$$

$$\left. [p_{00} + p_{10}t_1 + p_{01}t_2 + p_{11}t_1t_2]^{k - (y - i)} \right\}; \qquad (3.4.4)$$

hence,

$$\Pi_X(t \mid y) = [p_{0+} + p_{1+}t]^{n_1 - k} \sum_{i=0}^{y} \frac{\binom{n_2 - k}{i}\binom{k}{y - i}}{\binom{n_2}{y}}$$

$$\left(\frac{p_{01}}{p_{+1}} + \frac{p_{11}}{p_{+1}}t\right)^{y - i} \left(\frac{p_{00}}{p_{+0}} + \frac{p_{10}}{p_{+0}}t\right)^{k - (y - i)}. \qquad (3.4.5)$$

From (3.4.5) we see that the conditional distribution of X given Y = y is the convolution of V_1 and V_2, where

$$V_1 \sim B(n_1 - k, p_{1+})$$

and V_2 is a finite mixture of $(U_{1i} + U_{2i})$, i = 0, 1, ..., y. Here U_{1i}, U_{2i} are independent with

$$U_{1i} \sim B\left(y - i, \frac{p_{11}}{p_{+1}}\right), \qquad U_{2i} \sim B\left(k - y + i, \frac{p_{10}}{p_{+0}}\right).$$

The weight w(i) of the i-th component of the mixture is

$$w(i) = \frac{\binom{n_2 - k}{i}\binom{k}{y - i}}{\binom{n_2}{y}},$$

which is a hypergeometric distribution with parameters n_2, $n_2 - k$ and y.

Similarly, the conditional distribution of Y given X = x can be shown to have the pgf

$$\Pi_Y(t \mid x) = [p_{+0} + p_{+1}\, t]^{n_2 - k} \sum_{i=0}^{x} \frac{\binom{n_1 - k}{i}\binom{k}{x - i}}{\binom{n_1}{x}} \left(\frac{p_{10}}{p_{1+}} + \frac{p_{11}}{p_{1+}}\, t\right)^{x - i} \left(\frac{p_{00}}{p_{0+}} + \frac{p_{01}}{p_{0+}}\, t\right)^{k - (x - i)}. \qquad (3.4.6)$$

Hence the conditional distribution of Y given X = x is the convolution of W_1 and W_2, where

$$W_1 \sim B(n_2 - k, p_{+1})$$

and W_2 is a finite mixture of $(S_{1i} + S_{2i})$, i = 0, 1, ..., x with S_{1i}, S_{2i} being independent and

$$S_{1i} \sim B\left(x - i, \frac{p_{11}}{p_{1+}}\right), \qquad S_{2i} \sim B\left(k - x + i, \frac{p_{01}}{p_{0+}}\right).$$

The weight $w^*(i)$ of the i-th component of the mixture is

$$w^*(i) = \frac{\binom{n_1 - k}{i}\binom{k}{x - i}}{\binom{n_1}{x}},$$

which is a hypergeometric distribution with parameters n_1, $n_1 - k$, and x.

The regression of X on Y (and similarly that of Y on X) can be determined by viewing the appropriate conditional distribution as a convolution. Hence,

$$E(X \mid y) = E(V_1) + E(V_2)$$
$$= (n_1 - k)\, p_{1+} + E(V_2).$$

Now V_2 can be decomposed into a finite mixture; thus

$$E(V_2) = \sum_{i=0}^{y} w(i)\,[E(U_{1i}) + E(U_{2i})]$$
$$= \sum_{i=0}^{y} w(i)\,\left[\frac{p_{11}}{p_{+1}}(y - i) + \frac{p_{10}}{p_{+0}}(k - y + i)\right].$$

Recalling that $\sum_{i=0}^{y} w(i) = 1$ and that the expectation of i is $y(n_2 - k)/n_2$, we have

$$E(V_2) = \left(\frac{p_{11}}{p_{+1}} - \frac{p_{10}}{p_{+0}}\right) y + k\frac{p_{10}}{p_{+0}} - \left(\frac{p_{11}}{p_{+1}} - \frac{p_{10}}{p_{+0}}\right)\left(1 - \frac{k}{n_2}\right) y$$

Combining these results, the regression of X on Y is

$$E(X \mid y) = n_1 p_{1+} + k\left[\left\{\frac{p_{11}}{p_{+1}} - \frac{p_{10}}{p_{+0}}\right\}\frac{y}{n_2} + \left\{\frac{p_{10}}{p_{+0}} - p_{1+}\right\}\right]. \qquad (3.4.7)$$

Equation (3.4.7) shows that the regression of X on Y is linear in y. This expression can be shown to be equal to equation (2.8) in Hamdan and Jensen (1976).

Similarly, the conditional distribution of Y given X = x can be shown to be

$$E(Y \mid x) = n_2 p_{+1} + k\left[\left\{\frac{p_{11}}{p_{1+}} - \frac{p_{01}}{p_{0+}}\right\}\frac{x}{n_2} + \left\{\frac{p_{01}}{p_{0+}} - p_{+1}\right\}\right], \qquad (3.4.8)$$

which is linear in x.

Using similar arguments, we can show that Var [X | y] and

Var [Y | x] are linear in y and x, respectively.

3.5 Trinomial distribution

In section 3.1 we have seen that the trinomial distribution arises if we combine any two of the four cells in the 2 x 2 table or if we take one of the cell probabilities to be equal to zero. [See Hamdan (1975).] Thus, taking $p_{11} = 0$, we have

$$p_{10} = P\{I_1 = 1,\ I_2 = 0\} = p_1,$$
$$p_{01} = P\{I_1 = 0,\ I_2 = 1\} = p_2,$$
$$p_{00} = P\{I_1 = 0,\ I_2 = 0\} = 1 - p_1 - p_2.$$

Here we are considering the random variables X and Y, where X counts the number of times that $\{I_1 = 1,\ I_2 = 0\}$ and Y counts the number of times that $\{I_1 = 0,\ I_2 = 1\}$ in n independent trials. Then the joint pgf of (X, Y) is given by

$$\Pi(t_1, t_2) = [1 + p_1 (t_1 - 1) + p_2 (t_2 - 1)]^n \qquad (3.5.1)$$

with pf

$$f(x, y) = \frac{n!}{x!\ y!\ (n - x - y)!}\ p_1^x\ p_2^y\ (1 - p_1 - p_2)^{n-x-y}. \qquad (3.5.2)$$

From (3.5.1) we can easily see that the marginal distributions are

$$X \sim B(n, p_1) \quad \text{and} \quad Y \sim B(n, p_2).$$

Factorial moment and factorial cumulant generating functions

The fmgf is obtained from (3.5.1) as

$$G(t_1, t_2) = \Pi(t_1 + 1, t_2 + 1)$$
$$= [1 + p_1 t_1 + p_2 t_2]^n, \qquad (3.5.3)$$

while the fcgf is

$$H(t_1, t_2) = \log G(t_1, t_2)$$
$$= n \log [1 + p_1 t_1 + p_2 t_2]. \qquad (3.5.4)$$

Factorial moments

The (r, s)th factorial moment of (X, Y) is obtained by taking the coefficient of $t_1^r\, t_2^s / r!\, s!$ in the expansion of (3.5.3) in powers of t_1 and t_2:

$$\mu_{[r,s]} = \frac{n!}{(n - r - s)!}\, p_1^r\, p_2^s.$$

In particular for r =1, s = 1

$$\mu_{[1,1]} = E\,(X\,Y) = n\,(n - 1)p_1\, p_2$$

and

$$\text{Cov}\,(n_{10}, n_{01}) = -\,n\, p_1\, p_2, \tag{3.5.5}$$

which is always negative.

Conditional distribution and regression

The conditional distribution of X given Y (and *vice versa*) can be found using Theorem 1.3.1 or directly from the definition of the conditional distribution. To use Theorem 1.3.1, we need

$$\Pi^{(0,y)}\,(t_1, t_2) = n^{[y]}\,[1 + p_1\,(t_1 - 1) + p_2\,(t_2 - 1)]^{\,n - y}\, p_2^{\,y}\, p_{01}^{\,n_{01}};$$

hence,

$$\Pi_X(t \mid y) = \left[\frac{1 - p_1 - p_2}{1 - p_2} + \frac{p_1 t}{1 - p_2}\right]^{n - y}, \tag{3.5.6}$$

which is B $\left(n - y, \dfrac{p_1}{1 - p_2}\right)$. The same result can be obtained directly by taking the ratio of (3.5.2) to the marginal distribution of Y,

$$g(y) = \binom{n}{y} p_2^y\,(1 - p_2)^{n - y},$$

yielding

$$h(x|y) = \binom{n-y}{x} \left(\frac{p_1}{1-p_2}\right)^x \left(1 - \frac{p_1}{1-p_2}\right)^{n-y-x}. \qquad (3.5.7)$$

From (3.5.7) the conditional moments can be found:

$$E(X|y) = \frac{n p_1}{1-p_2} - \frac{p_1}{1-p_2} y$$

and

$$\mathrm{Var}(X|y) = (n-y)\left(\frac{p_1}{1-p_2}\right)\left(1 - \frac{p_1}{1-p_2}\right),$$

which are both linear in y as in the BVB Type I distribution.

3.6 Canonical representation

As discussed in section 1.5, it is of interest to obtain the canonical expansions and canonical variables in bivariate distributions. Aitken and Gonin (1935) have shown that the joint pf of the BVB Type I can be expressed as a function of the marginal pf's and Krawtchouk polynomials.

Definition 3.6.1 Krawtchouk polynomials

The rth Krawtchouk polynomial in x is defined by

$$G_r(x; n, p)\, b(x; n, p) = (p-1)^r \Delta^r [x^{(r)}\, b(x; n, p)],$$

where $\Delta f(x) = f(x+1) - f(x)$, $x^{(r)} = x!/(x-r)!$ and $b(x; n, p)$ is the binomial probability function.

From this definition, it follows that

$$G_r(x; n, p) = \sum_{j=0}^{r} (-1)^j \binom{r}{j} \frac{(n-r+j)!}{(n-r)!} p^j x^{(r-j)} \qquad (3.6.1)$$

(*Encyclopedia of Statistical Sciences*, *4*, p. 410). It is also known that Krawtchouk polynomials are orthogonal with respect to the weight function $b(x; n, p)$. [See *e. g.* Szego (1959, p. 36).]

Aitken and Gonin (1935) show that the factorial moment generating function of $b(x; n, p)$ $G_r^*(x; n, p)$ is

$$t^r (1 + p\,t)^{n-r} \left\{\binom{n}{r} p^r (1-p)^r\right\}^{\frac{1}{2}}, \tag{3.6.2}$$

where

$$G_r^*(x; n, p) = \frac{G_r(x; n, p)}{\left\{\binom{n}{r} p^r (1-p)^r\right\}^{\frac{1}{2}}}$$

are *orthonormal* Krawtchouk polynomials.

We will use (3.6.2) to express the general BVB Type II pf as a function of its marginal pf's and the Krawtchouk polynomials. Recall that the pgf of the BVB Type II is given by (3.4.1) with the corresponding factorial moment generating function

$$G(t_1, t_2) = [1 + p_{1+}t_1]^{n_1 - k} [1 + p_{+1}t_2]^{n_2 - k} \; [1 + p_{1+}t_1 + p_{+1}t_2 + p_{11}t_1t_2]^k. \tag{3.6.3}$$

In this case the marginal distributions of X and Y are B (n_1, p_{1+}) and B (n_2, p_{+1}), respectively, with fmgf's $[1 + p_{1+}t_1]^{n_1}$ and $[1 + p_{+1}t_2]^{n_2}$.

In order to use (3.6.2), we will need to rewrite the fmgf given in (3.6.3). Note that the last term

$$[1 + p_{1+}t_1 + p_{+1}t_2 + p_{11}t_1t_2]^k$$

can be written as

$$[1 + p_{1+}t_1]^k \; [1 + p_{+1}t_2]^k \left\{1 + \frac{[p_{11} - p_{1+}p_{+1}]\, t_1t_2}{[1 + p_{1+}t_1]\;[1 + p_{+1}t_2]}\right\}^k$$

$$= [1 + p_{1+}t_1]^k \; [1 + p_{+1}t_2]^k \sum_{i=0}^{k} \binom{k}{i} \frac{[p_{11} - p_{1+}p_{+1}]^i\, t_1^i t_2^i}{[1 + p_{1+}t_1]^i \; [1 + p_{+1}t_2]^i}.$$

Substituting the above expansion into $G(t_1, t_2)$ given in (3.6.3), we have

$$G(t_1, t_2) = [1 + p_{1+}t_1]^{n_1} \; [1 + p_{+1}t_2]^{n_2}$$

$$+\sum_{i=1}^{k} \binom{k}{i} [p_{11} - p_{1+}p_{+1}]^i t_1^i t_2^i \; [1+ p_{1+}t_1]^{n_1 - i} [1+ p_{+1}t_2]^{n_2 - i}.$$

(3.6.4)

Using the one-to-one relationship between the fmgf and the probability function of a random variable, we can invert $G(t_1, t_2)$ in (3.6.4) to yield $f(x, y)$. Proceeding term-by-term, we see that the first term on the right hand side of (3.6.4) inverts to

$$b\,(x;\, n_1,\, p_{1+})\; b\,(y;\, n_2,\, p_{+1}).$$

The ith term in the summation of (3.6.4) is

$$\binom{k}{i} [p_{11} - p_{1+}p_{+1}]^i \{t_1^i \; [1+ p_{1+}t_1]^{n_1 - i}\} \{t_2^i \; [1+ p_{+1}t_2]^{n_2 - i}\}.$$

The inverse of this term is

$$\binom{k}{i} [p_{11} - p_{1+}p_{+1}]^i \left\{ \frac{G_i^*(x;\, n_1,\, p_{1+})\; b(x;\, n_1,\, p_{1+})}{\left[\binom{n_1}{i} p_{1+}^i \,(1 - p_{1+})^i\right]^{\frac{1}{2}}} \right\} \cdot$$

$$\left\{ \frac{G_i^*(y;\, n_2,\, p_{+1})\; b(y;\, n_2,\, p_{+1})}{\left[\binom{n_2}{i} p_{+1}^i \,(1 - p_{+1})^i\right]^{\frac{1}{2}}} \right\},$$

which can be written as

$$\eta_i\, b(x;\, n_1,\, p_{1+})\; b(y;\, n_2,\, p_{+1})\; G_i^*(x;\, n_1,\, p_{1+})\; G_i^*(y;\, n_2,\, p_{+1})$$

with

$$\eta_i = \frac{k^{(i)}}{\left[n_1^{(i)}\, n_2^{(i)}\right]^{1/2}} \frac{[p_{11} - p_{1+}p_{+1}]^i}{\left[p_{1+}\,(1 - p_{1+})\, p_{+1}\,(1 - p_{+1})\right]^{\frac{i}{2}}}.$$

Combining these results, we have

$$f(x, y) = b(x; n_1, p_{1+})\, b(y; n_2, p_{+1}) \{1 + \sum_{i=1}^{k} \eta_i\, G_i^*(x; n_1, p_{1+})\, G_i^*(y; n_2, p_{+1})\} \tag{3.6.5}$$

Lancaster (1958, 1963) refers to the normalized functions $G_i^*(x; n_1, p_{1+})$ and $G_i^*(y; n_2, p_{+1})$ as the ith canonical variables of the representation given in (3.6.5). The coefficient η_i is the correlation between the two variables and is referred to as the canonical correlation. Krishnamoorthy (1951) has extended the representation given in (3.6.5) to the case of the multivariate binomial.

From the expansion given in (3.6.5), we see that the random variables X and Y are independent if and only if $p_{11} = p_{1+}p_{+1}$, which is equivalent to requiring $\eta_i = 0$. It should also be noted that in a 2 x 2 table this is the necessary and sufficient condition for independence.

In section 3.3 we obtained the conditional distribution of Y given X = x using the pgf. This conditional distribution can also be expressed as a canonical expansion determined from (3.6.5). Thus

$$f(x \mid y) = b(x; n_1, p_{1+}) \{1 + \sum_{i=1}^{k} \eta_i\, G_i^*(x; n_1, p_{1+})\, G_i^*(y; n_2, p_{+1})\}$$

and

$$f(y \mid x) = b(y; n_2, p_{+1}) \{1 + \sum_{i=1}^{k} \eta_i\, G_i^*(x; n_1, p_{1+})\, G_i^*(y; n_2, p_{+1})\}.$$

The corresponding regression equations are then given by

$$E(X \mid y) = n_1 p_{1+} + \eta_1 \, [n_1 p_{1+} (1 - p_{1+}) / n_2 p_{+1}(1 - p_{+1})]^{\frac{1}{2}} (y - n_2 p_{+1})$$

and

$$E(Y \mid x) = n_2 p_{+1} + \eta_2 \, [n_2 p_{+1} (1 - p_{+1}) / n_1 p_{1+}(1 - p_{1+})]^{\frac{1}{2}} (x - n_1 p_{1+})$$

which are equivalent to the results given in (3.4.7) and (3.4.8), respectively.

Canonical expansions can also be obtained for the trinomial distribution and the BVB Type I distribution. The difference between these two distributions has been pointed out in earlier sections. Hamdan (1972) has studied this difference in terms of the canonical representation. Taking $i = n_{11}$, $r = n_{1+}$ and $s = n_{+1}$, the pf of the BVB Type I given in (3.3.2) can be rewritten as

$$f(r, s) = n! \sum_i \frac{p_{00}^{n-r-s+i}\, p_{10}^{r-i}\, p_{01}^{s-i}\, p_{11}^{i}}{(n-r-s+i)!\,(r-i)!\,(s-i)!\,i!}, \qquad (3.6.6)$$

while the trinomial pf given in (3.5.2) is

$$f^*(r, s) = \frac{n!\, p_{1+}^{r}\, p_{+1}^{s}\,(1 - p_{1+} - p_{+1})^{n-s-r}}{(n-r-s)!\, r!\, s!}. \qquad (3.6.7)$$

Aitken and Gonin (1935) have shown the canonical representation of (3.6.6) to be

$$f(r, s) = b(r; n, p_{1+})\, b(s; n, p_{+1}) \cdot \left\{1 + \sum_{k=1}^{n} \rho^k G_k^*(r; n, p_{1+})\, G_k^*(s; n, p_{+1})\right\}, \qquad (3.6.8)$$

where

$$\rho = \frac{p_{11} - p_{1+}p_{+1}}{\{p_{1+}(1 - p_{1+})\, p_{+1}(1 - p_{+1})\}^{\frac{1}{2}}}.$$

On the other hand, the trinomial pf given in (3.6.7) can be similarly expanded with ρ replaced by

$$\rho^* = \frac{-p_{1+}p_{+1}}{\{(1 - p_{1+})(1 - p_{+1})\}^{\frac{1}{2}}}.$$

A comparison of ρ and ρ^* shows that, although ρ can be positive or negative, ρ^* must always be negative. If $\rho = 0$, X and Y are independent; however, if $\rho^* = 0$, then p_{1+} and/or p_{+1} must be zero. In this case the trinomial reduces to a univariate binomial (or a single-point degenerate distribution); hence, X and Y cannot be independent in the trinomial distribution.

3.7 Estimation of the parameters

The estimation of the parameters is mainly attacked from the maximum likelihood principle. The complexity of the problem is very much dependent upon the form of the distribution under consideration. Estimation for each of the cases will be considered individually.

3.7.1 Trinomial distribution

Let (x_1, x_2, x_3) be the observations from the trinomial distribution with $n = x_1 + x_2 + x_3$. (See section 3.5.) The likelihood function for p_1, p_2 is given by

$$L = C\, p_1^{x_1}\, p_2^{x_2}\, (1 - p_1 - p_2)^{n - x_1 - x_2}$$

with

$$C = \frac{n!}{x_1!\, x_2!\, (n - x_1 - x_2)!}.$$

Maximizing the log of the likelihood L with respect to p_1, p_2 as in section 2.1 is equivalent to solving for p_1 and p_2 in the equations

$$\frac{\partial}{\partial p_i}\left\{(x_{1+} \log p_1 + x_{2+} \log p_2 + (n - x_1 - x_2) \log (1 - p_1 - p_2)\right\} = 0$$

for i = 1, 2; or

$$\frac{x_1}{\hat{p}_1} - \frac{n - x_1 - x_2}{1 - \hat{p}_1 - \hat{p}_2} = 0$$

$$\frac{x_1}{\hat{p}_2} - \frac{n - x_1 - x_2}{1 - \hat{p}_1 - \hat{p}_2} = 0,$$

which yield

$$\hat{p}_1 = \frac{x_1}{n}, \quad \hat{p}_2 = \frac{x_2}{n} \quad \text{and} \quad \hat{p}_3 = 1 - \hat{p}_1 - \hat{p}_2 = \frac{x_3}{n}.$$

The variance matrix of these maximum likelihood estimators can be determined directly from the properties of the sample means; thus,

$$\mathrm{Var}(\hat{p}_1) = \frac{p_1(1 - p_1)}{n}, \quad \mathrm{Var}(\hat{p}_2) = \frac{p_2(1 - p_2)}{n}$$

and

$$\mathrm{Cov}(\hat{p}_1, \hat{p}_2) = -\frac{p_1 p_2}{n}.$$

It can easily be seen that the maximum likelihood estimators are also the moment estimators, which are obtained by equating the first two sample moments with their parametric equivalents.

The asymptotic distribution of $(\hat{p}_1, \hat{p}_2)$ is bivariate normal with the mean vector (p_1, p_2) and variance matrix

$$\begin{bmatrix} \frac{p_1(1 - p_1)}{n} & -\frac{p_1 p_2}{n} \\ \cdots & \frac{p_2(1 - p_2)}{n} \end{bmatrix}.$$

The following example taken from Elandt-Johnson (1971, p. 296) illustrates the application of the trinomial distribution in the area of genetics.

Example 3.7.1

The main blood group locus in rats is called Ag-B. Several alleles are associated with this locus. It appears that in certain crosses significant deviation from the Mendelian ratios are observed. In 200 serological tests from offspring of intercrosses of two strains of rats, one Lewis ($Ag\text{-}B^1$, $Ag\text{-}B^1$) and the other DA ($Ag\text{-}B^4$, $Ag\text{-}B^4$), the following results were obtained. For simplicity the crosses are depicted as $B^1 B^4 \times B^1 B^4$ with the symbol Ag omitted.

Genotype	$B^1 B^1$	$B^1 B^4$	$B^4 B^4$
Observed Frequency	58	129	13

The expected Mendelian ratios are 1: 2: 1. The deviation can be explained by a differential embryonic survival ratios of the zygotes leading to the genotypic ratios $B^1 B^1$: $B^1 B^4$: $B^4 B^4$ as 1: 2: ϕ with $0 < \phi < 1$. Here ϕ is unknown and has to be estimated from the data. Under this assumption, the expected frequencies are

Genotype	$B^1 B^1$	$B^1 B^4$	$B^4 B^4$
Observed Frequency	$\frac{n}{3+\phi}$	$\frac{2n}{3+\phi}$	$\frac{n\phi}{3+\phi}$

Prior to the estimation of ϕ, let us estimate p_1 and p_2. Since

$$x_1 = 58, \quad x_2 = 129, \quad x_3 = 13, \quad n = 200,$$

$$\hat{p}_1 = \frac{58}{200} = 0.290, \quad \hat{p}_2 = \frac{129}{200} = 0.645, \quad \hat{p}_3 = \frac{13}{200} = 0.065.$$

The variance matrix is estimated by

$$\frac{1}{200}\begin{bmatrix} (.29)(.71) & -(.29)(.645) \\ \cdots & (.645)(.355) \end{bmatrix} = 10^{-4}\begin{bmatrix} 10.2950 & -9.3525 \\ \cdots & 11.4488 \end{bmatrix}$$

with the estimated correlation coefficient between $\hat{p}_1$ and $\hat{p}_2$ equal to -0.8614.

To estimate ϕ from the data, the trinomial probabilities are expressed in terms of ϕ and the resulting likelihood function is maximized with respect to ϕ ; that is,

$$L(\phi; \underline{x}) = C\left(\frac{1}{3+\phi}\right)^{x_1}\left(\frac{2}{3+\phi}\right)^{x_2}\left(\frac{\phi}{3+\phi}\right)^{x_3},$$

which yields the likelihood equation

$$\frac{x_{3+}}{\hat{\phi}} - \frac{n}{3+\hat{\phi}} = 0$$

or

$$\hat{\phi} = \frac{3\ x_3}{n - x_3} = \frac{39}{187} = 0.2086.$$

The asymptotic variance of $\hat{\phi}$ is

$$\left\{E\left[-\frac{\partial^2 \log L}{\partial\phi^2}\right]\right\}^{-1} = \frac{\phi\,(3+\phi)^2}{3n};$$

hence, the asymptotic variance can be estimated by

$$\hat{Var}(\hat{\phi}) = \frac{\hat{\phi}(3+\hat{\phi})^2}{3n} = 0.003579.$$

3.7.2 Type I bivariate binomial distribution

A large part of the literature dealing with 2 x 2 contingency tables has been concerned with testing independence of the two characteristics under study and in measuring the degree of association between them. As early as 1900 Pearson introduced the tetrachoric correlation as an estimate of the correlation coefficient in a bivariate normal model underlying the 2 x 2 table. The question arises as to the form of this correlation when the variables are truly discrete and the assumption of bivariate normality is not plausible. Tests of independence are based on the distribution of the table entries, which are jointly quadrinomial. We are interested in estimating the parameters of the Type I BVB and determining their variances.

Data from the 2 x 2 contingency table can be of two types: (i) only the marginal totals n_{1+} and n_{+1} or (ii) $\{n_{rs}\}$, all the entries from the 2 x 2 table. The two cases are discussed below:

Case I

Estimation of the parameters is not possible if information is available from only one table (i. e., a single sample of size n). In order to estimate the parameters, p_{1+}, p_{+1}, p_{11}, we will assume that marginal information is available from k independent samples. This type of data is discussed in detail in Hamdan and Nasro (1986). Now the data consist only of pairs of observations (n_{1+}, n_{+1}) rather than the entire 2 x 2 table. We shall consider a sample of k such independent pairs. Writing $x_i = n_{1+}$ and $y_i = n_{+1}$ for the ith sample with i = 1, 2, ..., k, we have the likelihood function given by

$$L = \prod_{i=1}^{k} g(x_i, y_i; n).$$

Recalling the form of $g(x_i, y_i; n)$ given in (3.3.2) and differentiating with respect to the parameters p_{1+}, p_{+1} and p_{11}, we have

$$\frac{1 - \hat{p}_{+1}}{\hat{p}_{1+} - \hat{p}_{11}} x_+ + y_+ - nk - \frac{1 - \hat{p}_{+1}}{\hat{p}_{1+} - \hat{p}_{11}} S = 0$$

$$\frac{1 - \hat{p}_{1+}}{\hat{p}_{+1} - \hat{p}_{11}} y_+ + x_+ - nk - \frac{1 - \hat{p}_{1+}}{\hat{p}_{+1} - \hat{p}_{11}} S = 0$$

$$x_+ + y_+ - nk + \frac{1 - \hat{p}_{1+} - \hat{p}_{+1}}{\hat{p}_{11}} S = 0,$$

where

$$x_+ = \sum_{i=1}^{k} x_i, \qquad y_+ = \sum_{i=1}^{k} y_i$$

and

$$S = \sum_{i=1}^{k} \left\{ \frac{\sum_{z_i} z_i \, h(x_i, y_i, z_i; n)}{g(x_i, y_i; n)} \right\}$$

with

$$h(x_i, y_i, z_i; n) = \frac{n! \, p_{11}^{z_i} (p_{1+} - p_{11})^{x_i - z_i} (p_{+1} - p_{11})^{y_i - z_i}}{z_i! \, (x_i - z_i)! \, (y_i - z_i)! \, (n - x_i - y_i + z_i)!} \cdot$$

$$(1 - p_{+1} - p_{1+} + p_{11})^{n - x_i - y_i + z_i}.$$

After some algebraic manipulation, it can be seen that

$$\hat{p}_{1+} = \frac{x_+}{nk}, \quad \hat{p}_{+1} = \frac{y_+}{nk}$$

and $\hat{p}_{11}$ can be obtained as the solution to the equation $\bar{R} = 1$, where

$$\bar{R} = \frac{1}{k} \sum_{i=1}^{k} R_i$$

with $R_i = g(x_i - 1, y_i - 1; n - 1) / g(x_i, y_i; n)$. The equation, $\bar{R} = 1$, has to be solved iteratively, using the Newton-Raphson technique.

To solve this equation we need

$$\frac{\partial R_i}{\partial p_{11}} = \frac{1}{g(x_i, y_i; n)} \frac{\partial g(x_i - 1, y_i - 1; n - 1)}{\partial p_{11}}$$

$$- \frac{g(x_i - 1, y_i - 1; n - 1)}{g^2(x_i, y_i; n)} \frac{\partial g(x_i, y_i; n)}{\partial p_{11}},$$

where

$$\frac{\partial g(x_i, y_i; n)}{\partial p_{11}} =$$

$$\sum_{n_{11}} \left\{ \frac{n_{11}}{p_{11}} - \frac{x_i - n_{11}}{p_{1+} - p_{11}} - \frac{y_i - n_{11}}{p_{+1} - p_{11}} + \frac{n - x_i - y_i + n_{11}}{1 - p_{+1} - p_{1+} + p_{11}} \right\} \cdot$$

$$\frac{n!\, p_{11}^{\,n_{11}} (p_{1+} - p_{11})^{x_i - n_{11}} (p_{+1} - p_{11})^{y_i - n_{11}}}{n_{11}!\, (x_i - n_{11})!\, (y_i - n_{11})!\, (n - x_i - y_i + n_{11})!} \cdot$$

$$(1 - p_{+1} - p_{1+} + p_{11})^{n - x_i - y_i + n_{11}}.$$

We can obtain similar expressions in terms of $g(x_i - 1, y_i - 1; n - 1)$.

A preliminary estimate for p_{11} can be obtained using the method of moments. Since

$$E(XY) = n(p_{11} - p_{1+}p_{+1}) + n^2 p_{1+} p_{+1},$$

$$\tilde{p}_{11} = \frac{1}{nk} \sum_{i=1}^{k} x_i y_i + \bar{x}\bar{y} - n\bar{x}\bar{y}$$

with $\bar{x} = \frac{x_+}{nk}$ and $\bar{y} = \frac{y_+}{nk}$.

Hamdan and Nasro (1986) construct an example using computer simulated data. Their simulation procedure is based on the joint distribution of $\{n_{rs}\}$, which is quadrinomial with parameters p_{11}, p_{01} , p_{10}, p_{00}. They generate a sample of size n from this population and classify it into the four cells. This gives rise to *one* pair of observations (x_i, y_i) and a repetition of the process k times provides k independent pairs.

It is also possible to generate observations directly from the Type I BVB using the conditional distribution of X given y and the marginal distribution of Y. From (3.3.5) $Y \sim B(n, p_{+1})$ and from (3.3.13) X given y is the convolution of V_1 with V_2, where

$$V_1 \sim B(n - y, P^*) \quad \text{and} \quad V_2 \sim B(y, p^*).$$

Hence to simulate a realization from the Type I BVB,

(1) generate a value of y using IMSL subroutine GGBN
(2) using this value of y as a parameter, generate v_1 and v_2 using GGBN
(3) form $x = v_1 + v_2$.

The resulting pair (x, y) is a realization from the Type I BVB. To obtain a sample of k pairs, this procedure is repeated k times.

In the technique used by Hamdan and Nasro, it is necessary to generate n obervations from the quadrinomial in order to specify a single pair (x, y). Using the conditional distribution approach, only three binomial variates need be generated regardless of the value of n.

Example 3.7.2

For $n = 10$, $p_{1+} = 0.23$, $p_{+1} = 0.32$, $p_{11} = 0.1$, $k = 25$ pairs (x_i, y_i) were generated using the technique based on the conditional distribution of X given y:

x_i	3	3	2	1	2	1	3	3	3	5	3	1	1	2	0	2	0	1	1	3	3	1	3	2	4
y_i	4	1	3	2	5	3	3	4	4	5	3	5	1	3	4	3	4	2	2	3	2	4	2	7	3

Hence, $\hat{p}_{1+} = 0.212$, $\hat{p}_{+1} = 0.328$ and $\hat{p}_{11}$ (determined iteratively) = 0.0807, where the method of moments estimate $\tilde{p}_{11} = 0.0782$ was used as the initial value for $\hat{p}_{11}$.

Case II

Consider a 2 x 2 table with neither of the margins fixed; that is, the form discussed in section 3.3. In the literature [Kendall and Stuart (1979, p. 580)] this type of table is referred to as a 'double dichotomy'. We assume, that for a sample of size n, we have data $\{n_{rs}\}$ available from the 2 x 2 table. The joint distribution of the $\{n_{rs}\}$ is quadrinomial and the corresponding likelihood can be written as

$$L \propto p_{11}^{\;n_{11}} \, p_{10}^{\;n_{10}} \, p_{01}^{\;n_{01}} \, p_{00}^{\;n_{00}}.$$

This form of the likelihood is the basis for the analysis of contingency table leading to tests of independence for the two characteristics of study. Here, however, we are interested in the Type I BVB which arises as the joint distribution of the margins (n_{1+}, n_{+1}). Recall from equation (3.3.3) that this distribution is a function of p_{1+}, p_{+1} and p_{11}. We can then reparameterize the above likelihood function in terms of the parameters of interest p_{1+}, p_{+1} and the covariance $\delta = (p_{11} - p_{1+} p_{+1})$:

$$L \propto (\delta + p_{1+} p_{+1})^{n_{11}} (p_{1+} - \delta - p_{1+}p_{+1})^{n_{10}} (p_{+1} - \delta - p_{1+}p_{+1})^{n_{01}} \cdot$$
$$(1 - p_{1+} - p_{+1} + p_{1+} p_{+1} + \delta)^{n_{00}}.$$

With this parameterization the likelihood equations are

$$\frac{\partial \log L}{\partial p_{1+}} = \frac{n_{11}p_{+1}}{\delta + p_{1+}p_{+1}} + \frac{n_{10}q_{+1}}{p_{1+}q_{+1} - \delta} - \frac{n_{01}p_{+1}}{p_{+1}q_{1+} - \delta} - \frac{n_{00}q_{+1}}{q_{1+}q_{+1} + \delta}$$

$$\frac{\partial \log L}{\partial p_{+1}} = \frac{n_{11}p_{1+}}{\delta + p_{1+}p_{+1}} - \frac{n_{10}p_{1+}}{p_{1+}q_{+1} - \delta} + \frac{n_{01}q_{1+}}{p_{+1}q_{1+} - \delta} - \frac{n_{00}q_{1+}}{q_{1+}q_{+1} + \delta}$$

$$\frac{\partial \log L}{\partial \delta} = \frac{n_{11}}{\delta + p_{1+}p_{+1}} - \frac{n_{10}}{p_{1+}q_{+1} - \delta} - \frac{n_{01}}{p_{+1}q_{1+} - \delta} + \frac{n_{00}}{q_{1+}q_{+1} + \delta}$$

with $q_{1+} = 1 - p_{1+}$, $q_{+1} = 1 - p_{+1}$. Solving these equations,

$$\hat{p}_{1+} = \frac{n_{1+}}{n}, \quad \hat{p}_{+1} = \frac{n_{+1}}{n} \quad \text{and} \quad \hat{\delta} = \frac{(n_{00}\, n_{11} - n_{01}\, n_{10})}{n^2}. \quad (3.7.1)$$

Here we see that, as expected, $\hat{p}_{1+}$ and $\hat{p}_{+1}$ are the observed proportions of the occurrences of the characteristics A and B, respectively. It can also be seen [Hamdan and Martinson (1971)] that the estimated correlation coefficient $\hat{\rho}$ is identical to Pearson's (1904) $\varphi = \sqrt{\chi^2/n}$, where χ^2 is the usual chi-square statistic for the 2 x 2 table. Pearson had introduced φ^2 as the "mean square contingency" for a bivariate distribution in order to derive a measure of association independent of the sample size n.

The variance matrix of $(\hat{p}_{1+}, \hat{p}_{+1}, \hat{\delta})$ for large n is seen to be the inverse of

$$\Gamma = \left\{ - E\left(\frac{\partial^2 \log L}{\partial^2\theta_i \, \partial^2\theta_j} \right) \right\},$$

where θ_i and θ_j stand for the parameters p_{1+}, p_{+1} and δ. Recalling that $E(n_{ij}) = n\, p_{ij}$, we have

$$\frac{\gamma_{11}}{n} = p_{11}\left(\frac{p_{+1}}{\delta + p_{1+}p_{+1}}\right)^2 + p_{10}\left(\frac{q_{+1}}{p_{1+}q_{+1} - \delta}\right)^2 - p_{01}\left(\frac{p_{+1}}{p_{+1}q_{1+} - \delta}\right)^2 - p_{00}\left(\frac{q_{+1}}{q_{1+}q_{+1} + \delta}\right)^2$$

$$\frac{\gamma_{22}}{n} = p_{11}\left(\frac{p_{1+}}{\delta + p_{1+}p_{+1}}\right)^2 - p_{10}\left(\frac{p_{1+}}{p_{1+}q_{+1} - \delta}\right)^2 + p_{01}\left(\frac{q_{1+}}{p_{+1}q_{1+} - \delta}\right)^2 - p_{00}\left(\frac{q_{1+}}{q_{1+}q_{+1} + \delta}\right)^2$$

$$\frac{\gamma_{33}}{n} = \frac{p_{11}}{(\delta + p_{1+}p_{+1})^2} - \frac{p_{10}}{(p_{1+}q_{+1} - \delta)^2} - \frac{p_{01}}{(p_{+1}q_{1+} - \delta)^2} + \frac{p_{00}}{(q_{1+}q_{+1} + \delta)^2}$$

$$\frac{\gamma_{12}}{n} = -\frac{p_{11}}{(\delta + p_{1+}p_{+1})} + \frac{p_{11}p_{1+}p_{+1}}{(\delta + p_{1+}p_{+1})^2} + \frac{p_{10}}{(p_{1+}q_{+1} - \delta)} - \frac{p_{10}p_{1+}q_{+1}}{(p_{1+}q_{+1} - \delta)^2} + \frac{p_{01}}{(p_{+1}q_{1+} - \delta)} - \frac{p_{01}p_{+1}q_{1+}}{(p_{+1}q_{1+} - \delta)^2} - \frac{p_{00}}{(q_{1+}q_{+1} + \delta)} + \frac{p_{00}q_{1+}q_{+1}}{(q_{1+}q_{+1} + \delta)^2}$$

$$\frac{\gamma_{13}}{n} = \frac{p_{11}p_{+1}}{(\delta + p_{1+}p_{+1})^2} - \frac{p_{10}q_{+1}}{(p_{1+}q_{+1} - \delta)^2} + \frac{p_{01}p_{+1}}{(p_{+1}q_{1+} - \delta)^2} - \frac{p_{00}q_{+1}}{(q_{1+}q_{+1} + \delta)^2}$$

$$\frac{\gamma_{23}}{n} = \frac{p_{11}p_{+1}}{(\delta + p_{1+}p_{+1})^2} + \frac{p_{10}p_{1+}}{(p_{1+}q_{+1} - \delta)^2} - \frac{p_{01}q_{1+}}{(p_{+1}q_{1+} - \delta)^2} - \frac{p_{00}q_{1+}}{(q_{1+}q_{+1} + \delta)^2}.$$

These quantities are simplified if we replace the parameters by their maximum likelihood estimates given in (3.7.1):

$$\hat{\gamma}_{11} = \frac{n_{+0}^2}{n_{00}n_{10}} + \frac{n_{+1}^2}{n_{01}n_{11}},$$

$$\hat{\gamma}_{22} = \frac{n_{0+}^2}{n_{00}n_{01}} + \frac{n_{1+}^2}{n_{10}n_{11}},$$

$$\hat{\gamma}_{33} = n^2\left[\frac{1}{n_{00}} + \frac{1}{n_{10}} + \frac{1}{n_{01}} + \frac{1}{n_{11}}\right],$$

$$\hat{\gamma}_{12} = \frac{n_{0+}n_{+0}}{n_{00}} - \frac{n_{0+}n_{+1}}{n_{01}} - \frac{n_{1+}n_{+0}}{n_{10}} + \frac{n_{1+}n_{+1}}{n_{11}},$$

$$\hat{\gamma}_{13} = n\left[\frac{n_{+1}^2}{n_{01}n_{11}} - \frac{n_{+0}^2}{n_{00}n_{10}}\right],$$

$$\hat{\gamma}_{23} = n\left[\frac{n_{1+}^2}{n_{10}n_{11}} - \frac{n_{0+}^2}{n_{00}n_{10}}\right].$$

Hamdan and Martinson (1971) have also considered the estimation of p_{1+}, p_{+1} and δ using minimum chi-square, modified minimum chi-square and minimum Kullback-Liebler separator methods.

Example 3.7.3

The following data are quoted in Kendall and Stuart (1979, p. 582). Forty-two children were classified according to the nature of their teeth and the type of feeding they received:

	Teeth		
	Normal	Mal-occluded	
Breast-fed	4	16	20
Bottle-fed	1	21	22
	5	37	42

Using the estimators in (3.7.1), we have

$$\hat{p}_{1+} = P\text{ (breast-fed child)} = \frac{20}{42} = 0.4762,$$

$$\hat{p}_{+1} = P\text{ (normal teeth)} = \frac{5}{42} = 0.1190,$$

$$\hat{\delta} = \frac{(4)(21) - (16)(1)}{(42)(42)} = 0.0385$$

and

$$\hat{\rho} = \frac{0.0385}{[(.4762)(.5238)(.1190)(.8810)]^{\frac{1}{2}}} = 0.2381.$$

4

BIVARIATE POISSON DISTRIBUTION

4.1 Introduction

In the univariate case, the Poisson distribution arises in several ways. One is as the limit of the binomial distribution when the index parameter is allowed to go to infinity and the probability of success tends to become small with a restriction on the mean being fixed. Perhaps, a more realistic way of generating the distribution is to consider physical processes in which two conditions hold: changes are homogeneous in time and the future changes are independent of the past experience. In other words, the process is such that the probability of the event of interest remains constant over intervals of the same length, unaffected by the location of the interval and the past history of the process. Under these assumptions the probability distribution of the random variable measuring the number of times the event of interest has occurred in an interval of specified length can be shown to have the Poisson distribution. For an excellent discussion of the various situations in which the distribution arises we refer to Feller (1959).

4.2 Models

In the bivariate case the distribution is seen to arise in several ways, somewhat similar in nature to the ones in the univariate case.

(i) Campbell (1934) derives the distribution as the limit of the bivariate binomial distribution with probability generating function

$$\Pi(t_1, t_2) = [1 + p_{1+}(t_1 - 1) + p_{+1}(t_2 - 1) + p_{11}(t_1 - 1)(t_2 - 1)]^n. \quad (4.2.1)$$

Now let λ_1, λ_2 and λ_{11} be positive constants independent of n. Put

$$p_{1+} = \frac{\lambda_1}{n}, \quad p_{+1} = \frac{\lambda_2}{n}, \quad p_{11} = \frac{\lambda_{11}}{n}$$

and write

$$\Pi_n(t_1, t_2) = [1 + \frac{\lambda_1}{n}(t_1 - 1) + \frac{\lambda_2}{n}(t_2 - 1) + \frac{\lambda_{11}}{n}(t_1 - 1)(t_2 - 1)]^n. \quad (4.2.2)$$

Taking limits as $n \to \infty$, we have

$$\Pi(t_1, t_2) = \exp[\lambda_1 (t_1 - 1) + \lambda_2 (t_2 - 1) + \lambda_{11} (t_1 - 1)(t_2 - 1)]. \quad (4.2.3)$$

The random variables (X, Y) are said to have the bivariate Poisson distribution with the parameters λ_1, λ_2 and λ_{11}.

(ii) The trivariate reduction method introduced in Chapter 1 can be used to construct the distribution. Let W_1, W_2 and W_3 be independent Poisson random variables with the parameters λ_1, λ_2 and λ_3, respectively. Let $X = W_1 + W_3$, and $Y = W_2 + W_3$. Then the joint pgf of (X, Y) is

$$\Pi(t_1, t_2) = E[t_1^X t_2^Y]$$

$$= E[t_1^{W_1} t_2^{W_2} (t_1 t_2)^{W_3}],$$

which gives

$$\Pi(t_1, t_2) = \exp[\lambda_1^* (t_1 - 1) + \lambda_2^* (t_2 - 1) + \lambda_3(t_1 - 1)(t_2 - 1)], \quad (4.2.4)$$

where $\lambda_1^* = \lambda_1 + \lambda_3$ and $\lambda_2^* = \lambda_2 + \lambda_3$.

This method is discussed by Holgate (1964) for providing a practical basis for the distribution.

(iii) An alternative method of arriving at the bivariate Poisson distribution is by the process of compounding a bivariate binomial with a Poisson distribution. Thus, given the parameter n, let (X, Y) have the joint pgf

$$\Pi(t_1, t_2 | n) = [1 + p_{1+}(t_1 - 1) + p_{+1}(t_2 - 1) + p_{11}(t_1 - 1)(t_2 - 1)]^n.$$

Let the parameter n have the Poisson distribution with the parameter λ.

Then the unconditional distribution of (X, Y) has the pgf

$$\Pi(t_1, t_2) = \sum_{n=0}^{\infty} \frac{e^{-\lambda} \lambda^n}{n!} \Pi(t_1, t_2| n).$$

Upon summing, the right hand side yields

$$\Pi(t_1, t_2) = \exp [\lambda_1 (t_1 - 1) + \lambda_2 (t_2 - 1) + \lambda_3(t_1 - 1)(t_2 - 1)], \quad (4.2.5)$$

where $\lambda_1 = \lambda\, p_{1+}$, $\lambda_2 = \lambda\, p_{+1}$ and $\lambda_3 = \lambda\, p_{11}$.

(iv) The most interesting development of the bivariate Poisson distribution is due to McKendrick (1926) who foreshadowed many of the modern ideas. Unfortunately, his paper failed to receive the attention that is due such an original and thought provoking piece of work. Irwin (1963) brought it out of the past and has given an excellent discussion. We reproduce the model here and derive the distribution.

Consider a grid in which events take place along the coordinate axes and the diagonal of each square. Let $p_{x,y}$ be the joint probability of $\{X = x, Y = y\}$. For instance, X could represent the number of accidents sustained by an individual in this quarter and Y the accidents in the previous quarter. Let the probability of transition over the x-axis or parallel to it be f which is assumed constant over time or space. Similarly, transition along the y-axis or parallel to it has probability g again assumed constant. In addition, transition along the diagonal has a constant probability h.

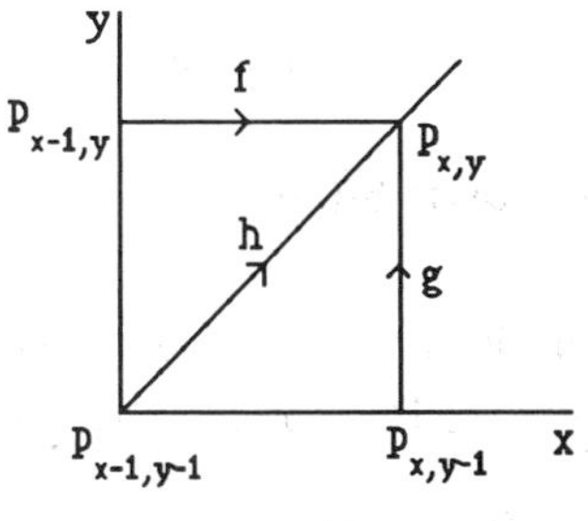

Then a change in the probability at (X=x, Y=y) in the time interval (t, t+dt) is

$$dp_{x,y} = dt\,[(f\,p_{x-1,y} - f\,p_{x,y}) + (g\,p_{x,y-1} - g\,p_{x,y}) + (h\,p_{x-1,y-1} - h\,p_{x,y})], \tag{4.2.6}$$

or

$$\frac{dp_{x,y}}{dt} = f\,(p_{x-1,y} - p_{x,y}) + g\,(p_{x,y-1} - p_{x,y}) + h\,(p_{x-1,y-1} - p_{x,y}), \tag{4.2.7}$$

which can be seen to be a differential-difference equation. To solve the equation, introduce the pgf $\Pi(t_1, t_2)$ and assume that the order of differentiation and addition can be interchanged. This gives

$$\begin{aligned}\frac{d}{dt}\Pi(t_1, t_2) &= f\sum_x\sum_y (p_{x-1,y} - p_{x,y})t_1^{\,x}\,t_2^{\,y} + g\sum_x\sum_y (p_{x,y-1} - p_{x,y})t_1^{\,x}\,t_2^{\,y} \\ &\qquad + h\sum_x\sum_y (p_{x-1,y-1} - p_{x,y})t_1^{\,x}\,t_2^{\,y} \\ &= (t_1 - 1)\,f\,\Pi(t_1, t_2) + (t_2 - 1)\,g\,\Pi(t_1, t_2) + (t_1\,t_2 - 1)\,h\,\Pi(t_1, t_2).\end{aligned} \tag{4.2.8}$$

The solution for equation (4.2.8) is

$$\Pi(t_1, t_2) = \exp[(t_1 - 1)\,(f + h)t + (t_2 - 1)\,(g + h)t + (t_1 - 1)(t_2 - 1)\,ht]. \tag{4.2.9}$$

Once again, this is a bivariate Poisson distribution with the parameters $\lambda_1 = (f + h)t$, $\lambda_2 = (g + h)t$ and $\lambda_3 = ht$.

When the probability h is zero then the transition takes place only along the axes, or parallel to them, and the random variables are readily seen to be independent. This stands to reason, as in this case the joint transition involving X and Y coordinates along the diagonal is precluded.

4.3 Properties of the distribution

As has been seen in the preceding models the probability generating function of the bivariate Poisson distribution can be written as

$$\Pi(t_1, t_2) = \exp[\lambda_1\,(t_1 - 1) + \lambda_2\,(t_2 - 1) + \lambda_3\,(t_1\,t_2 - 1)], \tag{4.3.1}$$

which is obtained from equation (4.2.3) by a rearrangement of the terms and a reparameterization. In what follows we will be taking (4.3.1) to be the joint pgf of the random variables (X, Y). However, we may sometimes refer to the pgf in the form

$$\Pi(t_1, t_2) = \exp[\lambda_1^* (t_1 - 1) + \lambda_2^* (t_2 - 1) + \lambda_3 (t_1 - 1)(t_2 - 1)], \qquad (4.3.2)$$

where $\lambda_i^* = \lambda_i + \lambda_3$, for i = 1, 2. This form may be found to be more appropriate in some instances for the derivation of the properties of the distribution, as will be seen in later sections.

Marginal distributions

The marginal generating functions are readily found from joint pgf (4.3.1) to be

$$\Pi_i(t) = \exp[(\lambda_i + \lambda_3)(t - 1)], \qquad i = 1, 2 \qquad (4.3.3)$$

from which the marginal distributions are each seen to be

$$X \sim P(\lambda_1 + \lambda_3) \text{ and } Y \sim P(\lambda_2 + \lambda_3).$$

An examination of the relationship of the joint pgf to the marginal pgf's shows that a necessary and sufficient condition for the independence of the random variables X and Y is that the parameter λ_3 be equal to zero.

Probability function

The joint probability function can be determined by expanding the joint pgf (4.3.1) in powers of t_1 and t_2 as

$$\Pi(t_1, t_2) = e^{-(\lambda_1+\lambda_2+\lambda_3)} \sum_{i=0}^{\infty} \frac{\lambda_1^i t_1^i}{i!} \sum_{j=0}^{\infty} \frac{\lambda_2^j t_2^j}{j!} \sum_{k=0}^{\infty} \frac{\lambda_3^k t_1^k t_2^k}{k!}$$

$$= e^{-(\lambda_1+\lambda_2+\lambda_3)} \sum_{r,s} \sum_{i} \left\{ \frac{\lambda_1^{r-i} \lambda_2^{s-i} \lambda_3^i}{(r-i)!\,(s-i)!\,i!} \right\} t_1^r t_2^s.$$

Hence the coefficient of $t_1^r t_2^s$ is

$$f(r, s) = e^{-(\lambda_1+\lambda_2+\lambda_3)} \sum_{i=0}^{\min(r,s)} \frac{\lambda_1^{r-i}\lambda_2^{s-i}\lambda_3^{i}}{(r-i)!\,(s-i)!\,i!}. \tag{4.3.4}$$

Replacing the parameters λ_i in terms of λ_i^*, for i = 1, 2, we have the representation in the form

$$f(r, s) = e^{-(\lambda_1^*+\lambda_2^* - \lambda_3)} \sum_{i=0}^{\min(r,s)} \frac{(\lambda_1^* - \lambda_3)^{r-i}(\lambda_2^* - \lambda_3)^{s-i}\lambda_3^{i}}{(r-i)!\,(s-i)!\,i!}. \tag{4.3.5}$$

Holgate (1964) gives the probability function in this form. It is obtained from the trivariate reduction method by examining the probability of the event {X = r, Y = s}. This event can be written equivalently as the union of the mutually exclusive events $\{W_1 = r - i, W_2 = s - i, W_3 = i\}$ for i = 0, 1, ..., min(r, s). Hence, the representation for the probability f(r, s) as in equation (4.3.5).

An alternative way of writing the probability function in (4.3.3) is

$$f(r, s) = \frac{e^{-\lambda_1}\lambda_1^r}{r!}\frac{e^{-\lambda_2}\lambda_2^s}{s!}e^{-\lambda_3}\sum_{i=0}^{\min(r,s)}\binom{r}{i}\binom{s}{i}i!\left\{\frac{\lambda_3}{\lambda_1\lambda_2}\right\}^i. \tag{4.3.6}$$

For $\lambda_3 = 0$, this reduces to

$$f(r, s) = \frac{e^{-\lambda_1}\lambda_1^r}{r!}\frac{e^{-\lambda_2}\lambda_2^s}{s!}, \tag{4.3.7}$$

showing that in this case $X \sim P(\lambda_1)$, $Y \sim P(\lambda_2)$, independently of each other. In the literature this distribution has been referred to as the double Poisson distribution.

Recurrence relations

Teicher (1954) gives the recurrence relations for the probability function (4.3.4) as

$$\left.\begin{aligned} r\,f(r, s) &= \lambda_1 f(r-1, s) + \lambda_3 f(r-1, s-1) \\ s\,f(r, s) &= \lambda_2 f(r, s-1) + \lambda_3 f(r-1, s-1) \end{aligned}\right\}. \tag{4.3.8}$$

The probabilities for $r \geq 1$, $s \geq 1$ can be computed recursively given the probabilities in the first row and column. The terms in the first row and column can be computed using the tables of the univariate Poisson distribution, as is seen from:

$$\left.\begin{aligned} f(0, 0) &= e^{-(\lambda_1 + \lambda_2 + \lambda_3)} \\ f(r, 0) &= e^{-(\lambda_2 + \lambda_3)} p(r; \lambda_1) \qquad r > 0 \\ f(0, s) &= e^{-(\lambda_1 + \lambda_3)} p(s; \lambda_2) \qquad s > 0 \end{aligned}\right\}.$$

Factorial moments and cumulants

The factorial moment generating function is given by

$$G(t_1, t_2) = \exp[(\lambda_1 + \lambda_3)t_1 + (\lambda_2 + \lambda_3)t_2 + \lambda_3 t_1 t_2]$$

$$= \sum_r \sum_s \{ \sum_i \frac{(\lambda_1 + \lambda_3)^{r-i} (\lambda_2 + \lambda_3)^{s-i} \lambda_3^{\,i}}{(r-i)!\,(s-i)!\,i!} \} t_1^r t_2^s .$$

The (r, s)th factorial moment is given as the coefficient of $\frac{t_1^r}{r!} \frac{t_2^s}{s!}$ in this expansion. Thus

$$\mu_{[r,s]} = \sum_{i=o}^{\min(r,s)} \binom{r}{i} \binom{s}{i} (\lambda_1 + \lambda_3)^{r-i} (\lambda_2 + \lambda_3)^{s-i}\, i!\, \lambda_3^i$$

$$= (\lambda_1 + \lambda_3)^r (\lambda_2 + \lambda_3)^s \sum_{i=o}^{\min(r,s)} \binom{r}{i} \binom{s}{i} i!\, \tau^i , \qquad (4.3.9)$$

where

$$\tau = \frac{\lambda_3}{[(\lambda_1 + \lambda_3)(\lambda_2 + \lambda_3)]} .$$

In particular, if $r = s = 1$

$$\mu'_{1,1} = (\lambda_1 + \lambda_3)(\lambda_2 + \lambda_3)[1 + \tau],$$

or

$$\text{Cov}(X, Y) = \lambda_3.$$

Hence

$$\rho_{XY} = \frac{\lambda_3}{[(\lambda_1 + \lambda_3)(\lambda_2 + \lambda_3)]^{\frac{1}{2}}} . \tag{4.3.10}$$

Since $\lambda_3 \geq 0$, it follows that for this model $\rho_{XY} \geq 0$. This shows that the condition of zero correlation is a necessary and sufficient condition for the independence of the random variables X and Y. Other forms of the distribution have been suggested in the literature that result in negative correlation for the random variables (X, Y). These will be mentioned in a later section.

The factorial cumulants generating function is

$$\begin{aligned} H(t_1, t_2) &= \log G(t_1, t_2) \\ &= (\lambda_1 + \lambda_3)t_1 + (\lambda_2 + \lambda_3)t_2 + \lambda_3 t_1 t_2, \end{aligned} \tag{4.3.11}$$

which shows that the factorial cumulants are

$$\kappa_{[1,0]} = \lambda_1 + \lambda_3 , \quad \kappa_{[0,1]} = \lambda_2 + \lambda_3 \quad \text{and} \quad \kappa_{[1,1]} = \lambda_3$$

while all other factorial cumulants are zero. This shows that the ordinary cumulants of these orders have these values while cumulants of higher order are zero.

Conditional distribution and regression

The conditional pgf of X given Y=y is from Theorem 1.3.1

$$\Pi_X(t|\, y) = e^{\lambda_1(t-1)} \left\{\frac{\lambda_2 + \lambda_3 t}{\lambda_2 + \lambda_3}\right\}^y , \tag{4.3.12}$$

which shows that the conditional distribution is the convolution of

$$X_1 \sim P(\lambda_1) \quad \text{and} \quad X_2 \sim B\left(y, \frac{\lambda_3}{\lambda_2 + \lambda_3}\right).$$

The regression of X on Y is thus seen to be

$$E[X|\, y] = \lambda_1 + \frac{\lambda_3}{\lambda_2 + \lambda_3}\, y, \tag{4.3.13}$$

which is linear in y. By the same argument the conditional variance is

$$\mathrm{Var}[X|\,y] = \mathrm{Var}[X_1] + \mathrm{Var}[X_2]$$

$$= \lambda_1 + [\frac{\lambda_3}{\lambda_2 + \lambda_3}][\frac{\lambda_2}{\lambda_2 + \lambda_3}]\,y. \qquad (4.3.14)$$

Similarly, the conditional distribution of Y given X = x is the convolution of $Y_1 \sim P(\lambda_2)$ and $Y_2 \sim B\left(x, \frac{\lambda_3}{\lambda_1 + \lambda_3}\right)$. From this the conditional expectation and variance are seen to be

$$E[Y|\,x] = \lambda_2 + \frac{\lambda_3}{\lambda_1 + \lambda_3}\,x, \qquad (4.3.15)$$

$$\mathrm{Var}[Y|\,x] = \lambda_2 + [\frac{\lambda_3}{\lambda_1 + \lambda_3}][\frac{\lambda_1}{\lambda_1 + \lambda_3}]\,x. \qquad (4.3.16)$$

From the linearity of the regressions and the regression coefficients, it can be verified that the coefficient of correlation is as given in equation (4.3.10).

4.4 Related distributions

Several related distributions have been considered in the literature. These are connected with the variety of operations that can be performed on the probability function of the distribution.

Distribution of the sum

Setting $t_1 = t_2$ in equation (4.3.1) we obtain the pgf of the sum Z = X + Y as

$$\Pi_Z(t) = \exp[(\lambda_1 + \lambda_2)(t - 1) + \lambda_3(t^2 - 1)]. \qquad (4.4.1)$$

Kemp and Kemp (1965) refer to the distribution of the random variable Z in this case as the Hermite distribution.

If the parameter λ_3 is zero, that is the random variables X and Y are independent, then the sum Z is once again distributed as a Poisson random variable with the parameter $\lambda_1 + \lambda_2$.

Compound distributions

Kocherlakota (1988) considers a class of distributions that arise out of compounding with the bivariate Poisson distribution. Details are provided in a later chapter. He looks at the joint pgf

$$\Pi(t_1, t_2|\tau) = \exp[\tau\{\lambda_1 (t_1 - 1) + \lambda_2 (t_2 - 1) + \lambda_3 (t_1 t_2 - 1)\}], \qquad (4.4.2)$$

where τ is a random component with the moment generating function M(t) and the probability density function g(s). Then the marginal pgf of the random variables (X, Y) is given by

$$\Pi(t_1, t_2) = \int \Pi(t_1, t_2|s)\, g(s)\, ds.$$

Upon substituting from (4.3.1) for the pgf in the integral, we have

$$\begin{aligned}\Pi(t_1, t_2) &= \int \exp[s\{\lambda_1 (t_1 - 1) + \lambda_2 (t_2 - 1) + \lambda_3 (t_1 t_2 - 1)\}]\, g(s)\, ds \\ &= M[\{\lambda_1 (t_1 - 1) + \lambda_2 (t_2 - 1) + \lambda_3 (t_1 t_2 - 1)\}]. \qquad (4.4.3)\end{aligned}$$

In this development τ has been taken to be a continuous random variable. If, however, the random variable is a discrete random variable with the probability function g(s) the same result is obtained for the unconditional pf.

Bivariate distributions with fixed marginals

Griffiths *et al.* (1979) have shown that a general class of bivariate distributions can be generated with fixed Poisson marginals. This is done by taking

$$\begin{aligned}\Pi(t_1, t_2|\mu) &= \exp[\lambda_1 (t_1 - 1) + \lambda_2 (t_2 - 1) + \mu (t_1 t_2 - 1)] \\ &= \exp[(\lambda_1 + \mu)(t_1 - 1) + (\lambda_2 + \mu)(t_2 - 1) + \mu(t_1 - 1)(t_2 - 1)]\end{aligned}$$

Writing $\lambda_i + \mu = \tau_i$, i=1, 2 and letting K be any distribution function over $[0, \min(\tau_1, \tau_2)]$, the unconditional distribution of (X, Y) is seen to have the pgf

$$\Pi(t_1, t_2) = \int \Pi(t_1, t_2|\mu)\, dK(\mu)$$

or, putting $\tau = \min(\tau_1, \tau_2)$,

$$\Pi(t_1, t_2) = \exp[\tau_1(t_1 - 1) + \tau_2(t_2 - 1)] \int_0^{\tau} \exp[\mu(t_1 - 1)(t_2 - 1)]\, dK(\mu) \tag{4.4.4}$$

It is clear that the pgf (4.4.4) gives rise to Poisson marginals while giving different "bivariate Poisson" distributions depending upon the choice of K. The term "bivariate Poisson" distribution is used by Griffiths and others to connote a general class of distributions having Poisson marginals. If the distribution function K is degenerate the pgf (4.4.4) reduces to form (4.3.2). We will have the opportunity to revisit this distribution later in connection with the correlation in the bivariate Poisson distribution.

Truncated distributions

The truncated distributions arise in practice through the nonobservability of certain values of the random variables. This could be caused by the differential viability of the organisms.

For example, when the distribution is truncated at (0, 0), (0, 1) and (1, 0) the probability function is given by

$$f_T(r, s) = \frac{f(r, s)}{[1 - \{f(0, 0) + f(1, 0) + f(0, 1)\}]}, \tag{4.4.5}$$

for $r = 0, 1, 2, \ldots$; $s = 0, 1, 2, \ldots$ but not (0, 0), (1, 0) and (0, 1).

The problem of estimation has received considerable attention in the literature. This will be examined in a later section.

4.5 Polynomial representation

As discussed in section 1.5 orthogonal polynomials have been used to represent bivariate probability functions. In the univariate case Charlier polynomials are related to the Poisson probability function being orthogonal with respect to this probability function. This relationship has been exploited to construct an orthogonal polynomial representation for the bivariate Poisson distribution.

Charlier polynomials

The Charlier polynomials are defined as

$$K_r(x;\lambda) = \left[x^{[r]} - r\lambda x^{[r-1]} + \binom{r}{2}\lambda^2 x^{[r-2]} - \dots + (-1)^r \lambda^r\right] / \lambda^r \qquad (4.5.1)$$

These polynomials are orthogonal with respect to the Poisson probability function $p(x; \lambda)$. Thus it is readily shown that

$$\sum_{x=0}^{\infty} K_r(x;\lambda) K_s(x;\lambda) p(x;\lambda) = \frac{r!\,\delta_{rs}}{\lambda^r} \qquad (4.5.2)$$

where δ_{rs} is the Kronecker δ assuming the value 1 if r=s and zero otherwise.

Campbell (1934) has shown that the probability function (4.3.4) can be represented in terms of these polynomials as

$$f(x, y) = f_1(x)\, f_2(y) \left[1 + \sum_{r=1}^{\infty} \frac{\lambda_3^r}{r!} K_r(x; \lambda_1 + \lambda_3) K_r(y; \lambda_2 + \lambda_3)\right], \qquad (4.5.3)$$

where the functions f_1 and f_2 are, respectively, the Poisson probability functions $p(x; \lambda_1 + \lambda_3)$ and $p(y; \lambda_2 + \lambda_3)$.

To normalize the polynomials we redefine them as

$$K_r^*(x;\lambda) = K_r(x;\lambda) \left[\frac{\lambda^r}{r!}\right]^{\frac{1}{2}}.$$

Replacing the polynomials in equation (4.5.3) by the orthonormal versions, we have

$$f(x, y) = f_1(x)\, f_2(y) \left[1 + \sum_{r=1}^{\infty} \rho^r K_r^*(x; \lambda_1 + \lambda_3) K_r^*(y; \lambda_2 + \lambda_3)\right], \qquad (4.5.4)$$

where the parameter ρ is the coefficient of correlation defined in equation (4.3.10). From the discussion in section 1.5 it is clear that the canonical correlation of order r is ρ^r. The regression of $K_r^*(x; \lambda_1+\lambda_3)$ on $K_r^*(y; \lambda_2+\lambda_3)$ is seen to be

$$E[K_r^*(x; \lambda_1 + \lambda_3) \mid K_r^*(y; \lambda_2 + \lambda_3)] = \rho^r K_r^*(y; \lambda_2 + \lambda_3), \qquad (4.5.5)$$

which is linear in the conditioning polynomial. In particular for r=1, since $K_1^*(x; \lambda) = (x - \lambda) / \lambda^{\frac{1}{2}}$, we have

$$E[\{X - (\lambda_1 + \lambda_3)\}/(\lambda_1 + \lambda_3)^{\frac{1}{2}} \mid \{y - (\lambda_2 + \lambda_3)\}/(\lambda_2 + \lambda_3)^{\frac{1}{2}}] = \frac{\rho}{(\lambda_2 + \lambda_3)^{\frac{1}{2}}} \{y - (\lambda_2 + \lambda_3)\}.$$

Or, substituting for the value of ρ and simplifying,

$$E[X \mid \{y - (\lambda_2 + \lambda_3)\}/(\lambda_2 + \lambda_3)^{\frac{1}{2}}] = \frac{\lambda_3 \, y}{(\lambda_2 + \lambda_3)} + \lambda_1$$
$$= E[X \mid y].$$

4.6 A note on the correlation: Infinite divisibility

It will be noted that the coefficient of correlation for the probability function in (4.3.1) is always nonnegative. This is obviously a requirement resulting from the nonnegative nature of λ_3. Also it is a necessary condition for the pgf

$$\Pi(t_1, t_2) = \exp[\lambda_1 (t_1 - 1) + \lambda_2 (t_2 - 1) + \lambda_3 (t_1 - 1)(t_2 - 1)] \qquad (4.6.1)$$

to be a proper bivariate probability generating function.

As pointed out earlier Dwass and Teicher (1957) have established that the only infinitely divisible bivariate Poisson distributions are those for which the pgf's are of the form (4.6.1). They further establish that for any pair of random variables (X, Y) with the joint pgf (4.6.1) there exists a decomposition of X and Y as X = U + W and Y = V + W, where U, V, W are independent Poisson random variables. From this decomposition it follows that the Cov(X, Y) = Var(W) ≥ 0. This once again reasserts the fact that only nonnegative correlation is permissible in the distributions of which the joint pgf is (4.6.1).

Further, we have seen that a class of bivariate Poisson distributions can be constructed by considering

$$\Pi(t_1, t_2) = \exp[\tau_1(t_1 -1) + \tau_2(t_2 -1)] \int_0^{\tau} \exp[\mu(t_1 -1)(t_2 -1)]\, dK(\mu) \tag{4.6.2}$$

where $\tau = \min(\tau_1, \tau_2)$ and K is any distribution function on $(0, \tau)$. When K is degenerate then this yields the bivariate Poisson distribution of the form (4.6.1). From the Dwass-Teicher result it follows that this distribution is not an infinitely divisible form of the bivariate Poisson distribution *except when* K *is degenerate.* However, from the construction of the pgf in (4.6.2) it is obvious that the distributions so generated will all have a nonnegative correlation. Griffiths *et al.* (1979) quote an unpublished result of Wood (1970) in which a study of the canonical representation of the bivariate distribution of the type in (4.6.2) is made. The canonical representation leads Wood to state that any bivariate Poisson distribution of which the matrix of canonical correlations with respect to the Charlier polynomials is diagonal must have the pgf of the form in (4.6.2). As pointed out by Griffiths *et al.* implicit in this result is the fact that *in the bivariate Poisson distributions of which the matrix of canonical correlations with respect to the Charlier polynomials is diagonal only nonnegative correlation is permissible.*

The question one would like to answer at this point is whether it is possible to construct non-infinitely divisible distributions with *negative correlation.* Indeed it is possible to do so as is shown in the following examples taken from Griffiths *et al.* (1979) :

Example 4.6.1

For $\{(r, s) : r = 0, 1, 2, \dots ; s = 0, 1, 2, \dots\}$ let

$$f(r, s) = p(r; 1)\, p(s; 1)\, \{1 + \alpha\, \psi(r)\, \psi(s)\} \tag{4.6.3}$$

where α is a constant and $\psi(.)$ maps (0, 1, 2, ...) to the real line. Choose ψ so that $|\psi(x)| \leq x$, x = 0, 1, 2, ... and $E[\psi(X)] = 0$. Then $f(r, s) \geq 0$ for values of $\alpha \in [-1, 1]$. Also each of the marginals in the joint distribution

(4.6.3) is Poisson(1). Now

$$\text{Cov}(X, Y) = \alpha\{\sum_{r=0}^{\infty} r\,\psi(r)\,p(r; 1)\}^2,$$

which shows that the covariance is negative if we choose $\alpha < 0$. This makes (4.6.3) a bivariate Poisson distribution with negative correlation. Referring to the canonical representation it is readily seen that (4.6.3) has a single pair of canonical variables.

Example 4.6.2

A bivariate distribution $\{f(x, y)\colon\ x = 0, 1, 2, \ldots ; y = 0, 1, 2, \ldots\}$ is constructed with specified marginals $\{g(i)\colon\ i = 0, 1, 2, \ldots\}$ as follows:

(i) for $x \geq 3$ and $y \geq 3$ let

$$f(x, y) = g^*(x)\,g^*(y) \quad \text{where } g^*(i) = \frac{g(i)}{[\sum_{k=3}^{\infty} g(i)]^{\frac{1}{2}}}.$$

(ii) $f(0, 2) = f(2, 0) = g(0)$, $f(1, 1) = g(1)$, $f(2, 2) = g(2) - g(0)$.

(iii) $f(x, y) = 0$ for all other (x, y).

In the particular case when the marginal specification is the univariate Poisson distribution, the joint distribution $f(x, y)$ is the bivariate Poisson as each marginal is univariate Poisson distribution. The coefficient of correlation is negative, the covariance being equal to

$$\frac{(13 - 5e)}{(2e^2 - 5e)} < 0.$$

Griffiths *et al.* (1979) give a general technique for constructing distributions with negative correlation from independent, and hence zero-correlated, random variables by moving the probability mass away from the *diagonal*. This results in a reduction of the correlation. The reader interested in this and the more general topic of construction of bivariate distributions with specified marginals is referred to the papers of Griffiths and his colleagues and of Nelsen (1987).

4.7 Estimation

The problem of estimation of the parameters of complete and truncated bivariate Poisson distributions has received much attention in the literature. The problems encountered in estimating the parameters of the truncated distribution are complex. In this respect they are distinct from those arising in the case of the complete distribution. Also, when the parameter λ_3 is zero the distribution is that of two independent Poisson random variables. In this case the estimation is very much simplified. We will consider various aspects of the techniques available for the estimation of the parameters of the complete distribution. The truncated distribution will be examined briefly.

4.7.1 Complete distribution

Let (x_i, y_i), i= 1, 2, ... n, be a random sample of size n from the population (4.7.1). We will assume that the frequency of the pair (r, s) is n_{rs} for r= 0, 1, 2, ..., s= 0, 1, 2, ... We recall that $\sum_{r,s} n_{rs} = n$. Also

$$\left.\begin{aligned} \bar{x} &= \frac{1}{n}\sum_{i=1}^{n} x_i = \frac{1}{n}\sum_{r=0}\sum_{s=0} r\, n_{rs} \\ \bar{y} &= \frac{1}{n}\sum_{i=1}^{n} y_i = \frac{1}{n}\sum_{r=0}\sum_{s=0} s\, n_{rs} \\ m_{1,1} &= \frac{1}{n}\sum_{i=1}^{n} \{x_i - \bar{x}\}\{y_i - \bar{y}\} = \frac{1}{n}\sum_{r=0}\sum_{s=0} r\, s\, n_{rs} - \bar{x}\,\bar{y} \end{aligned}\right\}. \tag{4.7.1}$$

Method of moments

Since $E[X] = \lambda_1^*$, $E[Y] = \lambda_2^*$ and $Cov(X, Y) = \lambda_3$, the moment estimators are readily seen to be

$$\tilde{\lambda}_1^* = \bar{x}, \quad \tilde{\lambda}_2^* = \bar{y}, \quad \tilde{\lambda}_3 = m_{1,1}.$$

The variance matrix of the estimators can be found from the results given in section 2.1.1. This is

$$\Sigma_{MM} = \frac{1}{n}\begin{bmatrix} \lambda_1^* & \lambda_3 & \lambda_3 \\ \cdots & \lambda_2^* & \lambda_3 \\ \cdots & \cdots & \lambda_1^*\lambda_2^*+\lambda_3(1+\lambda_3) \end{bmatrix}. \tag{4.7.2}$$

Zero-zero cell frequency method

The (0, 0) cell frequency combined with the two marginal moments has sometimes been used in the literature to give a quick set of estimators for the parameters. It is clear that the estimates for the first two parameters remain unchanged while parameter λ_3 is determined from the equation

$$\log\left[\frac{n_{00}}{n}\right] = -(\tilde{\lambda}_1^* + \tilde{\lambda}_2^* - \approx{\lambda}_3), \tag{4.7.3}$$

or

$$\approx{\lambda}_3 = \log\left[\frac{n_{00}}{n}\right] + \tilde{\lambda}_1^* + \tilde{\lambda}_2^*. \tag{4.7.4}$$

The variance matrix of the estimators is found as before from the results in section 2.1.3. Thus for the vector $(\tilde{\lambda}_1^*, \tilde{\lambda}_2^*, \approx{\lambda}_3)$ the variance matrix is seen to be

$$\Sigma_{DZ} = \frac{1}{n}\begin{bmatrix} \lambda_1^* & \lambda_3 & -\lambda_1^*\delta \\ \cdots & \lambda_2^* & -\lambda_2^*\delta \\ \cdots & \cdots & \delta(1-\delta) \end{bmatrix}, \tag{4.7.5}$$

where $\delta = \exp - (\lambda_1^* + \lambda_2^* - \lambda_3)$. It should be recalled that the 2x2 upper left hand submatrix is the same as was found in the method of moments.

Method of maximum likelihood

Writing the joint probability function in terms of the parameters λ_1^*

and λ_2^* we have

$$f(r, s) = e^{-(\lambda_1^*+\lambda_2^*-\lambda_3)} \sum_{i=0}^{\min(r,s)} \frac{(\lambda_1^*-\lambda_3)^{r-i}(\lambda_2^*-\lambda_3)^{s-i}\, \lambda_3^i}{(r-i)!\,(s-i)!\,i!}. \qquad (4.7.6)$$

From this the recurrence relations (4.3.7) take the form

$$\left.\begin{aligned} r\, f(r, s) &= [\lambda_1^* - \lambda_3]f(r-1, s) + \lambda_3 f(r-1, s-1) \\ s\, f(r, s) &= [\lambda_2^* - \lambda_3]f(r, s-1) + \lambda_3 f(r-1, s-1) \end{aligned}\right\}. \qquad (4.7.7)$$

Upon differentiating the probability function with respect to the parameters, we obtain

$$\left.\begin{aligned} \frac{\partial f(r, s)}{\partial \lambda_1^*} &= f(r-1, s) - f(r, s) \\ \frac{\partial f(r, s)}{\partial \lambda_2^*} &= f(r, s-1) - f(r, s) \\ \frac{\partial f(r, s)}{\partial \lambda_3} &= f(r, s) - f(r-1, s) - f(r, s-1) + f(r-1, s-1) \end{aligned}\right\}. \qquad (4.7.8)$$

Using these recurrence relations in the partial derivatives of the logarithm of the likelihood with respect to the parameters, the following likelihood equations are obtained by Holgate (1964):

$$\left.\begin{aligned} \frac{\bar{x}}{[\hat{\lambda}_1^* - \hat{\lambda}_3]} - \frac{\hat{\lambda}_3 \bar{R}}{[\hat{\lambda}_1^* - \hat{\lambda}_3]} - 1 &= 0 \\ \frac{\bar{y}}{[\hat{\lambda}_2^* - \hat{\lambda}_3]} - \frac{\hat{\lambda}_3 \bar{R}}{[\hat{\lambda}_2^* - \hat{\lambda}_3]} - 1 &= 0 \end{aligned}\right\} \qquad (4.7.9a)$$

$$\frac{\bar{x}}{[\hat{\lambda}_1^* - \hat{\lambda}_3]} + \frac{\bar{y}}{[\hat{\lambda}_2^* - \hat{\lambda}_3]} - [1 + \frac{\hat{\lambda}_3}{[\hat{\lambda}_1^* - \hat{\lambda}_3]} + \frac{\hat{\lambda}_3}{[\hat{\lambda}_2^* - \hat{\lambda}_3]}]\bar{R} - 1 = 0, \tag{4.7.9b}$$

where

$$\bar{R} = \frac{1}{n}\sum_{i=1}^{n} \frac{f(x_i - 1,\ y_i - 1)}{f(x_i,\ y_i)}$$
$$= \frac{1}{n}\sum_{r,s} n_{rs} \frac{f(r - 1,\ s - 1)}{f(r,\ s)}. \tag{4.7.10}$$

Solving the equations (4.7.9a and b) we get $\hat{\lambda}_1^* = \bar{x}$, $\hat{\lambda}_2^* = \bar{y}$ while for λ_3

the equation becomes $\bar{R} = 1$ which has to be solved by an iterative

technique. For these techniques $\frac{\partial \bar{R}}{\partial \lambda_3}$ is needed:

$$\frac{\partial \bar{R}}{\partial \lambda_3} = \frac{1}{n}\sum_{r,s} n_{rs} \frac{1}{f(r,\ s)}\frac{\partial f(r - 1,\ s - 1)}{\partial \lambda_3} - \frac{f(r - 1,\ s - 1)}{f^2(r,\ s)}\frac{\partial f(r,\ s)}{\partial \lambda_3}$$

$$= \sum_{r,s} n_{rs} \Big\{ -\frac{f(r - 2,\ s - 1)}{f^2(r,\ s)} - \frac{f(r - 1,\ s - 2)}{f(r,\ s)}$$
$$+ \frac{f(r - 1,\ s - 1)\, f(r - 1,\ s)}{f^2(r,\ s)} + \frac{f(r - 2,\ s - 2)}{f(r,\ s)}$$
$$+ \frac{f(r - 1,\ s - 1)\, f(r,\ s - 1)}{f^2(r,\ s)} - \frac{f^2(r - 1,\ s - 1)}{f^2(r,\ s)}\Big\}.$$

For large n the asymptotic covariance matrix is found as the inverse of the information matrix. With

$$\tau = \sum_{r,s=1}^{\infty} [\frac{f^2(r-1,\ s-1)}{f(r,\ s)}]$$

$$\delta_1 = \lambda_3[1 - \lambda_3(\tau-1)]$$

$$\delta_2 = -(\lambda_1 + \lambda_2) + [\lambda_1^* \lambda_2^* - \lambda_3^2](\tau - 1)$$

$$\delta_3 = [\lambda_1^* \lambda_2^* - \lambda_3^2]\{\tau - 1 - (\lambda_1 + \lambda_2)\}$$

the information matrix can be written as

$$\Gamma = n \begin{bmatrix} \dfrac{\lambda_1 - \delta_1}{\lambda_1^2} & -\dfrac{\delta_1}{\lambda_1\lambda_2} & -\dfrac{\delta_2\lambda_3}{\lambda_1^2\lambda_2} \\ \cdots & \dfrac{\lambda_2 - \delta_1}{\lambda_2^2} & -\dfrac{\delta_2\lambda_3}{\lambda_1\lambda_2^2} \\ \cdots & \cdots & \dfrac{\delta_3}{\lambda_1^2\lambda_2^2} \end{bmatrix},$$

where $\lambda_1 = \lambda_1^* - \lambda_3$ and $\lambda_2 = \lambda_2^* - \lambda_3$.

The variance matrix is the inverse of this matrix and is given by

$$\frac{1}{n}\begin{bmatrix} \lambda_1^* & \lambda_3 & \lambda_3 \\ \cdots & \lambda_2^* & \lambda_3 \\ \cdots & \cdots & \dfrac{\lambda_1\lambda_2 + \lambda_3(\lambda_1 + \lambda_2)[\lambda_3(\tau - 1) - 1]}{\delta_2} \end{bmatrix}. \tag{4.7.11}$$

It should be noted that Holgate's paper has some errors which have been corrected in the above.

Even point estimation

The even point method of estimation was applied to the bivariate Poisson distribution in Loukas *et al.* (1986). In Papageorgiou and Loukas (1988) the method is modified by the use of the conditional probability generating function in place of the pgf.

A. In Loukas *et al.* (1986) the pgf is taken as

$$\Pi(t_1, t_2) = \exp[\lambda_1^* (t_1 -1) + \lambda_2^* (t_2 -1) + \lambda_3 (t_1 -1) (t_2 -1)].$$

As discussed in Chapter 2 the even points method consists of determining the value of

$$\Pi(1,1) + \Pi(-1,-1) = 1 + \exp[-2(\lambda_1^* + \lambda_2^* - 2\lambda_3)]$$

$$= 2 [\sum f(2r, 2s) + \sum f(2r+1, 2s+1)].$$

Upon replacing the probabilities in the last equation by the observed relative frequencies and using the maximum likelihood estimates for the parameters λ_1^* and λ_2^*, the estimator of λ_3 is given by

$$\hat{\hat{\lambda}}_3 = \frac{1}{2}[\bar{x} + \bar{y}] + \frac{1}{4}\log [2 A/n - 1], \qquad (4.7.12)$$

where A is the total of the observed frequencies at the points (2x, 2y) and (2x+1, 2y+1) for the data set. Here it is assumed that the quantity in the square brackets on the right hand side of (4.7.12) is positive; that is $A > n/2$.

B. For using the conditional even points method we recall that the conditional pgf of X given Y=y is

$$\Pi_X(t|\,y) = e^{[\lambda_1^* - \lambda_3](t-1)} \left\{\frac{\lambda_2^* + \lambda_3 [t - 1]}{\lambda_2^*}\right\}^y, \qquad (4.7.13)$$

while that of Y given X=x is

$$\Pi_Y(t|\,x) = e^{[\lambda_2^* - \lambda_3](t-1)} \left\{\frac{\lambda_1^* + \lambda_3 [t - 1]}{\lambda_1^*}\right\}^x. \qquad (4.7.14)$$

Setting y = 0 in (4.7.13) then t =1 and - 1, successively thereafter, we get

$$\Pi_X(1|0) + \Pi_X(-1|0) = 1 + \exp[-2\{\lambda_1^* - \lambda_3\}]$$

$$= \frac{2}{f_1(0)} \sum_{s=0}^{\infty} f(0, 2s), \qquad (4.7.15)$$

where $f_1(0)$ is the marginal probability of $X = 0$. Replacing the parameter values by their sample equivalents, and using the maximum likelihood estimators where necessary, we have for the conditional even points estimator of λ_3

$$\hat{\hat{\lambda}}_3 = \bar{y} + \frac{1}{2} \log \left[\frac{2A}{A+B} - 1\right]. \qquad (4.7.16)$$

The quantities A and B on the right hand side are the sample relative frequencies in the conditional distribution of X at $y = 0$.

It is possible to use the conditional pgf of Y given $x = 0$ to determine the estimator for λ_3 as

$$\hat{\hat{\lambda}}_3 = \bar{x} + \frac{1}{2} \log \left[\frac{2A'}{A'+B'} - 1\right] \qquad (4.7.17)$$

where the quantities A' and B' on the right hand side are the sample relative frequencies in the conditional distribution of Y at $x = 0$.

Asymptotic relative efficiencies

As usual the performance of an estimator is evaluated in terms of the ratio of the generalized variance of the ML estimator to that of the estimator under consideration. As given in Holgate (1964), Loukas *et al.* (1986), and Paul and Ho (1989) the expressions for the generalized variances of the estimators discussed above are:

$$|\Sigma_{ML}| = \frac{1}{n^3 \delta_2} [\lambda_1 \lambda_2]^2,$$

$$|\Sigma_{MM}| = \frac{1}{n^3} [(\lambda_1^* \lambda_2^* - \lambda_3^2)(\lambda_1^* \lambda_2^* + \lambda_3^2 + \lambda_3) - (\lambda_1^* + \lambda_2^* - 2\lambda_3)\lambda_3^2],$$

$$|\Sigma_{DZ}| = \frac{1}{n^3} [(\lambda_1^* \lambda_2^* - \lambda_3^2)(e^{\lambda_1^* + \lambda_2^* - \lambda_3} - 1) - \lambda_1^* \lambda_2^* (\lambda_1^* + \lambda_2^* - 2\lambda_3)],$$

$$|\Sigma_{EP}| = \frac{1}{16n^3} [(\lambda_1^* \lambda_2^* - \lambda_3^2)(e^{4(\lambda_1+\lambda_2)} - 1) + 4(2\lambda_3\lambda_1\lambda_2 - \lambda_1^* \lambda_2^2 - \lambda_1^2 \lambda_2^*)].$$

From these the ARE's of the MM, DZ and EP estimators relative to the ML estimators are determined for various parameter combinations with $\lambda_1^* = \lambda_2^*$ as indicated by the following graphs. In the graphs $a = \lambda_1^*$, $b = \lambda_2^*$ and $d = \lambda^3$ with $r = d/a$.

Figure 4.7.1 ARE's for Method of Moments

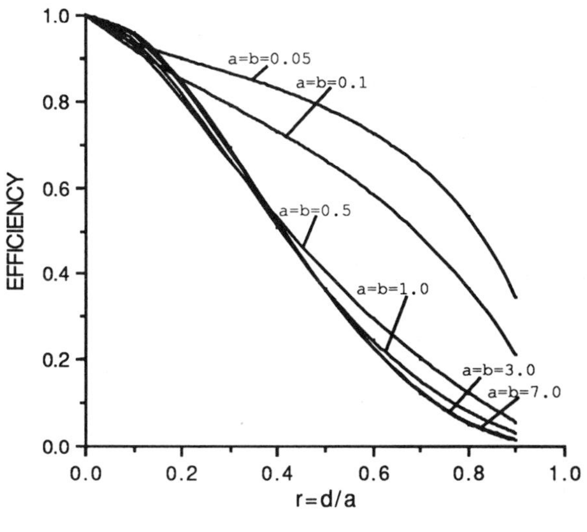

Figure 4.7.2 ARE's for Method of Double-zero Frequency

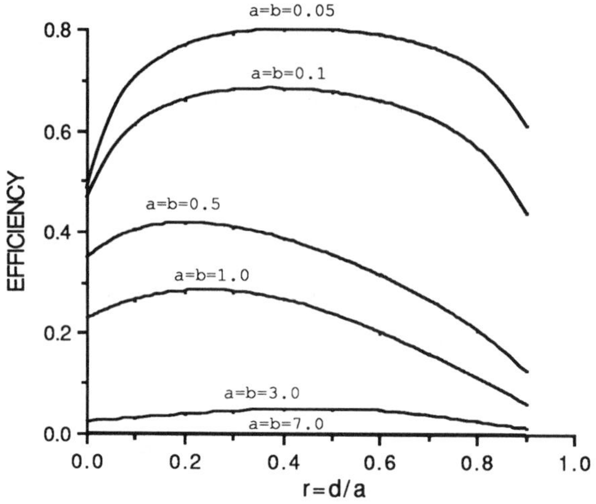

Figure 4.7.3 ARE's for Method of Even Points

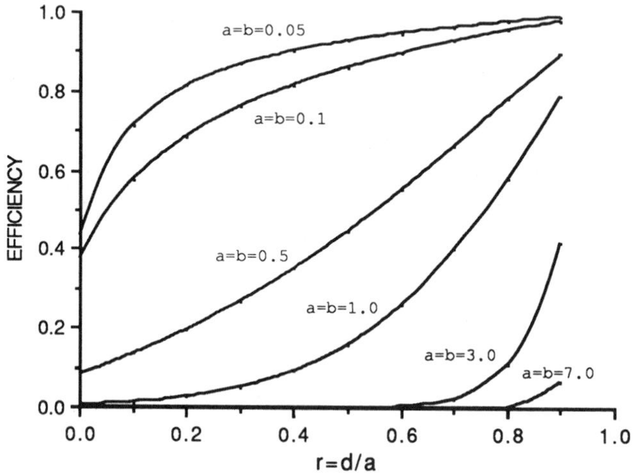

4.7.2 Truncated distribution

In this section we will consider the zero truncated distribution. Hamdan (1972) gives the pf of this distribution as

$$\left.\begin{aligned} f_T(r, s) &= \frac{f(r, s)}{V} \quad r = 1, 2, \dots \; s = 1, 2, \dots \\ &= 0 \qquad \text{otherwise} \end{aligned}\right\}, \tag{4.7.18}$$

where V is the 'volume' of the region over which the pf $f_T(r, s)$ is positive. It should be noted that the Charlier polynomial representation of the joint pf in terms of λ_1^* and λ_2^* for the untruncated case is

$$f(x, y) = f_1(x)\, f_2(y) \left[1 + \sum_{r=1}^{\infty} \frac{\lambda_3^r}{r!} K_r(x; \lambda_1^*) K_r(y; \lambda_2^*)\right] \tag{4.7.19}$$

and $f_1(x)$, $f_2(y)$ are the Poisson probability functions with parameters λ_1^*, λ_2^*, respectively. The respective orthogonality properties are satisfied by each of the functions $K_r(x; \lambda_1^*)$ and $K_r(y; \lambda_2^*)$, in this case.

Moments

The moments of the distribution can be derived using the orthogonality properties of the polynomials. The following special cases are easily verified:

(i) $\sum_{x=1}^{\infty} K_r(x; \lambda_1^*) f(x) = (-1)^{r+1} \exp(-\lambda_1^*)$

(ii) Substituting for x in terms of $K_1(x; \lambda_1^*)$, we have

$$\begin{aligned} \sum_{x=1}^{\infty} x\, K_r(x; \lambda_1^*) f(x) &= 1 \quad \text{for } r = 1 \\ &= 0 \quad \text{for } r \neq 1 \end{aligned}$$

(iii) Writing x^2 in terms of $K_2(x; \lambda_1^*)$ and $K_1(x; \lambda_1^*)$, we have

$$\sum_{x=1}^{\infty} x^2 K_r(x; \lambda_1^*)f(x) = 2\lambda_1^* + 1 \quad \text{for } r = 1$$

$$= 2 \quad \text{for } r = 2$$

$$= 0 \quad \text{for } r > 2$$

with similar identities holding for the variable y.

From these identities and the orthogonality of the polynomials we can determine the moments of the truncated distribution in terms of the parameters λ_1^*, λ_2^* and λ_3 as:

(i) $V = [1 - e^{-\lambda_1^*}][1 - e^{-\lambda_2^*}] + e^{-\lambda_1^* - \lambda_2^*}[e^{\lambda_3} - 1]$

(ii) $V\mu'_{1,0} = \lambda_1^*[1 - e^{-\lambda_2^*}] + \lambda_3 e^{-\lambda_2^*}$ and

$V\mu'_{0,1} = \lambda_2^*[1 - e^{-\lambda_1^*}] + \lambda_3 e^{-\lambda_1^*}$

(iii) $V\mu'_{2,0} = \lambda_1^* + \lambda_1^{*2} - (\lambda_1^* - \lambda_3)(\lambda_1^* + \lambda_3 + 1)e^{-\lambda_2^*}$ and

$V\mu'_{0,2} = \lambda_2^* + \lambda_2^{*2} - (\lambda_2^* - \lambda_3)(\lambda_2^* + \lambda_3 + 1)e^{-\lambda_1^*}$

(iv) $V\mu'_{1,1} = \lambda_1^*\lambda_2^* + \lambda_3$

Moment estimators

Representing the sample moments by replacing the Greek letters by the corresponding Roman letters, we have the following three equations as suggested by Hamdan (1972) :

$$\left.\begin{aligned} \tilde{V} m'_{1,0} &= \tilde{\lambda}_1^*[1 - e^{-\tilde{\lambda}_2^*}] + \tilde{\lambda}_3 e^{-\tilde{\lambda}_2^*} \\ \tilde{V} m'_{0,1} &= \tilde{\lambda}_2^*[1 - e^{-\tilde{\lambda}_1^*}] + \tilde{\lambda}_3 e^{-\tilde{\lambda}_1^*} \\ \tilde{V} m'_{1,1} &= \tilde{\lambda}_1^* \tilde{\lambda}_2^* + \tilde{\lambda}_3 \end{aligned}\right\} \quad (4.7.20)$$

where $\tilde{V}$ is the value of V with the estimates substituted for the parameters. Some special cases are examined now.

$\lambda_3 = 0$

If λ_3 is zero then the first two equations yield, since in this case the random variables are independent, the estimates as the solutions to the equations

$$m'_{1,0} = \frac{\tilde{\lambda}_1^*}{[1 - e^{-\tilde{\lambda}_1^*}]}, \quad m'_{0,1} = \frac{\tilde{\lambda}_2^*}{[1 - e^{-\tilde{\lambda}_2^*}]}$$

which are the maximum likelihood equations obtained by Cohen (1960) for the zero-truncated univariate Poisson distribution. The properties of the estimators are studied by him.

Symmetric case

If $\lambda_1^* = \lambda_2^*$ the distribution becomes symmetric. Writing $a_1 = m'_{10}/m'_{11}$ and $a_2 = m'_{20}/m'_{11}$ the estimating equations for ρ and the common value λ of $\lambda_1^* = \lambda_2^*$ are

$$\tilde{\rho} = [1 - a_1\tilde{\lambda} - \exp(-\tilde{\lambda})]/[a_1 - \exp(-\tilde{\lambda})] \tag{4.7.21}$$

$$a_2 = [1 + \tilde{\lambda} - (1 - \tilde{\rho})(1 + \tilde{\lambda} + \tilde{\lambda}\tilde{\rho})\exp(-\tilde{\lambda})]/(\tilde{\lambda} + \tilde{\rho}) \tag{4.7.22}$$

These equations have to be solved iteratively. Further discussion of the truncated bivariate Poisson distribution can be found in Charalambides (1984), Dahiya (1977) and Patil *et al.* (1977).

4.8 Tests of hypotheses

In this section we will consider tests of fit and tests for the independence of random variables X and Y. There are several tests suggested in the literature for the purpose.

In Chapter 2 tests for goodness-of-fit were developed using Pearson's chi-square statistic and empirical pgf. In the case of the bivariate

Poisson distribution specific tests have been developed: (i) a 'quick' test based on method of moment estimators [Crockett (1979)] and (ii) an index of dispersion test [Loukas and Kemp (1986)].

Tests of independence are equivalent to testing the hypothesis $H_0 : \lambda_3 = 0$ against the alternative $H_a : \lambda_3 > 0$. Reference may be made to Kocherlakota and Singh (1982), Kocherlakota and Kocherlakota (1985) and Kocherlakota and Kocherlakota (1991) for a detailed discussion of the problems relating to the test of this hypothesis. Neyman (1959) gives the $C(\alpha)$ test in this case. More recently Paul and Ho (1989) have proposed the likelihood ratio and Wald's test.

4.8.1 Tests for goodness-of-fit

Recall that the pf is given by

$$f(x, y) = e^{-(\lambda_1+\lambda_2+\lambda_3)} \sum_{i=0}^{\min(x,y)} \frac{\lambda_1^{x-i} \lambda_2^{y-i} \lambda_3^{i}}{(x-i)!\,(y-i)!\,i!}. \tag{4.8.1}$$

In using the traditional Pearson's χ^2 statistic,

$$\underline{X}^2 = \sum_{r,s} \frac{[n_{rs} - n\hat{f}(r, s)]^2}{n\hat{f}(r, s)},$$

it has been pointed out in Chapter 2 that we are faced with the problem of combining the expected cell frequencies so that the expectations are at least one. This difficulty does not arise in using the other suggested tests.

Crockett's 'quick' test

This test is based on a bivariate extension of the dispersion test for a univariate Poisson which uses that property that the mean and variance for the Poisson distribution are equal. The test statistic is given by

$$T = n\{\bar{y}^2 Z_x^2 - 2 m_{1,1}^2 Z_x Z_y + \bar{x}^2 Z_y^2\} / 2(\bar{x}^2 \bar{y}^2 - m_{1,1}^4),$$

where $Z_x = m_{2,0} - \bar{x}$ and $Z_y = m_{0,2} - \bar{y}$. Here T is approximately distributed as a χ^2 on two degrees of freedom.

Index of dispersion test

Loukas and Kemp (1986) also developed this test as an extension of the univariate dispersion test. They are interested in testing for departures from the bivariate Poisson against alternatives which involve an increase in the generalized variance $\mu_{2,0}\,\mu_{0,2} - \mu_{1,1}^2$. Here the statistic is

$$I_B = n\{\bar{y}\, m_{2,0} - 2\, m_{1,1}^2 + \bar{x}\, m_{0,2}\} / \{\bar{x}\,\bar{y} - m_{1,1}^2\},$$

which is approximately distributed as a χ^2 with 2n - 3 degrees of freedom, since three parameters have been estimated. Loukas and Kemp have shown that the statistic I_B has greater power than either T or $\underline{\overline{X}^2}$.

Tests based on the probability generating function

From section 2.2 this test based on the statistic

$$Z = \frac{P(t_1, t_2) - \Pi(t_1, t_2)}{\sigma} \sim N(0, 1)$$

where $P(t_1, t_2)$ and $\Pi(t_1, t_2)$ are the empirical and the population pgf's, respectively. Also, the standard deviation is determined from the equation

$$\sigma^2 = \frac{1}{n}\,[\Pi(t_1^2, t_2^2) - \Pi^2(t_1, t_2)] - \sum_{r,s=1}^{3} \sigma_{rs}\, \frac{\partial\Pi(t_1, t_2)}{\partial\theta_r}\, \frac{\partial\Pi(t_1, t_2)}{\partial\theta_s},$$

where $\{\sigma_{rs}\}$ is the variance matrix given in (4.7.11). The partial derivatives of the pgf are given by

$$\frac{\partial\Pi(t_1, t_2)}{\lambda_1} = (t_1 - 1)\,\Pi(t_1, t_2)$$

$$\frac{\partial\Pi(t_1, t_2)}{\lambda_2} = (t_2 - 1)\,\Pi(t_1, t_2)$$

$$\frac{\partial \Pi(t_1, t_2)}{\lambda_3} = (t_1 t_2 - 1)\, \Pi(t_1, t_2).$$

To perform the test, all parameters are replaced by their maximum likelihood estimates.

4.8.2 Tests of independence

Here we are considering tests of the null hypothesis, $H_0 : \lambda_3 = 0$ against the alternative $H_a : \lambda_3 > 0$.

Efficient score test

Consider the pf given in (4.8.1) and let the random sample of size n be classified so that (x_r, y_s) occurs with the frequency n_{rs}. Then the log of the likelihood of the parameter set, given the sample, is

$$\log L = \sum_{r,s} n_{rs} \log f(x_r, y_s)$$

and hence, under the null hypothesis using the maximum likelihood estimate for the nuisance parameters, we have

$$\left.\begin{aligned} \frac{1}{\sqrt{n}} \frac{\partial \log L}{\partial \lambda_1} &= \sqrt{n}\left[\frac{\bar{x}}{\lambda_1} - 1\right] \\ \frac{1}{\sqrt{n}} \frac{\partial \log L}{\partial \lambda_2} &= \sqrt{n}\left[\frac{\bar{y}}{\lambda_2} - 1\right] \\ \frac{1}{\sqrt{n}} \frac{\partial \log L}{\partial \lambda_3} &= \sqrt{n}\left[\frac{m'_{1,1}}{\lambda_1 \lambda_2} - 1\right] \end{aligned}\right\}. \tag{4.8.2}$$

The information matrix for a single observation is

$$V = \begin{bmatrix} \frac{1}{\lambda_1} & 0 & \frac{1}{\lambda_1} \\ 0 & \frac{1}{\lambda_2} & \frac{1}{\lambda_2} \\ \frac{1}{\lambda_1} & \frac{1}{\lambda_2} & \frac{1}{\lambda_1} + \frac{1}{\lambda_2} + \frac{1}{\lambda_1 \lambda_2} \end{bmatrix}. \tag{4.8.3}$$

The efficient score statistic from equation (2.2.9) is given by

$$S_c = \frac{1}{n \bar{x} \bar{y}} \{\sum_{r,s} n_{rs} (x_r - \bar{x}) (y_s - \bar{y})\}^2. \tag{4.8.4}$$

Under H_0 the statistic S_c has approximately the χ^2 distribution on one degree of freedom.

Likelihood ratio test

The likelihood ratio test for testing the hypothesis is based on the statistic

$$2 \left[n \hat{\lambda}_3 + \sum_{r,s} n_{r,s} \log \left\{ \sum_{i=0}^{\min(x_r, y_s)} \frac{\hat{\lambda}_1^{x_r - i} \hat{\lambda}_2^{y_s - i} \hat{\lambda}_3}{(x_r - i)!\,(y_s - i)!\,i!} \right\} \right.$$

$$\left. - n \bar{x} \log \bar{x} - n \bar{y} \log \bar{y} + \sum_{r,s} n_{r,s} \log (x_r!\, y_s!) \right] \tag{4.8.5}$$

which, once again under H_0, has asymptotically the χ^2 distribution on one degree of freedom.

Wald test

The Wald test is performed on the basis of the statistic $W = \hat{\lambda}_3 / \sqrt{\mathrm{Var}(\hat{\lambda}_3)}$ of which the asymptotic distribution is N(0, 1) when the hypothesis H_0 is true.

Transforms of the sample correlation

Kocherlakota and Kocherlakota (1985) have studied Fisher's z-transform and Student's t for tests of independence by examining the maintenance of the level of significance.

The behavior of the transforms of the sample coefficient of correlation for testing the hypothesis of independence has also been studied by Kocherlakota and Singh (1982). Their development is based on the behavior of the moments as discussed in Subrahmaniam and Gajjar (1980). In addition they have also carried out an empirical examination of the behavior of the transforms by the density estimation technique of Tarter *et al.* (1967).

Discussion of the results

The general consensus emerging from the studies of Kocherlakota and Kocherlakota (1985) as well as that of Paul and Ho (1989) is that Fisher's z-transform performs just as well as the efficient score test in terms of the power when the level is maintained at the prescribed value. This, in addition to the simplicity of the computation, leads them to recommend the use of Fisher's z-transform in practice.

4.9 Computer generation of the distribution

The bivariate Poisson distribution can be simulated using the trivariate reduction method: Consider three independent random variables with $Z_i \sim P(\lambda_i)$ for i = 1, 2, 3. Let

$$X = Z_1 + Z_3$$
$$Y = Z_2 + Z_3,$$

then (X, Y) will be distributed as bivariate Poisson with the pf (4.3.5). The following table has 100 observations generated from the bivariate Poisson distribution with the parameters $\lambda_1^* = 1.5$, $\lambda_2^* = 1.0$ and $\lambda_3 = 0.5$. The IMSL subroutine GGPON was used for simulating the univariate Poisson random variates.

x \ **y**	**0**	**1**	**2**	**3**	**4**	**5**	**6**	**Total**
0	13	4	3	1	0	0	0	21
1	12	16	3	2	1	0	0	34
2	5	16	8	0	1	0	0	30
3	2	3	1	1	1	0	1	9
4	0	3	2	1	0	0	0	6
Total	32	42	17	5	3	0	1	100

Estimation

The summary statistics for this data set are:

$$\bar{x} = 1.45, \quad \bar{y} = 1.09, \quad m_{2,0} = 1.2197, \quad m_{0,2} = 1.2140, \quad m_{1,1} = .403535.$$

Based on these statistics, the estimates for λ_1^* and λ_2^* are 1.45 and 1.09 for all of the procedures. The estimates for λ_3 can be summarized as:

ML: 0.4658, MM: 0.3995, DZ: 0.4998, EP: 0.6386.

Tests of goodness-of-fit

For each of the testing procedures discussed in section 4.8, we conclude that the data do fit the bivariate Poisson distribution. For the $\overline{\underline{X}}^2$ test and pgf technique the ML estimates were used. For the dispersion tests based on T and I_B the MM estimates were used. The results are summarized below:

Statistic	Value	Degrees of Freedom	Prob-value
$\overline{\underline{X}}^2$	11.56	11	0.3976
T	2.117	2	0.3469
I_B	194.9637	197	0.5276

For the pgf technique, the value of the test statistic Z depends on the choice of t_1 and t_2. The minimum value of Z = 0.0145 occurred with t_1 = 0.1, t_2 = 0.1 (Prob-value = 0.9884) and the maximum value was Z = - 1.0306 for t_1 = 0.5, t_2 = 0.9 (Prob-value = 0.3027).

Tests of independence

Here we present the efficient score test given in (4.8.4), the LR test given in (4.8.5) and Fisher's Z given by

$$Z = \frac{\frac{1}{2}\left\{ \log \frac{1+r}{1-r} - \log \frac{1+\rho}{1-\rho} \right\}}{\frac{1}{\sqrt{n-3}}}$$

with $\rho = 0$ under the null hypothesis. The results given below indicate a rejection of the null hypothesis for each testing procedure:

Statistic	Value	Prob-value
S_c	10.098 on 1 d.f.	0.000692
Fisher's Z	3.194	0.000344
LR	3.528 on 1 d.f.	0.000419

5

BIVARIATE NEGATIVE BINOMIAL DISTRIBUTION

5.1 Introduction

The genesis of the bivariate negative binomial, as does that of its univariate counterpart, depends upon the underlying chance mechanism. The simplest derivation of the univariate negative binomial distribution involves waiting times. Recall that in Chapter 3 we discussed Bernoulli trials. Now consider an *infinite* sequence of Bernoulli trials with p as the probability of success and (1 - p) as the probability of a failure. Let the random variable X count the number of failures preceding the rth success; then X has the univariate negative binomial distribution with pf

$$P\{X = x\} = \frac{(r + x - 1)!}{(r - 1)!\, x!} p^r (1 - p)^x \quad x = 0, 1, \ldots \tag{5.1.1}$$

and pgf

$$\begin{aligned} \Pi_X(t) &= p^r [1 - (1 - p)t]^{-r} \\ &= [1 - \frac{q}{p}(t - 1)]^{-r}. \end{aligned} \tag{5.1.2}$$

This will be denoted by

$$X \sim NB(r, p).$$

In the case in which r = 1, we are looking for the distribution of the number of failures preceding the *first* success. Then X has a geometric distribution with pf

$$P\{X = x\} = p\,(1 - p)^x \quad x = 0, 1, \ldots \tag{5.1.3}$$

Several other chance mechanisms give rise to the univariate negative binomial. A summary of these is given in *Encyclopedia of Statistical Sciences* (Volume 6, pp. 169-177). We shall now develop the bivariate analog of (5.1.1) under various forms of chance mechanism.

5.2 Inverse sampling

Consider a sequence of independent trials resulting in one of three possible outcomes, A, B or C with $P(A) = p_1$, $P(B) = p_2$ and $P(C) = 1 - p_1 - p_2$. Suppose we are interested in fixing the number of occurrences of C at r. Let X count the number of times that A occurs and Y the number of times that B occurs at the rth occurrence of C. Then the joint pf of X and Y is

$$f(x, y) = \frac{(r + x + y - 1)!}{(r - 1)!\, x!\, y!} p_1^x p_2^y (1 - p_1 - p_2)^r \tag{5.2.1}$$

with x = 0, 1, ... and y = 0, 1, ... The joint distribution in (5.2.1) can be derived by recalling that, with the sampling scheme under consideration, the first (r + x + y -1) trials constitute a trinomial distribution with the index parameter (r + x + y - 1) and probabilities p_1, p_2 and $(1 - p_1 - p_2)$. Here the event {X = x, Y = y} has the probability

$$\frac{(r + x + y - 1)!}{(r - 1)!\, x!\, y!} p_1^x p_2^y (1 - p_1 - p_2)^{r-1} .$$

In this case the last trial has to lead to C; hence , we have the pf given in (5.2.1). This development for the bivariate negative binomial distribution has been given by Guldberg (1934).

Probability generating function and probability function

The pgf of this distribution is given by

$$\Pi_{X,Y}(t_1, t_2) = \sum_{x=0}^{\infty} \sum_{y=0}^{\infty} \frac{(r + x + y - 1)!}{(r - 1)!\, x!\, y!} (p_1t_1)^x (p_2t_2)^y (1 - p_1 - p_2)^r$$

$$= (1 - p_1 - p_2)^r (1 - p_1t_1 - p_2t_2)^{-r}. \tag{5.2.2}$$

The corresponding fmgf is

$$G(t_1, t_2) = (1 - p_1 - p_2)^r [1 - p_1(t_1 + 1) - p_2(t_2 + 1)]^{-r}$$

$$= \left[1 - \frac{p_1 t_1}{1 - p_1 - p_2} - \frac{p_2 t_2}{1 - p_1 - p_2}\right]^{-r}. \qquad (5.2.3)$$

Marginal distributions

From (5.2.2) we can easily determine the pgf of X and of Y:

$$\left.\begin{aligned} \Pi_X(t) &= \Pi(t, 1) = \left[1 - \frac{p_1}{1 - p_1 - p_2}(t - 1)\right]^{-r} \\ \Pi_Y(t) &= \Pi(1, t) = \left[1 - \frac{p_2}{1 - p_1 - p_2}(t - 1)\right]^{-r} \end{aligned}\right\}. \qquad (5.2.4)$$

Each of the pgf's in (5.2.4) is readily seen to be a univariate negative binomial. From these the marginal means and variances are found to be

$$E(X) = \frac{r\,p_1}{1 - p_1 - p_2}, \qquad E(Y) = \frac{r\,p_2}{1 - p_1 - p_2},$$

$$\mathrm{Var}(X) = \left(\frac{r\,p_1}{1 - p_1 - p_2}\right)\left(1 + \frac{p_1}{1 - p_1 - p_2}\right),$$

$$\mathrm{Var}(Y) = \left(\frac{r\,p_2}{1 - p_1 - p_2}\right)\left(1 + \frac{p_2}{1 - p_1 - p_2}\right).$$

Correlation, conditional distributions and regressions

From the joint pgf in (5.2.2), we have

$$\mathrm{Cov}(X, Y) = \frac{r\,p_1\,p_2}{(1 - p_1 - p_2)^2}$$

and

$$\rho_{X,Y} = \left\{\frac{p_1\,p_2}{(1 - p_1)(1 - p_2)}\right\}^{\frac{1}{2}},$$

which is always positive.

Using Theorem 1.3.1, the pgf of the conditional distribution of X given Y = y is

$$\Pi_X(t \mid y) = \Pi^{(0,y)}(t, 0) \,/\, \Pi^{(0,y)}(1, 0).$$

Since

$$\Pi^{(0,y)}(t_1, t_2) = (1 - p_1 - p_2)^r p_2^y \frac{\Gamma(r+y)}{\Gamma(r)} (1 - p_1 t_1 - p_2 t_2)^{-(r+y)},$$

$$\Pi_X(t \mid y) = (1 - p_1)^{(r+y)} (1 - p_1 t)^{-(r+y)},$$

which is NB(r+y, 1 - p_1). Hence, the regression of X on Y is

$$E(X \mid y) = r \frac{p_1}{1 - p_1} + y \frac{p_1}{1 - p_1},$$

which is linear in y with a positive slope. Similar results can be obtained for the regression of Y on X.

It should be noted that the random variables X and Y cannot be independent in the bivariate negative binomial distribution since it is not possible for

$$\Pi_{X,Y}(t_1, t_2) = \Pi_{X,Y}(t_1, 1)\, \Pi_{X,Y}(1, t_2)$$

to hold.

5.3 Bivariate shock model

Downton (1970) describes a model for developing the bivariate geometric distribution which arises when r = 1 in (5.2.1). Thus the pgf in (5.2.2) becomes

$$\Pi_{X,Y}(t_1, t_2) = (1 - p_1 - p_2)(1 - p_1 t_1 - p_2 t_2)^{-1}. \qquad (5.3.1)$$

The bivariate geometric distribution given in (5.3.1) arises in a shock model with two components. Downton (1970) describes this model in the following way: "Suppose that the number of shocks suffered by each component before failure can be represented by a population in which proportions p_1 and p_2 affected the first and second component, respectively, without failure and a proportion 1 - p_1 - p_2 of the shocks lead to failure of both components." This can be summarized by letting

$$\bar{F}_1 \text{ and } \bar{F}_2$$

represent the events that components 1 and 2, respectively, do *not* fail. Then

$$P(\bar{F}_1) = p_1, \quad P(\bar{F}_2) = p_2$$

and

$$P(\bar{F}_1 \cup \bar{F}_2) = p_1 + p_2, \quad P(F_1 F_2) = 1 - P(\bar{F}_1 \cup \bar{F}_2) = 1 - p_1 - p_2.$$

It should be noted that under these assumptions $P(\bar{F}_1 \cap \bar{F}_2) = 0$. If we sample successively and observe

X = number of shocks to component 1 *prior* to the first failure

Y = number of shocks to component 2 *prior* to the first failure,

then the joint pgf of (X, Y) is given by (5.3.1) with corresponding pf

$$f(x, y) = \binom{x+y}{x} p_1^x p_2^y (1 - p_1 - p_2) \tag{5.3.2}$$

with x = 0, 1, ... and y = 0, 1, ...

Downton (1970) consider the random variables

$$X^* = X + 1, \quad Y^* = Y + 1,$$

which measure the number of shocks to component 1 and component 2, respectively, leading (rather than prior to) the first failure. In this case

$$f(x^*, y^*) = \binom{x^* + y^* - 2}{x^* - 1} p_1^{x^*-1} p_2^{y^*-1} (1 - p_1 - p_2) \tag{5.3.3}$$

with x* = 1, 2, ... and y* = 1, 2, ...

This shock model can be generalized to the joint distribution of (X, Y) when they measure the number of shocks to the components before the rth failure of the system (i. e., they both fail). Then we have the joint pf given in (5.2.1) and the corresponding pgf given in (5.2.2).

As in the univariate case, the bivariate negative binomial can be derived as the convolution of r independent identical bivariate geometric random variables. Let (X_i, Y_i) be the number of shocks to the two components before the ith failure of the system, but after the (i-1)th failure. If (X, Y) are the number of shocks to the components for the rth failure to occur, then

$$X = \sum_{i=1}^{r} X_i \text{ and } Y = \sum_{i=1}^{r} Y_i,$$

where (X_i, Y_i) for $i = 1, 2, \ldots, r$ are independent random variables with the joint pgf

$$\Pi_{X_i,Y_i}(t_1, t_2) = (1 - p_1 - p_2)(1 - p_1t_1 - p_2t_2)^{-1}$$

as in (5.3.1). Since (X_i, Y_i) are independent,

$$\begin{aligned}\Pi_{X,Y}(t_1, t_2) &= E\{t_1^X t_2^Y\} \\ &= \prod_{i=1}^{r} E\{t_1^{X_i} t_2^{Y_i}\} \\ &= (1 - p_1 - p_2)^r (1 - p_1t_1 - p_2t_2)^{-r},\end{aligned}$$

which is the pgf given in (5.2.2). Thus we can consider (X, Y) as the convolution of r independent identical bivariate geometric random variables.

We can also consider a translated model

$$X^* = X + r, \quad Y^* = Y + r,$$

which measure the number of shocks to each of the components for the system to fail for the rth time, then

$$f(x^*, y^*) = \frac{(x^* + y^* - r - 1)!}{(r - 1)!\,(x^* - r)!\,(y^* - r)!} p_1^{x^*-r} p_2^{y^*-r} (1 - p_1 - p_2)^r \tag{5.3.4}$$

with $x^* = r, r + 1, \ldots$ and $y^* = r, r + 1 \ldots$. Here the pgf is

$$\begin{aligned}\Pi_{X^*,Y^*}(t_1, t_2) &= E[t_1^{X^*} t_2^{Y^*}] \\ &= t_1^r t_2^r E[t_1^X t_2^Y] \\ &= t_1^r t_2^r \Pi_{X,Y}(t_1, t_2).\end{aligned} \tag{5.3.5}$$

5.4 Compounding

The univariate negative binomial distribution arises from the Poisson distribution by a process of 'compounding'. Greenwood and Yule (1920) developed this model in their study of accident proneness. They assumed that the number of accidents sustained by α individuals in a time interval has a Poisson distribution with parameter λ. To account for the heterogeneity, or variation from individual to individual, they allowed λ to be a random variable with a gamma distribution. The resulting distribution is a univariate negative binomial distribution.

Let the random variable X given λ be a Poisson random variable with the pf

$$h(x|\lambda) = \exp(-\alpha\lambda)\,\frac{(\alpha\lambda)^x}{x!}, \qquad x = 0, 1, 2, \ldots$$

If we let λ itself be a random variable have the gamma distribution

$$g(\lambda) = \frac{\beta^\nu}{\Gamma(\nu)}\,\lambda^{\nu-1} \exp(-\beta\lambda), \qquad \lambda > 0, \qquad (5.4.1)$$

then the unconditional distribution of X is

$$f(x) = \int_0^\infty h(x|\lambda)\, g(\lambda)\, d\lambda$$

$$= \frac{\Gamma(\nu + x)}{\Gamma(\nu)\, x!}\left(\frac{\beta}{\alpha + \beta}\right)^{\nu}\left(\frac{\alpha}{\alpha + \beta}\right)^{x}, \qquad x = 0, 1, 2, \ldots, \qquad (5.4.2)$$

which is a univariate negative binomial with the index parameter ν and p $= \frac{\beta}{\alpha + \beta}$. It should be noted here that ν in this case can be any real number such that $\Gamma(\nu)$ is defined. On the other hand, in the model based on inverse sampling, the shock model or convolution techniques, the parameter ν is taken to be a positive integer.

During the 1950's this model received considerable attention, particularly in problems concerning accident proneness. A review of the model as applied to accident proneness can be found in Kemp (1970). The univariate model has been extended to the bivariate case by Arbous and Kerrich (1951) and to the multivariate case by Bates and Neyman (1952a). The latter authors credit Greenwood and Yule (1920) and Newbold (1926) with the bivariate extension using the following assumptions:

(i) Let X and Y represent the number of accidents incurred by the same individual within the same period of time or in two non-overlapping consecutive periods. The random variables X and Y are assumed to have Poisson distributions with parameters λ and μ, respectively. Conditional on λ and μ, X and Y are *independent.* Here λ and μ characterize the proneness of the individual to the two types of accidents.

(ii) The parameter μ is taken to be proportional to λ; that is, $\mu = a\lambda$, where a is a constant.

(iii) λ has the gamma distribution given in (5.4.1).

Under these assumptions, the conditional pgf of X and Y given λ is given by

$$\Pi^*_{X,Y}(t_1, t_2|\lambda) = \exp[\lambda\{(t_1 - 1) + a(t_2 - 1)\}].$$

Combining this with (5.4.1), we have the conditional pgf of X and Y

$$\Pi^*_{X,Y}(t_1, t_2) = [1 - (\frac{1}{\beta})\{(t_1-1) + a(t_2-1)\}]^{-\nu} \qquad (5.4.3)$$

and the corresponding pf

$$g^*(x, y) = \frac{\beta^\nu a^y \Gamma(\nu + x + y)}{\Gamma(x + 1)\Gamma(y + 1)\Gamma(\nu)[\beta + a + 1]^{\nu+x+y}}. \qquad (5.4.4)$$

[The (*) in the expressions for the pgf's and pf denotes the assumption of independence.]

The assumption of independence given in assumption (i) was relaxed by Edwards and Gurland (1961) and Subrahmaniam (1966); that is, for a given individual the two types of accidents are assumed to be positively *correlated.* Let (X, Y) have the bivariate Poisson distribution, given a parameter λ, with joint pgf

$$\Pi_{X,Y}(t_1, t_2|\lambda) = \exp[\lambda\{\alpha_1(t_1 - 1) + \alpha_2(t_2 - 1) + \alpha_3(t_1t_2 - 1)\}] \qquad (5.4.5)$$

and let λ have a Pearson Type III distribution with the pdf given in (5.4.1).

Then the unconditional distribution of X and Y has the joint pgf

$$\Pi_{X,Y}(t_1, t_2) = \int_0^\infty \Pi_{X,Y}(t_1, t_2|\lambda)\, g(\lambda)\, d\lambda$$

$$= \beta^\nu [\beta - \{\alpha_1(t_1 - 1) + \alpha_2(t_2 - 1) + \alpha_3(t_1t_2 - 1)\}]^{-\nu}$$

$$= [1 - \frac{\alpha_1}{\beta}(t_1 - 1) - \frac{\alpha_2}{\beta}(t_2 - 1) - \frac{\alpha_3}{\beta}(t_1t_2 - 1)]^{-\nu}, \qquad (5.4.6)$$

which can also be written as

$$\Pi_{X,Y}(t_1, t_2) = [1 + \frac{\alpha_1}{\beta} + \frac{\alpha_2}{\beta} + \frac{\alpha_3}{\beta} - \frac{\alpha_1}{\beta}t_1 - \frac{\alpha_2}{\beta}t_2 - \frac{\alpha_3}{\beta}t_1t_2]^{-\nu}$$

$$= q^\nu [1 - p_1t_1 - p_2t_2 - p_3t_1t_2]^{-\nu}, \qquad (5.4.7)$$

where

$$p_i = \alpha_i/(\alpha_1 + \alpha_2 + \alpha_3 + \beta), \quad i = 1, 2, 3$$

and

$$q = \beta/(\alpha_1 + \alpha_2 + \alpha_3 + \beta) = 1 - (p_1 + p_2 + p_3).$$

The above general formulation was first given by Edwards and Gurland (1961), who unfortunately referred to the distribution as the compound correlated bivariate Poisson (CCBP) distribution. Independently Subrahmaniam (1966), working along the same lines, developed the distribution, giving it the more appropriate terminology of the bivariate negative binomial distribution. This name seems more appropriate in as much as the identical univariate development results in the negative binomial or Pascal distribution. Several authors seem to have accepted the latter terminology.

A simpler model based on compounding in the independent bivariate Poisson distribution is obtained from (5.4.6) or (5.4.7) by setting $\alpha_3 = 0$. In this case the pgf is given by

$$\Pi^*_{X,Y}(t_1, t_2) = \left(\frac{\beta}{\alpha_1 + \alpha_2 + \beta}\right)^\nu [1 - p_1t_1 - p_2t_2]^{-\nu}, \qquad (5.4.8)$$

which is equivalent to (5.4.3) with $\alpha_1 = 1$ and $\alpha_2 = a$.

We shall study the distribution given in (5.4.6). Much of the following material was developed by Subrahmaniam (1966).

Probability function

Expanding (5.4.7) in powers of t_1 and t_2 and identifying the term involving $t_1^r\, t_2^s$, we obtain the joint pf $P\{X = r, Y = s\}$ as

$$g(r, s) = q^{\nu} \sum_{i=0}^{\min(r,s)} \frac{\Gamma(\nu + r + s - i)}{\Gamma(\nu)\, i!\, (r - i)!\, (s - i)!} p_1^{r-i}\, p_2^{s-i}\, p_3^{i}, \qquad (5.4.9)$$

which is given in Subrahmaniam (1966), equation (2.4).

Marginal distributions

The marginal distributions of X and Y can be obtained from the pgf given in (5.4.7):

$$\Pi_X(t) = \Pi_{X,Y}(t, 1) = q^{\nu}\, [(1 - p_2) - (p_1 + p_3)\, t\,]^{-\nu},$$

$$= [1 - \frac{P_1}{1 - P_1}(t - 1)]^{-\nu},$$

where $P_1 = (p_1 + p_3)/(1 - p_2)$; that is,

$$X \sim NB(\nu, 1 - P_1).$$

Similarly,

$$Y \sim NB(\nu, 1 - P_2)$$

with $P_2 = (p_2 + p_3)/(1 - p_1)$.

Factorial moments and correlation

From (5.4.7) the fmgf is seen to be

$$G(t_1, t_2) = \Pi(1+t_1, 1+t_2)$$

$$= [1 - \frac{p_1 + p_3}{q} t_1 - \frac{p_2 + p_3}{q} t_2 - \frac{p_3}{q} t_1\, t_2]^{-\nu} \qquad (5.4.10)$$

Expanding (5.4.10) in powers of t_1, t_2 and identifying the coefficient of $t_1^r t_2^s / r!\, s!$, we have

$$\mu_{[r,s]} = r!\, s! \left(\frac{p_1+p_3}{q}\right)^r \left(\frac{p_2+p_3}{q}\right)^s \cdot \sum_{i=0}^{\min(r,s)} \frac{\Gamma(\nu+r+s-i)}{\Gamma(\nu)\, i!\, (r-i)!\, (s-i)!} \left\{\frac{q\, p_3}{(p_1+p_3)(p_2+p_3)}\right\}^i. \qquad (5.4.11)$$

Since the marginal factorial moments of X and of Y, of order r and s, respectively, are

$$\mu_{[r]}^{(x)} = \frac{\Gamma(\nu+r)}{\Gamma(\nu)} \left(\frac{p_1+p_3}{q}\right)^r$$

$$\mu_{[s]}^{(y)} = \frac{\Gamma(\nu+s)}{\Gamma(\nu)} \left(\frac{p_2+p_3}{q}\right)^s,$$

(5.4.11) can be written as

$$\mu_{[r,s]} = \left\{\frac{\Gamma(\nu)\, r!}{\Gamma(\nu+r)} \mu_{[r]}^{(x)}\right\} \left\{\frac{\Gamma(\nu)\, s!}{\Gamma(\nu+s)} \mu_{[s]}^{(y)}\right\} \cdot \sum_{i=0}^{\min(r,s)} \frac{\Gamma(\nu+r+s-i)}{\Gamma(\nu)\, i!\, (r-i)!\, (s-i)!} \left\{\frac{q\, p_3}{(p_1+p_3)(p_2+p_3)}\right\}^i. \qquad (5.4.12)$$

Using the notation $\begin{bmatrix} \nu \\ i \end{bmatrix}$ to represent the ratio $\frac{\Gamma(\nu)}{i!\, \Gamma(\nu - i)}$, (5.4.12) can be expressed as

$$\mu_{[r,s]} = \mu_{[r]}^{(x)} \mu_{[s]}^{(y)} \sum_{i=0}^{\min(r,s)} \frac{\begin{bmatrix} \nu+r+s-i \\ \nu+r \end{bmatrix} \binom{r}{i}}{\begin{bmatrix} \nu+s \\ s \end{bmatrix}} \left\{\frac{q\, p_3}{(p_1+p_3)(p_2+p_3)}\right\}^i. \qquad (5.4.13)$$

If $p_3 = 0$,

$$\mu_{[r,s]} = \frac{\Gamma(\nu+r+s)\, \Gamma(\nu)}{\Gamma(\nu+r)\, \Gamma(\nu+s)} \mu_{[r]}^{(x)} \mu_{[s]}^{(y)}. \qquad (5.4.14)$$

From (5.4.13) and (5.4.14) it is clear that X and Y can never be independent in this model.

From the moments of the marginal distributions and (5.4.13),

$$\rho_{X,Y} = \frac{p_3 + p_1 p_2}{\left\{(1 - p_1)(1 - p_2)(p_1 + p_3)(p_2 + p_3)\right\}^{\frac{1}{2}}},$$

which is always positive. When $p_3 = 0$, the correlation coefficient has a simpler form

$$\rho^*_{X,Y} = \left\{\frac{p_1 p_2}{(1 - p_1)(1 - p_2)}\right\}^{\frac{1}{2}},$$

which is again always positive.

Conditional distributions and regressions

Using Theorem 1.3.1, the conditional distribution of X given Y = y is

$$\Pi_X(t \mid y) = \left\{\frac{p_2 + p_3 t}{p_2 + p_3}\right\}^y \left\{1 - \frac{p_1}{1 - p_1}(t - 1)\right\}^{-(\nu + y)} \qquad (5.4.15)$$

from which the conditional distribution of X given Y = y is found to be the convolution of

$$X_1 \sim B(y, \frac{p_3}{p_2 + p_3}) \quad \text{and} \quad X_2 \sim NB(\nu + y, 1 - p_1).$$

Hence,

$$\left.\begin{aligned} E(X \mid y) &= \frac{\nu p_1}{1 - p_1} + \left\{\frac{p_3}{p_2 + p_3} + \frac{p_1}{1 - p_1}\right\} y \\ Var(X \mid y) &= \frac{\nu}{(1 - p_1)^2} + \left\{\frac{p_2 p_3}{(p_2 + p_3)^2} + \frac{p_1}{(1 - p_1)^2}\right\} y \end{aligned}\right\}. \qquad (5.4.16)$$

Both of the expressions in (5.4.16) are linear in y.

A similar development for the conditional distribution of Y given X = x leads to the pgf

$$\Pi_Y(t \mid x) = \left\{\frac{p_1 + p_3 t}{p_1 + p_3}\right\}^x \left\{1 - \frac{p_2}{1 - p_2}(t - 1)\right\}^{-(\nu + x)}; \qquad (5.4.17)$$

hence, the conditional distribution is the convolution of

$$Y_1 \sim B(x, \frac{p_3}{p_1 + p_3}) \quad \text{and} \quad Y_2 \sim NB(\nu + x, 1 - p_2).$$

From these results it can be shown that $E(Y \mid x)$ and $Var(Y \mid x)$ are of the same form as (5.4.16) with p_1 and p_2 interchanged. Here again both the conditional expectation and variance are linear in x.

5.5 Canonical representation

In section 3.6 we saw that the bivariate binomial distribution can be expressed in terms of Krawtchouk polynomials. In this section we will develop a similar canonical expansion for the bivariate negative binomial distribution in terms of the canonical correlations and a set of orthogonal polynomials which are orthonormal with respect to the negative binomial probability function. A detailed discussion of this representation is given in Hamdan and Al-Bayyati (1971).

Consider the polynomial

$$G_r(x; k, p) = p^r \left(1 - \frac{q\Delta}{p}\right)^{k+r-1} x^{(r)}$$

$$= p^r \sum_{i=0}^{r} \binom{k+r-1}{i} \left(-\frac{q}{p}\right)^i \Delta^i x^{(r)},$$

where $\Delta f(x) = f(x+1) - f(x)$. Applying Δ^i to $x^{(r)}$, we have

$$\Delta^i x^{(r)} = \frac{r!}{(r-i)!} \frac{x!}{(x-r+i)!}$$

and

$$G_r(x; k, p) = p^r \sum_{i=0}^{r} \binom{k+r-1}{i} \left(-\frac{q}{p}\right)^i r^{(i)} x^{(r-i)}. \tag{5.5.1}$$

The generating function of the set $\{G_r(x; k, p)\}$ is

$$\gamma_k(w; x, p) = = (1+w)^x (1 + q w)^{-(x+k)}. \tag{5.5.2}$$

That this is the generating function can be seen by determining the coefficient of $w^r/r!$ in (5.5.2) and comparing it with $G_r(x; k, p)$.

Now

$$\sum_{x=0}^{\infty} G_s(x; k, p)\, G_r(x; k, p) \frac{\Gamma(x+k)}{\Gamma(k)\, x!} p^k q^x$$

$$= (r!)^2 q^r \frac{\Gamma(k+r)}{\Gamma(k)\, r!} \qquad r = s$$

$$= 0 \qquad r \neq s.$$

Therefore,

$$G_r^*(x; k, p) = \frac{G_r(x; k, p)}{\left\{ r!\, q^{\frac{r}{2}} \left(\frac{\Gamma(k+r)}{\Gamma(k)\, r!} \right)^{\frac{1}{2}} \right\}}$$

is an orthonormal set of polynomials with respect to the negative binomial weight function.

Let

$$f(x) = \frac{\Gamma(x+k)}{\Gamma(k)\, x!} p^k q^x, \qquad x = 0, 1, 2, \ldots$$

be the pf of a NB(k, p) with q = 1- p. The factorial moment generating of $G_r(x; k, p)f(x)$ is

$$\sum_{x=0}^{\infty} (1+x)^x G_r(x; k, p) f(x) = r! \binom{k+r-1}{r} \left(\frac{tq}{p}\right)^r \left(1 - \frac{tq}{p}\right)^{-(k+r)}. \qquad (5.5.3)$$

Recall from equation (5.4.7) that the joint fmgf of the bivariate negative binomial distribution is given by

$$G(t_1, t_2) = \left[1 - \frac{p_1 + p_3}{q} t_1 - \frac{p_2 + p_3}{q} t_2 - \frac{p_3}{q} t_1 t_2\right]^{-\nu}.$$

Since the marginal distributions of X and of Y are NB(ν, 1 - P_1) and NB(ν, 1 - P_2), respectively, then their fmgf's are

$$G_X(t) = \left[1 - \frac{P_1}{1 - P_1} t\right]^{-\nu}, \qquad P_1 = \frac{p_1 + p_3}{1 - p_2}$$

and

$$G_Y(t) = \left[1 - \frac{P}{1 - P_2} t_2\right]^{-\nu}, \qquad P_2 = \frac{p_2 + p_3}{1 - p_1}.$$

The procedure, outlined in section 3.6, developed by Aitken and Gonin for the (positive) bivariate binomial, consists of expressing $G(t_1, t_2)$ in terms of $G_X(t_1)$ and $G_Y(t_2)$ and then inverting the resulting expression, term by term. For notational simplicity, write

$$G(t_1, t_2) = [1 - A_1 t_1 - A_2 t_2 - A_3 t_1 t_2]^{-\nu}$$

$$= [(1 - A_1 t_1)(1 - A_2 t_2) - (A_3 + A_1 A_2) t_1 t_2]^{-\nu}$$

$$= (1 - A_1 t_1)^{-\nu} (1 - A_2 t_2)^{-\nu} \left\{1 - \frac{(A_3 + A_1 A_2) t_1 t_2}{(1 - A_1 t_1)(1 - A_2 t_2)}\right\}^{-\nu} \tag{5.5.4}$$

with

$$A_1 = \frac{p_1 + p_3}{q} = \frac{P_1}{1 - P_1}, \qquad A_2 = \frac{p_2 + p_3}{q} = \frac{P_2}{1 - P_2}, \qquad A_3 = \frac{p_3}{q}.$$

Now expand the last term in (5.5.4) as

$$\sum_{i=0}^{\infty} \frac{\Gamma(\nu+i)}{\Gamma(\nu)\, i!} \frac{(A_3 + A_1 A_2)^i (t_1 t_2)^i}{(1 - A_1 t_1)^i (1 - A_2 t_2)^i}. \tag{5.5.5}$$

Recalling that $\rho_{X,Y}$ can be written as

$$\rho_{X,Y} = \frac{q (A_3 + A_1 A_2)}{\{(1 - p_1)(1 - p_2) A_1 A_2\}^{\frac{1}{2}}}$$

or

$$A_3 + A_1 A_2 = \frac{\rho_{X,Y}}{q} \{(1 - p_1)(1 - p_2) A_1 A_2\}^{\frac{1}{2}},$$

(5.5.5) becomes

$$\sum_{i=0}^{\infty} \frac{\Gamma(\nu+i)}{\Gamma(\nu)\, i!} \frac{\rho_{X,Y}^i (1 - p_1)^{\frac{i}{2}} (1 - p_2)^{\frac{i}{2}}}{q^i (A_1 A_2)^{\frac{1}{2}}} \left(\frac{A_1 t_1}{1 - A_1 t_1}\right)^i \left(\frac{A_2 t_2}{1 - A_2 t_2}\right)^i.$$

Therefore,

$$G(t_1, t_2) = (1 - A_1 t_1)^{-\nu} (1 - A_2 t_2)^{-\nu} \cdot$$

$$\sum_{i=0}^{\infty} \frac{\Gamma(\nu+i)}{\Gamma(\nu)\, i!} \frac{\rho_{X,Y}^{i} (1-p_1)^{\frac{i}{2}} (1-p_2)^{\frac{i}{2}}}{q^i (A_1 A_2)^{\frac{1}{2}}} \left(\frac{A_1 t_1}{1 - A_1 t_1}\right)^i \left(\frac{A_2 t_2}{1 - A_2 t_2}\right)^i . \tag{5.5.6}$$

The general term of this expansion is of the form

$$(A_r t_r)^i (1 - A_r t_r)^{-(\nu+i)}, \qquad r = 1, 2;$$

hence, from (5.5.3) the inverse is

$$G_i(x; \nu, \theta_r)\, f(x; \nu, \theta_r) \frac{\Gamma(\nu)}{\Gamma(\nu+i)} = G_i(x; \nu, \theta_r) \frac{\Gamma(x+\nu)}{\Gamma(\nu+i)\, x!} \theta_r^{\nu} (1 - \theta_r)^x$$

with $\theta_r = (1 + A_r)^{-1}$. We can then determine the pf corresponding to the fmgf given in (5.5.4):

$$g(x, y) = g_1(x)\, g_2(y) \sum_{i=0}^{\infty} \left\{ \frac{\rho_{X,Y} [(1-p_1)(1-p_2)]^{\frac{1}{2}}}{q (A_1 A_2)^{\frac{1}{2}}} \right\}^i \cdot$$

$$\frac{\Gamma(\nu+i)}{\Gamma(\nu)\, i!} \frac{\Gamma^2(\nu)}{\Gamma^2(\nu+i)} G_i(x; \nu, \theta_1)\, G_i(y; \nu, \theta_2)$$

$$= g_1(x)\, g_2(y) \sum_{i=0}^{\infty} \rho_{X,Y}^{i} \left\{ \frac{(1-p_1)(1-p_2)}{q^2 A_1 A_2} \right\}^{\frac{i}{2}} \cdot$$

$$\frac{\Gamma(\nu)}{\Gamma(\nu+i)\, i!} G_i(x; \nu, \theta_1)\, G_i(y; \nu, \theta_2) \tag{5.5.7}$$

with

$$\theta_1 = \frac{q}{1 - p_2}, \quad \theta_2 = \frac{q}{1 - p_1}.$$

Using the relationship given above between the orthonormal polynomials $G^*_r(x; k, p)$ and the polynomials $G_r(x; k, p)$, we have

$$G^*_i(x; \nu, \frac{q}{1 - p_2}) = \left\{\frac{\Gamma(\nu)}{\Gamma(\nu+i)\, i!}\right\}^{\frac{1}{2}} \left\{\frac{1 - p_2}{p_1 + p_3}\right\}^{\frac{i}{2}} G_i(x; \nu, \frac{q}{1 - p_2})$$

$$G^*_i(y; \nu, \frac{q}{1 - p_1}) = \left\{\frac{\Gamma(\nu)}{\Gamma(\nu+i)\, i!}\right\}^{\frac{1}{2}} \left\{\frac{1 - p_1}{p_2 + p_3}\right\}^{\frac{i}{2}} G_i(y; \nu, \frac{q}{1 - p_1}).$$

Substituting these results in (5.5.7), the joint pf of X and Y can be written as

$$g(x, y) = g_1(x)\, g_2(y) \sum_{i=0}^{\infty} \rho^i_{X,Y}\, G^*_i(x; \nu, \frac{q}{1 - p_2})\, G^*_i(y; \nu, \frac{q}{1 - p_1}) \quad (5.5.8)$$

with

$$g_1(x) = \frac{\Gamma(\nu+x)}{\Gamma(\nu)\, x!} \left(\frac{q}{1 - p_2}\right)^{\nu} \left(\frac{p_1 + p_3}{1 - p_2}\right)^{x}$$

and

$$g_2(y) = \frac{\Gamma(\nu+y)}{\Gamma(\nu)\, y!} \left(\frac{q}{1 - p_1}\right)^{\nu} \left(\frac{p_2 + p_3}{1 - p_1}\right)^{y}.$$

These results can be modified for the model in which $p_3 = 0$.

5.6 Mitchell and Paulson model

The bivariate negative binomial distribution defined in the preceding sections, while arising from practical situations, suffers from two main drawbacks: (i) the coefficient of correlation between X and Y has to be positive and (ii) the regression of Y on X and of X on Y must be linear. Such restrictions are not always tenable. Mitchell and Paulson (1981) have introduced less restrictive models in which negative correlation as well as nonlinear regression is possible.

Basic to the understanding of their bivariate negative model is the construction of the bivariate geometric distribution defined by Paulson and Uppuluri (1972). In section 5.6.1 the bivariate geometric will be developed and then in 5.6.2 this distribution will be generalized to give a bivariate negative binomial model.

5.6.1 Bivariate geometric distribution

It is well known that bivariate analogs of univariate distributions can be developed by appropriately extending the generating functions of the univariate random variables. Consider the random variable X to be the number of successes preceding the first failure, the pgf is

$$\Pi_X(t) = \left[1 + \frac{p}{1 - p}(1 - t)\right]^{-1}. \tag{5.6.1}$$

A natural extension of this pgf to two dimensions would be

$$\Pi_{X,Y}(t_1, t_2) = \left[1 + \frac{p_1}{1 - p_1}(1 - t_1) + \frac{p_2}{1 - p_2}(1 - t_2)\right]^{-1} \tag{5.6.2}$$

Such extensions have led to the bivariate negative binomial distribution discussed earlier.

In this section we will consider a bivariate geometric distribution obtained from an extension of a characterization of the geometric distribution in the one dimensional case.

Theorem 5.6.1 [Paulson and Uppuluri (1972)]

Let $\Psi(t)$ be the characteristic function of a random variable and V a random variable with the distribution function G(v). Then the solution $\phi(t)$ of the characteristic functional equation

$$\phi(t) = \Psi(t)\, E[\phi(tV)] \tag{5.6.3}$$

is the characteristic function of the geometric distribution if and only if $\Psi(t)$ is the characteristic function of a geometric distribution and G(v) is such that

$$P\{V = 0\} = a, \quad P\{V = 1\} = b \quad \text{with } a + b = 1,\ 0 < a \le 1.$$

Proof:

For a proof of the theorem, we refer to the source paper. □

To define a bivariate extension of this theorem let V be a 2x2 matrix valued random variable with the value set

$$\left\{ \begin{pmatrix} 0 & 0 \\ 0 & 0 \end{pmatrix}, \begin{pmatrix} 1 & 0 \\ 0 & 0 \end{pmatrix}, \begin{pmatrix} 0 & 0 \\ 0 & 1 \end{pmatrix}, \begin{pmatrix} 1 & 0 \\ 0 & 1 \end{pmatrix} \right\}$$

and the respective probabilities $\alpha_1, \alpha_2, \alpha_3, \alpha_4$ with $\sum_{i=1}^{4} \alpha_i = 1$ and $\alpha_2 + \alpha_4 < 1$, $\alpha_3 + \alpha_4 < 1$. Let the characteristic function of (X_1, X_2) be

$$\Psi(t_1, t_2) = E[\exp\{it_1 X_1 + it_2 X_2\}]$$

and let

$$\phi(t_1, t_2) = E[\exp\{it_1 Y_1 + it_2 Y_2\}] .$$

Then the bivariate form of the characteristic functional equation is

$$\phi(t_1, t_2) = \Psi(t_1, t_2) \;\; E[\phi(\underline{t}' V)], \tag{5.6.4}$$

where $\underline{t}' = (t_1, t_2)$. Choosing

$$\Psi(t_1, t_2) = \Psi_1(t_1, 0)\, \Psi_2(0, t_2)$$

with

$$\Psi_1(t_1, 0) = \left[1 + \frac{p_1}{1 - p_1}(1 - e^{it_1})\right]^{-1}$$

$$\Psi_2(0, t_2) = \left[1 + \frac{p_2}{1 - p_2}(1 - e^{it_2})\right]^{-1}$$

and recalling the distribution of the matrix V, we have for the characteristic functional equation

$$\phi(t_1, t_2) = \Psi_1(t_1, 0)\, \Psi_2(0, t_2) \;\; [\alpha_1 + \alpha_2\, \phi(t_1, 0) + \alpha_3\, \phi(0, t_2) + \alpha_4\, \phi(t_1, t_2)]. \tag{5.6.5}$$

Now set $t_2 = 0$ in (5.6.5) and since $\phi(0, 0) = 1$, we have

$$\phi(t_1, 0) = \Psi_1(t_1, 0)\,[\alpha_1 + \alpha_2\,\phi(t_1, 0) + \alpha_3 + \alpha_4\,\phi(t_1, 0)]$$

$$= \frac{\Psi_1(t_1, 0)\,(\alpha_1 + \alpha_3)}{1 - (\alpha_2 + \alpha_4)\,\Psi_1(t_1, 0)}$$

$$= \left[1 + \theta_1\,(1 - e^{it_1})\right]^{-1} \tag{5.6.6}$$

with $\theta_1 = p_1 / [(1 - p_1)\,(\alpha_1 + \alpha_3)]$. Similarly,

$$\phi(0, t_2) = \left[1 + \theta_2\,(1 - e^{it_2})\right]^{-1} \tag{5.6.7}$$

with $\theta_2 = p_2 / [(1 - p_2)\,(\alpha_1 + \alpha_2)]$. Hence, the marginal distributions of each is a univariate geometric distribution.

From (5.6.5), (5.6.6) and (5.6.7), it can be readily seen that the random variables Y_1 and Y_2 have the moments

$$E(Y_i) = \theta_i, \quad Var(Y_i) = \theta_i\,(1 + \theta_i) \quad \text{for } i = 1, 2$$

and

$$Cov(Y_1, Y_2) = \frac{\alpha_1\,\alpha_4 - \alpha_2\,\alpha_3}{1 - \alpha_4}\,\theta_1\,\theta_2;$$

hence,

$$\rho_{Y_1, Y_2} = \frac{\alpha_1\,\alpha_4 - \alpha_2\,\alpha_3}{1 - \alpha_4} \left\{\frac{\theta_1\,\theta_2}{(1 + \theta_1)\,(1 + \theta_2)}\right\}^{\frac{1}{2}},$$

which is obviously positive or negative.

To determine the joint probability function of (Y_1, Y_2), we can equate the coefficients in (5.6.5). Now

$$\phi(t_1, t_2) = \sum_{\alpha=0}^{\infty} \sum_{\beta=0}^{\infty} f_{\alpha\beta}\ \exp(it_1\alpha + it_2\beta)$$

$$\phi(t_1, 0) = \sum_{\alpha=0}^{\infty} \left\{\sum_{\beta=0}^{\infty} f_{\alpha\beta}\right\} \exp(it_1\alpha)$$

$$\phi(0, t_2) = \sum_{\beta=0}^{\infty} \left\{\sum_{\alpha=0}^{\infty} f_{\alpha\beta}\right\} \exp(it_2\beta).$$

Let

$$f_{\alpha+} = \sum_{\beta=0}^{\infty} f_{\alpha\beta} \text{ and } f_{+\beta} = \sum_{\alpha=0}^{\infty} f_{\alpha\beta}$$

in (5.6.5). Equating coefficients of $\exp(it_1\alpha + it_2\beta)$, we have

$$f_{00} = A\,[\alpha_1 + \alpha_2 f_{0+} + \alpha_3 f_{+0}] / D$$

$$f_{\alpha 0} = [A\,\alpha_2 f_{(\alpha-1)+} + p_1 f_{(\alpha-1)0}] / D$$

$$f_{0\beta} = [A\,\alpha_3 f_{+(\beta-1)} + p_2 f_{0(\beta-1)}] / D$$

$$f_{\alpha\beta} = [p_1 f_{(\alpha-1)\beta} + p_2 f_{\alpha(\beta-1)} - p_1 p_2 f_{(\alpha-1)(\beta-1)}] / D,$$

where $A = (1 - p_1)(1 - p_2)$, $D = 1 - \alpha_4 A$ and $f_{\alpha+}$, $f_{+\beta}$ are determined from the marginal distributions given in (5.6.6) and (5.6.7), respectively.

It should be noted here that the bivariate geometric distribution defined by

$$\phi(t_1, t_2) = \left[1 + \frac{p_1}{1 - p_1}(1 - e^{it_1}) + \frac{p_2}{1 - p_2}(1 - e^{it_2})\right]$$

does not enjoy the property of independence. There is no choice of parameters that reduces $\phi(t_1, t_2)$ to the product $\phi(t_1, 0)\ \phi(0, t_2)$. However, the Paulson-Uppuluri model does give rise to independence in the special case when

$$\alpha_1\,\alpha_4 = \alpha_2\,\alpha_3.$$

This can be proved by noting that in (5.6.5) the condition for independence is

$$\phi(t_1, t_2) = \phi(t_1, 0)\ \phi(0, t_2)$$

or

$$\Psi_1(t_1, 0)\,\Psi_2(0, t_2)\,[\alpha_1 + \alpha_2\,\phi(t_1, 0) + \alpha_3\,\phi(0, t_2) + \alpha_4\,\phi(t_1, t_2)] = \phi(t_1, 0)\ \phi(0, t_2). \tag{5.6.8}$$

Noting that

$$\phi(t_1, 0) / \Psi_1(t_1, 0) = (\alpha_1 + \alpha_3) + (\alpha_2 + \alpha_4)\,\phi(t_1, 0)$$

and

$$\phi(0, t_2) / \Psi_2(0, t_2) = (\alpha_1 + \alpha_2) + (\alpha_3 + \alpha_4)\, \phi(0, t_2)$$

and substituting in (5.6.8) with $\alpha_1\, \alpha_4 = \alpha_2\, \alpha_3$, we have the relationship

$$\phi(t_1, t_2) = \phi(t_1, 0)\ \ \phi(0, t_2).$$

5.6.2 Bivariate negative binomial distribution

As an extension of section 5.6.1, Mitchell and Paulson (1981) suggest such a bivariate negative binomial distribution defined by the pgf

$$\Pi_{X,Y}(t_1, t_2) = \left[\frac{1 - d}{\{1 + \tau_1 (1 - t_1)\}\, \{1 + \tau_2(1 - t_2)\} - d}\right]^m \cdot$$

$$\left[\frac{\alpha}{1 - d} + \frac{\alpha_1}{(1 - d)\, [1 + \theta_1(1 - t_1)]} + \frac{\alpha_2}{(1 - d)\, [1 + \theta_2(1 - t_2)]}\right]^m, \tag{5.6.9}$$

where

$$\tau_i = \frac{p_i}{1 - p_i} \qquad \text{for } i = 1, 2;$$

$$\theta_1 = \frac{\tau_1}{\alpha + \alpha_2}, \qquad \theta_2 = \frac{\tau_2}{\alpha + \alpha_1}$$

and

$$\alpha_1 + \alpha_2 + \alpha + d = 1.$$

This model has correlation which may be positive or negative.

From (5.6.9) the marginal pgf's of X and of Y are given by

$$\Pi_X(t) = \left[\frac{1 - d}{1 + \tau_1\, (1 - t) - d}\right]^m \cdot$$

$$\left[\frac{\alpha + \alpha_2}{1 - d} + \frac{\alpha_1}{(1 - d)\, [1 + \theta_1(1 - t)]}\right]^m$$

$$= [1 + \theta_1\, (1 - t)]^{-m}$$

and

$$\Pi_Y(t) = [1 + \theta_2 (1 - t)]^{-m};$$

hence, $X \sim NB(m, \frac{1}{1 + \theta_1})$ and $Y \sim NB(m, \frac{1}{1 + \theta_2})$.

Setting

$$\lambda_1 = \alpha / (1 - d), \quad \lambda_2 = \alpha_1 / (1 - d), \quad \lambda_3 = \alpha_2 / (1 - d),$$

the probability function corresponding to the pgf in (5.6.9) is given by

$$g(x, y) = \sum_{\alpha,\beta,\gamma} \frac{m!}{\alpha!\, \beta!\, \gamma!} \lambda_1^{\alpha} \lambda_2^{\beta} \lambda_3^{\gamma} \sum_{i=0}^{x} \sum_{j=0}^{y} g_X(x - i)\, g_Y(y - j)\, g^*(i, j), \tag{5.6.10}$$

where the summation runs over all $\alpha, \beta, \gamma, \geq 0$ with $\alpha + \beta + \gamma = m$ and $\alpha_1 + \alpha_2 + \alpha + d = 1$ with each parameter ≥ 0 and $\alpha_1 + d < 1$, $\alpha_2 + d < 1$. The functions g_X and g_Y are the marginal pf's

$$g_X(x) = \frac{\Gamma(x+m)}{\Gamma(m)\, x!} \left(\frac{1}{1 + \theta_1}\right)^m \left(\frac{\theta_1}{1 + \theta_1}\right)^x,$$

$$g_Y(y) = \frac{\Gamma(y+m)}{\Gamma(m)\, y!} \left(\frac{1}{1 + \theta_2}\right)^m \left(\frac{\theta_2}{1 + \theta_2}\right)^y$$

and

$$g(i, j) = (1 - d)^m h_1(i)\, h_2(j)\, {}_2F_1[m+i, m+j; m; d(1 - p_1)(1 - p_2)]$$

with

$$h_r(i) = \frac{\Gamma(i+m)}{\Gamma(m)\, i!} (1-p_r)^m p_r^i.$$

Although the pf given in (5.6.10) is quite complicated, the fmgf can be found from the pgf given in (5.6.9):

$$G(t_1, t_2) = \left[(1 + \tau_1 t_1)(1 + \tau_2 t_2) - d\right]^{-m}.$$

$$\left[\alpha + \frac{\alpha_1}{1 + \theta_1 t_1} + \frac{\alpha_2}{1 + \theta_2 t_2}\right]^m. \qquad (5.6.11)$$

From (5.6.11) the correlation coefficient can be determined as

$$\rho_{X,Y} = \frac{\alpha d - \alpha_1 \alpha_2}{1 - d}\left\{\frac{\theta_1 \theta_2}{(1 + \theta_1)(1 + \theta_2)}\right\}^{\frac{1}{2}},$$

which can be positive or negative.

5.7 Marshall-Olkin model

Marshall and Olkin (1985) consider a modified version of the Guldberg (1934) model by considering X, the number of trials needed for the rth occurrence of an event A and Y, the number needed for the sth occurrence of an event B. If we are dealing with Bernoulli trials, then X and Y are each distributed as negative binomial random variables. Consider the bivariate Bernoulli situation where I_1 indicates the occurrence of the event A and I_2 indicates the occurrence of B with

$$P\{I_1 = i, I_2 = j\} = p_{ij}, \ i = 0, 1; j = 0,1.$$

Now consider the random variables X and Y defined by

X = the number of trials needed for I_1 to be '1' for the rth time
Y = the number of trials needed for I_2 to be '1' for the sth time.

Clearly X and Y are not independent. Marginally, X and Y have negative binomial distributions:

$$f_1(x) = \binom{x-1}{r-1} p_{1+}^r \, p_{0+}^{x-r}, \qquad x = r, r+1, \ldots$$

$$f_2(y) = \binom{y-1}{r-1} p_{+1}^r \, p_{+0}^{y-r}, \qquad y = r, r+1, \ldots .$$

Although the marginal pf's are of the usual form, the joint pf of (X, Y) is quite complicated. Reference may be made to Marshall and Olkin (1985, eq. 7.2) for details. Due to the complexity of the pgf, the following recurrence relation due to Marshall and Olkin for $\Pi_{r,s}(t_1, t_2)$, the pgf of X and Y, may be useful:

$$\Pi_{r,1} = p_{11}\left(\frac{p_{1+}}{1 - p_{0+}t_1}\right)^{r-1} + p_{01}\, t_1\left(\frac{p_{1+}}{1 - p_{0+}t_1}\right)^{r}$$

$$+ p_{10}\, t_2\, \Pi_{r-1,1} + p_{00}\, t_1\, t_2\, \Pi_{r,1}, \qquad r = 2, 3, \ldots \qquad (5.7.1)$$

and

$$\Pi_{r,s} = p_{11}\, \Pi_{r-1,s-1} + p_{01}\, t_1\, \Pi_{r,s-1} + p_{10}\, t_2\, \Pi_{r-1,s} + p_{00}\, t_1\, t_2\, \Pi_{r,s},$$

$$r, s = 2, 3, \ldots \qquad (5.7.2)$$

For the special case when $p_{11} = p_{00} = 0$ and $r = s = 2$, the correlation coefficient is given by

$$\rho_{X,Y} = -2\,(p_{01}\, p_{10})^{\frac{3}{2}}.$$

Unfortunately, the distribution is so complicated that it does not seem to lend itself to any deep study. As pointed out by Marshall and Olkin the distribution cannot be represented as a convolution of geometric random variables even for $r = s$.

5.8 Limiting forms

Several different limiting forms of the distribution arise depending upon the conditions placed on the parameters. The important results in this connection are:

(1) *Bivariate logarithmic series distribution.* The univariate logarithmic series distribution was developed by Fisher *et al.* (1943) as the limit of the negative binomial distribution truncated at the origin with the index parameter tending to zero. Subrahmaniam (1966) generalized this result to the bivariate case, considering the pgf of the truncated bivariate distribution

$$\Pi_T(t_1, t_2) = \frac{\Pi(t_1, t_2) - \Pi(0, 0)}{1 - \Pi(0, 0)},$$

which, in this case, from (5.4.7) becomes

$$\Pi_T(t_1, t_2) = \frac{q^{\nu}\,[1 - p_1t_1 - p_2t_2 - p_3t_1t_2]^{-\nu} - q^{\nu}}{1 - q^{\nu}}.$$

Taking the limit as $\nu \to 0$, we have

$$\Pi^*(t_1, t_2) = \frac{\log[1 - p_1t_1 - p_2t_2 - p_3t_1t_2]}{-\log[1 - p_1 - p_2 - p_3]},$$

the pgf of bivariate logarithmic series distribution, which will be studied in detail in Chapter 7.

(2) *Bivariate Poisson distribution.* Again consider the pgf given in (5.4.7)

$$\Pi(t_1, t_2) = q^{\nu}\,[1 - p_1t_1 - p_2t_2 - p_3t_1t_2]^{-\nu}.$$

Taking $p_i = \dfrac{m_i}{\nu}$ and letting $\nu \to \infty$, $p_i \to 0$ with m_i constant for i = 1, 2, 3, we have

$$\Pi(t_1, t_2) = \frac{\exp[m_1t_1 + m_2t_2 + m_3t_1t_2]}{\exp[m_1 + m_2 + m_3]}$$

$$= \exp[m_1(t_1-1) + m_2(t_2-1) + m_3(t_1t_2-1)],$$

which is the pgf of a bivariate Poisson distribution with parameters m_1, m_2 and m_3. If $p_3 = 0$, the limiting distribution is that of two independent Poisson random variables.

(3) *Bivariate normal distribution.* Subrahmaniam (1966) also mentions that the moment generating function of appropriately standardized variables

$$Z_i = \frac{X_i - E(X_i)}{\{\mathrm{Var}(X_i)\}^{\frac{1}{2}}}, \qquad i = 1, 2$$

tends to that of a standardized bivariate normal distribution with the correlation coefficient ρ.

5.9 Estimation

For the purpose of estimation the parameterization given in (5.4.7) is useful. We will discuss the estimation of the parameters in two variations of this model:

(i) $p_3 = 0$ (or $\alpha_3 = 0$)
(ii) $p_3 \neq 0$ (or $\alpha_3 \neq 0$).

5.9.1 $p_3 = 0$

In this case the model reduces to the ones with the pgf's given by (5.4.3) or (5.2.2). If we assume that the index parameter r is known as is usually the case in inverse sampling, then the maximum likelihood estimators for β and a are

$$\hat{\beta} = r/\bar{x} \quad \text{and} \quad \hat{a} = \bar{y}/\bar{x}. \tag{5.9.1}$$

In the case in which the index parameter ν is not necessarily an integer and is unknown, Bates and Neyman (1952a) have derived the maximum likelihood estimators:

$$\hat{\beta} = \hat{\nu}/\bar{x}, \quad \hat{a} = \bar{y}/\bar{x} \tag{5.9.2a}$$

and ν as the solution to the equation

$$\log\left(1 + \frac{\bar{x} + \bar{y}}{\nu}\right) = \sum_{t=0}^{\infty} \frac{1 - \sum_{i=0}^{t} \frac{q_i}{n}}{\nu + t}, \tag{5.9.2b}$$

where q_i represents the frequency of pairs (x, y) which have their sum $x + y = t$. This latter equation has to be solved iteratively.

5.9.2 $p_3 \neq 0$

This model is defined by the bivariate negative binomial distribution with pf given in (5.4.9). The form of the marginal moments $\mu'_{1,0}$, $\mu'_{0,1}$ and the mixed moment $\mu_{1,1}$ suggests a reparameterization of (5.4.9) in terms of

$$p_1 = \frac{\gamma_0 - \gamma_2}{1 + \gamma_0 + \gamma_1 + \gamma_2}, \qquad p_2 = \frac{\gamma_1 - \gamma_2}{1 + \gamma_0 + \gamma_1 + \gamma_2},$$

$$p_3 = \frac{\gamma_2}{1 + \gamma_0 + \gamma_1 + \gamma_2}, \qquad q = \frac{1}{1 + \gamma_0 + \gamma_1 + \gamma_2}.$$

With this parameterization (5.4.9) becomes

$$g(x, y) = \frac{(\gamma_0 - \gamma_2)^x (\gamma_1 - \gamma_2)^y \Gamma(\nu + x + y)}{x!\, y!\, \Gamma(\nu)\, (1 + \gamma_0 + \gamma_1 - \gamma_2)^{\nu+x+y}} S(x, y), \qquad (5.9.3)$$

where

$$S(x, y) = \sum_{i=0}^{\min(x,y)} \frac{\binom{x}{i}\binom{y}{i}\tau^i}{\binom{\nu+x+y-1}{i}} \qquad (5.9.4)$$

and

$$\tau = \frac{\gamma_2 (1 + \gamma_0 + \gamma_1 - \gamma_2)}{(\gamma_0 - \gamma_2)(\gamma_1 - \gamma_2)}.$$

Estimation using this parameterization (5.9.3) has been examined by Subrahmaniam and Subrahmaniam (1973). Three techniques are summarized: (1) method of moments, (2) zero-zero cell frequency and (3) maximum likelihood.

(1) *Method of moments.* The moments of the distribution in (5.9.3) are

$$\mu'_{1,0} = \nu\gamma_0, \quad \mu'_{0,1} = \nu\gamma_1, \quad \mu_{1,1} = \nu(\gamma_2 + \gamma_0\gamma_1),$$

$$\mu_{2,0} = \nu\gamma_0(1 + \gamma_0), \quad \mu_{0,2} = \nu\gamma_1(1 + \gamma_1),$$

$$\mu_{2,1} = \nu(2\gamma_0 + 1)(\gamma_2 + \gamma_0\gamma_1), \quad \mu_{1,2} = \nu(2\gamma_0 + 1)(\gamma_1 + \gamma_0\gamma_1),$$

$$\mu_{2,2} = \nu\,\{2(r+1)\gamma_2^2 + 2(\nu+2)\gamma_2\gamma_0\gamma_1 + 3(\nu+2)\gamma_0^2\gamma_1^2 + \gamma_2(2\gamma_0 + 2\gamma_1 + 1)$$
$$+ (\nu+1)\gamma_0\gamma_1(1+\gamma_0+\gamma_1) + \gamma_0\gamma_1(\gamma_0+\gamma_1)\}.$$

Using the sample marginal first moments $\bar{x}$, $\bar{y}$ and the mixed moment $m_{1,1}$, we have

$$\nu\,\tilde{\gamma}_0 = \bar{x}, \qquad \nu\,\tilde{\gamma}_1 = \bar{y}, \qquad \nu\,(\tilde{\gamma}_2 + \tilde{\gamma}_0\,\tilde{\gamma}_1) = m_{1,1}.$$

Hence the moment estimators are

$$\tilde{\gamma}_0 = \frac{\bar{x}}{\nu}, \qquad \tilde{\gamma}_1 = \frac{\bar{y}}{\nu}, \qquad \tilde{\gamma}_2 = \frac{m_{1,1}}{\nu} - \frac{\bar{x}\,\bar{y}}{\nu^2}. \tag{5.9.5}$$

The variance matrix of these estimators to $O(n^{-1})$ [Subrahmaniam and Subrahmaniam (1973)] is given by

$$\Sigma_{MM} = \frac{1}{n\nu^2}\begin{bmatrix} \mu_{2,0} & \mu_{1,1} & \mu_{2,1} \\ \cdot & \mu_{0,2} & \mu_{1,2} \\ \cdot & \cdot & \mu_{2,2} - \mu_{1,1}^2 \end{bmatrix}.$$

(2) *Zero-zero cell frequency.* From (5.9.3)

$$g(0,0) = (1 + \gamma_0 + \gamma_1 - \gamma_2)^{-\nu}.$$

A combination of $\bar{x}$, $\bar{y}$ and the observed zero-zero cell relative frequency $\frac{n_{00}}{n}$ can be used to estimate the parameters γ_0, γ_1 and γ_2:

$$\tilde{\tilde{\gamma}}_0 = \frac{\bar{x}}{\nu}, \qquad \tilde{\tilde{\gamma}}_1 = \frac{\bar{y}}{\nu}, \qquad \tilde{\tilde{\gamma}}_2 = 1 + \frac{\bar{x}}{\nu} + \frac{\bar{y}}{\nu} - \left\{\frac{n_{00}}{n}\right\}^{-\frac{1}{\nu}}. \tag{5.9.6}$$

Section 2.1.3 gives the requisite moments of the above sample quantities; hence,

$$\mathrm{Var}\left[\left\{\frac{n_{00}}{n}\right\}^{-\frac{1}{\nu}}\right] \cong \{\,g(0,0)\,\}^{-\frac{2}{\nu}}\ \mathrm{Var}\left\{\frac{n_{00}}{n}\right\} / \nu^2$$

$$= \left\{ \frac{1}{g(0, 0)} - 1 \right\} (1 + \gamma_0 + \gamma_1 - \gamma_2)^2 / n\, \nu^2,$$

$$\mathrm{Cov}\ \left[\bar{x}, \left\{ \frac{n_{00}}{n} \right\}^{-\frac{1}{\nu}} \right] \cong \frac{(1 + \gamma_0 + \gamma_1 - \gamma_2)\, \gamma_0}{n}$$

and

$$\mathrm{Cov}\ \left[\bar{y}, \left\{ \frac{n_{00}}{n} \right\}^{-\frac{1}{\nu}} \right] \cong \frac{(1 + \gamma_0 + \gamma_1 - \gamma_2)\, \gamma_1}{n}.$$

Using these results, we find

$$\mathrm{Cov}\ (\tilde{\tilde{\gamma}}_0, \tilde{\tilde{\gamma}}_2) \cong \gamma_2 (1 + \gamma_0) / n\, \nu$$

$$\mathrm{Cov}\ (\tilde{\tilde{\gamma}}_1, \tilde{\tilde{\gamma}}_2) \cong \gamma_2 (1 + \gamma_1) / n\, \nu$$

and

$$\mathrm{Var}\ (\tilde{\tilde{\gamma}}_2) \cong \frac{(1 + \gamma_0 + \gamma_1)\, (2\, \gamma_2 - \gamma_0 - \gamma_1)}{n\, \nu}$$

$$+ \frac{1}{n\, \nu^2} \left\{ \frac{1}{f(0, 0)} - 1 \right\} (1 + \gamma_0 + \gamma_1 - \gamma_2)^2 .$$

(3) *Maximum likelihood estimators.* Differentiating the pf given in (5.8.1) with respect to the parameters γ_0, γ_1 and γ_2, we have, after considerable algebraic manipulation, the following maximum likelihood equations:

$$\frac{\bar{x}}{\hat{\gamma}_0 - \hat{\gamma}_2} - \frac{\nu + \bar{x} + \bar{y}}{1 + \hat{\gamma}_0 + \hat{\gamma}_1 - \hat{\gamma}_2} - \frac{(1 + \hat{\gamma}_1)\, \bar{U}}{(\hat{\gamma}_0 - \hat{\gamma}_2)\, (1 + \hat{\gamma}_0 + \hat{\gamma}_1 - \hat{\gamma}_2)} = 0$$

$$- \frac{\nu + \bar{x} + \bar{y}}{1 + \hat{\gamma}_0 + \hat{\gamma}_1 - \hat{\gamma}_2} + \frac{\bar{y}}{\hat{\gamma}_1 - \hat{\gamma}_2} - \frac{(1 + \hat{\gamma}_0)\, \bar{U}}{(\hat{\gamma}_1 - \hat{\gamma}_2)\, (1 + \hat{\gamma}_0 + \hat{\gamma}_1 - \hat{\gamma}_2)} = 0$$

$$- \frac{\bar{x}}{\hat{\gamma}_0 - \hat{\gamma}_2} - \frac{\bar{y}}{\hat{\gamma}_1 - \hat{\gamma}_2} + \frac{\nu + \bar{x} + \bar{y}}{1 + \hat{\gamma}_0 + \hat{\gamma}_1 - \hat{\gamma}_2}$$

$$+\left\{\frac{1}{\hat\gamma_2} + \frac{1}{\hat\gamma_0 - \hat\gamma_2} + \frac{1}{\hat\gamma_1 - \hat\gamma_2} - \frac{1}{1 + \hat\gamma_0 + \hat\gamma_1 - \hat\gamma_2}\right\}\bar{U} = 0,$$

where

$$U(x, y) = S'(x, y) / S(x, y)$$

with

$$S'(x, y) = \sum_{i=0}^{\min(x,y)} \frac{i\,\tau^i \binom{x}{i}\binom{y}{i}}{\binom{\nu+x+y-1}{i}} \tag{5.9.7}$$

and $S(x, y)$ is as defined in (5.9.4). Here $\bar{U}$ is the sample mean of $U(x, y)$. Solving these equations leads to

$$\hat\gamma_0 = \frac{\bar{x}}{\nu}, \qquad \hat\gamma_1 = \frac{\bar{y}}{\nu},$$

while $\hat\gamma_2$ is the solution to the equation

$$\nu\,\hat\gamma_2 = \bar{U}. \tag{5.9.8}$$

As in the case of the bivariate Poisson distribution, this equation for $\hat\gamma_2$ can be solved iteratively using either the moment or zero-zero cell frequency estimate as an initial value.

The variance matrix of $(\hat\gamma_0, \hat\gamma_1, \hat\gamma_2)$ is obtained as the inverse of the information matrix given by

$$\left\{\sum_{x,y} Q_h(x, y)\, Q_i(x, y)\, g(x, y)\right\},$$

where

$$Q_1(x, y) = \frac{x}{\gamma_0 - \gamma_2} - \frac{\nu + x + y}{1 + \gamma_0 + \gamma_1 - \gamma_2} - \frac{(1 + \gamma_1)\,U(x, y)}{(\gamma_0 - \gamma_2)\,(1 + \gamma_0 + \gamma_1 - \gamma_2)}$$

$$Q_2(x, y) = \frac{y}{\gamma_1 - \gamma_2} - \frac{\nu + x + y}{1 + \gamma_0 + \gamma_1 - \gamma_2} - \frac{(1 + \gamma_0)\,U(x, y)}{(\gamma_1 - \gamma_2)\,(1 + \gamma_0 + \gamma_1 - \gamma_2)}$$

$$Q_3(x, y) = -\frac{x}{\gamma_0 - \gamma_2} - \frac{y}{\gamma_1 - \gamma_2} + \frac{\nu + x + y}{1 + \gamma_0 + \gamma_1 - \gamma_2} + \lambda\,U(x, y)$$

with

$$\lambda = \frac{1}{\gamma_2} + \frac{1}{\gamma_0 - \gamma_2} + \frac{1}{\gamma_1 - \gamma_2} - \frac{1}{1 + \gamma_0 + \gamma_1 - \gamma_2}.$$

The computation of the maximum likelihood estimate for γ_2 and the variance matrix of $(\hat{\gamma}_0, \hat{\gamma}_1, \hat{\gamma}_2)$ is simplified by using recurrence relationships developed by Subrahmaniam and Subrahmaniam (1973). Recall the functions S(x, y) and S'(x, y) as defined in (5.9.4) and (5.9.7), respectively. For the iterative solution of (5.9.8), a similar function

$$\frac{dS'(x, y)}{d\gamma_2} = \lambda\, S''(x, y)$$

with

$$S''(x, y) = \sum_{i=0}^{\min(x,y)} \frac{i^2\, \tau^i \binom{x}{i} \binom{y}{i}}{\binom{v+x+y-1}{i}} \tag{5.9.9}$$

is useful. The recurrence relationships are summarized as:

$$S(x, 0) = S(0, y) = 1, \qquad x \geq 0,\ y \geq 0,$$

$$S(x, y) = S(x, y-1) + S(x-1, y-1) \left\{ \frac{\gamma_2 (v+x-1)(1 + \gamma_0 + \gamma_1 - \gamma_2)}{(\gamma_0 - \gamma_2)(\gamma_1 - \gamma_2)(v+x+y-1)(v+x+y-2)} \right\}$$

$$S'(x, y) = S'(x, y-1) + \left\{ S(x-1, y-1) + S'(x-1, y-1) \right\} \left\{ \frac{x(v+x-1)\tau}{(v+x+y-1)(v+x+y-2)} \right\}$$

$$\begin{aligned} S''(x, y) = {} & S''(x, y-1) + S'(x, y) - S'(x, y-1) + \\ & \left\{ S'(x-1, y-1) + S''(x-1, y-1) \right\} \left\{ \frac{x(v+x-1)\tau}{(v+x+y-1)(v+x+y-2)} \right\} \end{aligned}$$

for $x \geq 1$, $y \geq 1$ and zero otherwise.

Using these recurrence relationships, it is quite easy to obtain the maximum likelihood solution for γ_2, iteratively, and the solutions for γ_0 and γ_1 can be obtained in a closed form. The method of moments or zero-zero cell frequency estimates for γ_2 can be used as an initial value in the iterative procedure.

Example 5.9.1

Estimation in the model with $\alpha_3 = 0$ will be illustrated using data taken from Arbous and Kerrich (1951). They studied accidents among 122 experienced shunters. Here x represents the number of accidents in the 6 year period 1937-42 and y the number of accidents in the 5 year period 1943-47.

	y								
x	**0**	**1**	**2**	**3**	**4**	**5**	**6**	**7**	**Total**
0	21	13	4	2	0	0	0	0	40
1	18	14	5	1	0	0	0	1	39
2	8	10	4	3	1	0	0	0	26
3	2	1	2	2	1	0	0	0	8
4	1	4	1	0	0	0	0	0	6
5	0	1	0	1	0	0	0	0	2
6	0	0	1	0	0	0	0	0	1
Total	50	43	17	9	2	0	0	1	122

The following summary statistics are useful for calculating the estimates:

$$\bar{x} = 1.27049, \quad \bar{y} = .975431$$

$$q_0 = 21,\ g_1 = 31,\ q_2 = 26,\ q_3 = 19,\ q_4 = 7,\ q_5 = 9,\ q_6 = 5,\ q_7 = 1,\ q_8 = 3.$$

Using these statistics, the maximum likelihood estimates are:

$$\hat{v} = 3.4201, \quad \hat{\beta} = 2.6919, \quad \hat{a} = .7677.$$

Example 5.9.2

Using the simulated data given in section 5.11, we can estimate the parameters of the model with $\alpha_3 \neq 0$. In this model ν is assumed to be known and the parameters γ_0, γ_1 and γ_2 can be estimated by method of moments, zero cell frequency technique or maximum likelihood. Using the summary data

$$\bar{x} = 1.5160, \quad \bar{y} = 1.6140, \quad m_{1,1} = .7980,$$

γ_0 and γ_1 for all three methods (with $\nu = 5$) are

$$\hat{\gamma}_0 = .3032 \quad \text{and} \quad \hat{\gamma}_1 = 0.3228$$

and the three estimates for γ_3 are

$$\tilde{\gamma}_2 = 0.06172, \qquad \approx{\gamma}_2 = 0.068, \qquad \hat{\gamma}_2 = 0.0572.$$

5.10 Testing the model

In section 5.4 two types of bivariate negative binomial distributions were developed by compounding bivariate Poisson random variables, which are (i) independent or (ii) correlated.

If we examine the conditional bivariate Poisson models, we see that the correlation coefficient in the case (i) is zero since X and Y are Independent.

In the correlated case (ii) as introduced by Subrahmaniam (1966) the correlation coefficient is

$$\rho_I = \frac{\alpha_3}{\{(\alpha_2 + \alpha_3)(\alpha_1 + \alpha_3)\}^{\frac{1}{2}}}.$$

Here ρ_I is the correlation between X and Y for a given individual in the population and it is not a function of the bivariate Poisson parameter λ. Subrahmaniam refers to ρ_I as the coefficient of 'intrinsic correlation' as distinct from the correlation arising from the unconditional distribution given in (5.4.7).

The intrinsic correlation coefficient ρ_I can be used for distinguishing between the two models. Testing for $\rho_I = 0$ is equivalent to testing for $\alpha_3 = 0$; thus, testing for the significance of $\rho_I = 0$ is actually testing for the adequacy of the model based on independent bivariate Poisson random variables. Subrahmaniam developed a test of this hypothesis using a locally asymptotic most powerful test developed by Neyman (1959) for composite hypotheses. This type of test, known as a $C(\alpha)$ test, is equivalent to Rao's efficient score test (1947). The equivalence of these two procedures is discussed in Kocherlakota and Kocherlakota (1991). We will develop the test in the context of the efficient score test.

Referring to section 2.2.2, we see that the efficient score test is based on the statistic

$$S_c = \underline{\phi}^{*'} \Gamma^{*-1} \underline{\phi}^*,$$

where

$$\phi^{*'} = \{u_1(\hat{\underline{\theta}}), \ldots, u_4(\hat{\underline{\theta}})\}$$

and Γ^* is the information matrix. S_c is asymptotically distributed as a χ^2 on one degree of freedom.

Here the score test is developed using the parameterization given in Subrahmaniam (1966). Writing

$$\beta = \frac{1}{\gamma_0 - \gamma_2}, \qquad \alpha_2 = \frac{\gamma_1 - \gamma_2}{\gamma_0 - \gamma_2}, \qquad \alpha_3 = \frac{\gamma_2}{\gamma_0 - \gamma_2},$$

the pf given in (5.9.3) becomes

$$g(x,y) = \frac{\Gamma(v + x + y)\, \beta^v\, \alpha_2^y}{x!\, y!\, \Gamma(v)\, (1 + \beta + \alpha_2 + \alpha_3)^{v+x+y}} S(x, y), \qquad (5.9.10)$$

where

$$\tau = \frac{\alpha_3 (1 + \beta + \alpha_2 + \alpha_3)}{\alpha_2}.$$

In this case elements of ϕ^* are obtained by differentiating the pf in (5.9.10) with respect to α_3, α_2, v and β . These partial derivatives are then evaluated under the null hypothesis $\alpha_3 = 0$. The maximum likelihood estimates obtained assuming the null hypothesis are substituted for the remaining parameters. Denote these partial derivatives by Q_i^* for i = 1, 2, 3, 4:

$$Q_1^*(x, y) = -\frac{v + x + y}{1 + \alpha_2 + \beta} + \frac{xy\,(1 + \alpha_2 + \beta)}{\alpha_2\,(v + x + y - 1)}$$

$$Q_2^* = \frac{y}{\alpha_2} - \frac{v + x + y}{1 + \alpha_2 + \beta}$$

$$Q_3^* = \frac{d}{dv} \log \Gamma(v + x + y) + \log \beta - \frac{d}{dv} \log\Gamma(v) - \log(1 + \alpha_2 + \beta)$$

$$Q_4^* = \frac{v}{\beta} - \frac{v + x + y}{1 + \alpha_2 + \beta}$$

and

$$u_i(\underline{\hat{\theta}}) = \frac{1}{\sqrt{n}} \sum_{x,y} Q_i^*(x, y) \qquad i = 1, 2, 3, 4$$

with α_2, β and v estimated by $\hat{a}$, $\hat{\beta}$, $\hat{v}$ as given in (5.9.2a, b). The information matrix Γ^* is given by

$$\Gamma^* = \left\{ \sum_{x,y} Q_i^*(x, y) \,) \, Q_h^*(x, y) \, g^*(x, y) \right\},$$

where $g^*(x, y)$ is defined in (5.4. 4).

A test of significance for $\alpha_3 = 0$ can be performed by evaluating the test statistic S_c. A summary of the test for examples 1 and 2 is given below.

	$\hat{a}$, $\hat{\beta}$, $\hat{v}$	S_c
Example 5.9.1	1.3025, 3.5062, 3.4201	0.3288
Example 5.9.2	0.7678, 2.4264, 3.6784	147.9704

These results indicate that, for shunter data in example 5.9.1, there is not sufficient evidence to reject the null hypothesis; hence, the model with $\alpha_3 = 0$ is adequate. In example 5.9.2 the hypothesis is clearly rejected, indicating that the model with the term α_3 is more appropriate.

5.11 Simulation

The distribution with $\alpha_3 \neq 0$ can be generated using the marginal and conditional distributions developed in section 5.4:

$$\left.\begin{array}{l} \text{(i) } Y \sim NB\left(v, \dfrac{p_2 + p_3}{1 - p_1}\right) \\ \text{(ii) } X \mid y \sim \text{convolution of } X_1 \text{ and } X_2 \end{array}\right\} \qquad (5.11.1)$$

with

$$X_1 \sim B(y, \frac{p_3}{p_2 + p_3}) \text{ and } X_2 \sim NB(\nu + y, 1 - p_1).$$

We have considered several parameterizations of the probability function. These can be found in (5.4.9), (5.9.4) and (5.9.10). The following relationships between the parameters may be useful in order to use the marginal and conditional distributions in the form given in (5.11.1)

Since the marginal distribution of Y is univariate negative binomial, a routine is needed for simulating this distribution. The technique used in the IMSL subroutine can only be used if the index parameter is a positive integer. For generating realizations from the bivariate negative binomial, we need a routine that allows any real value of the index parameter ν. We have used a routine UPAS (available from the authors), in which a univariate Poisson random variable is compounded with a gamma random variable (5.4.1) to give the univariate negative binomial given in (5.4.2).

The generating procedure can be summarized as: Generate a realization s of Y from the negative binomial distribution using subroutine UPAS described above. Generate X_1 from $B(s, \frac{p_3}{p_2 + p_3})$ using the subroutine GGBN and X_2 from $NB(\nu + s, 1 - p_1)$ using UPAS. The realized value of X is $x_1 + x_2$. The index parameter for each of these is a function of the value s that has been generated for the marginal distribution. This procedure is repeated N times to give a sample of size N from a bivariate negative binomial distribution.

For the model with $\alpha_3 = 0$ given in (5.4.4), the conditional distribution of X given Y = s reduces to a univariate negative binomial with index $\nu + s$ and parameter $\frac{1+\beta}{1+a+\beta}$.

With $\gamma_0 = .308$, $\gamma_1 = .323$, $\gamma_3 = .051$ and $\nu = 5.0$, N = 1000 observations were simulated from (5.9.4) and are displayed in the table given below. The expected values based on the parameters used for the simulation appear in parentheses. Two observations at (4, 11) and (6, 9) are not displayed, but they were used in calculating the goodness-of-fit. For this table the χ^2 goodness-of-fit statistic was 50.642 on 48 degrees of freedom with a prob-value of 0.37. A fit of the distribution with $\alpha_3 \neq 0$ was also attempted based on estimates of the parameters a, β, ν given in (5.9.2a, b); however, this distribution, as expected, did not fit adequately.

x \ y	0	1	2	3	4	5	6	7
0	109	74	55	17	7	5	0	0
	(101.56)	(87.42)	(45.15)	(18.14)	(6.24)		(2.67)	
1	90	97	75	36	10	6	1	0
	(82.6)	(101.7)	(68.33)	(33.80)	(13.82)	(4.96)	(2.12)	
2	38	63	58	32	9	8	2	1
	(40.30)	(64.57)	(54.31)	(32.46)	(15.61)	(6.45)	(2.38)	(1.06)
3	16	22	36	28	7	3	1	2
	(15.03)	(30.17)	(30.67)	(21.69)	(12.10)	(5.70)	(2.37)	(1.20)
4	3	9	17	11	6	2	1	2
	(4.98)	(11.66)	(13.93)	(11.43)	(7.31)	(3.89)	(2.86)	
5	0	0	8	3	8	1	3	2
	(1.46)	(3.95)	(5.44)	(5.09)	(3.68)	(2.20)	(1.57)	
6	0	0	1	2	2	0	1	0
	(1.61)		(1.90)	(2.00)	(1.62)	(1.07)	(1.06)	
7	0	1	1	2	0	1	0	0
				(6.14)				
8	0	2	1	0	0	0	0	0

6

SAMPLING FROM FINITE POPULATIONS

6.1 Introduction

In the univariate case, the hypergeometric distribution arises in sampling from a finite population with a dichotomy of items, provided that the sampling is without replacement. Thus, we may consider a population of N items consisting of N_1 of Type I and N_2 of Type II. A random sample of n items is drawn, without replacement, from this population. Let the random variable X be defined as

X = number of Type I items appearing in the sample.

Then it can be seen from the basic counting principle that the

$$P\{X = x\} = \frac{\binom{N_1}{x}\binom{N_2}{n-x}}{\binom{N}{n}}, \qquad \max[0, n - N_2] \le x \le \min[n, N_1]. \qquad (6.1.1)$$

The crucial desiderata in this development are the finite characteristic of the population and that the sampling is without replacement. These two criteria lead to the *dependence* of the trials or draws. If we relax the first of these two requirements, then the distribution in (6.1.1) *approaches* that of the binomial with the parameters n and $p = N_1/N$. On the other hand, if we sample with replacement from the finite population, (6.1.1) is replaced by the binomial distribution with the same parameters.

6.2 Bivariate generalizations

There are two distinct types of bivariate hypergeometric distributions that arise in practice. The first type, like the Type I bivariate binomial, is based on the double dichotomy. Wicksell (1923) showed that in a sample taken without replacement from a finite population the joint distribution of the marginal totals is a bivariate hypergeometric.

The second type of bivariate hypergeometric is an analog of the trinomial distribution and seems to have been first developed by Guldberg (1934).

6.2.1 Double dichotomy

Consider the following representation of a *population* of N items.

Characteristic II

		Present	Absent	
Characteristic I	Present	N_{11}	N_{12}	N_{1+}
	Absent	N_{21}	N_{22}	N_{2+}
		N_{+1}	N_{+2}	N

A random sample of size n is taken from this population *without replacement*. The corresponding representation of one observed *sample* using the same double dichotomy is:

Characteristic II

		Present	Absent	
Characteristic I	Present	x	r - x	r
	Absent	u - x	n - u - r + x	n - r
		u	n - u	n

Let the random variables corresponding to the above table be defined as follows:

Characteristic II

		Present	Absent	
Characteristic I	Present	X_{11}	X_{12}	R
	Absent	X_{21}	X_{22}	n - R
		U	n - U	n

Then the

$$P\{X_{11} = x_{11}, X_{12} = x_{12}, X_{21} = x_{21}, X_{22} = x_{22}\}$$

$$= \frac{\binom{N_{11}}{x_{11}}\binom{N_{12}}{x_{12}}\binom{N_{21}}{x_{21}}\binom{N_{22}}{x_{22}}}{\binom{N}{n}}, \qquad (6.2.1)$$

where

$$x_{11} = 0, 1, \ldots, \min(n, N_{11}), \quad x_{12} = 0, 1, \ldots, \min(n, N_{12}),$$
$$x_{21} = 0, 1, \ldots, \min(n, N_{21}), \quad \max(0, n - N_{22}) \le x_{11} + x_{12} + x_{21} \le n.$$

With this sampling scheme, we can now examine the joint probability function of the random variables R and U:

$$P\{R = r, U = u\} = \sum_x P\{X_{11} = x, X_{12} = r - x, X_{21} = u - x, X_{22} = n - r - u + x\}. \qquad (6.2.2)$$

Since the event $\{R = r, U = u\}$ arises from all the mutually exclusive configurations of the *sample* tables, given that the marginal totals are fixed by this event, the summation will be over all x such that

$$\max(0, r + u - n) \le x \le \min(r,u).$$

Substituting from equation (6.2.1) into (6.2.2), we have

$$P\{R = r, U = u\} = \sum_x \frac{\binom{N_{11}}{x}\binom{N_{12}}{r-x}\binom{N_{21}}{u-x}\binom{N_{22}}{n-r-u+x}}{\binom{N}{n}}. \qquad (6.2.3)$$

The random variables (R, U) are said to have the bivariate hypergeometric distribution. Marshall and Olkin (1985) credit Wicksell (1923) with the development of this model.

6.2.2 Trichotomous populations

Consider a population containing three types of items I, II, III of sizes N_1, N_2 and N_3, respectively. Let a random sample of size n be taken without replacement. Define, for i = I, II, III,

$$X_i = \text{number of items of Type i in the sample.}$$

Then, since $X_1 + X_2 + X_3 = n$, we can write

$$P\{X_1 = x_1, X_2 = x_2\} = \frac{\binom{N_1}{x_1}\binom{N_2}{x_2}\binom{N_3}{n-x_1-x_2}}{\binom{N}{n}} \tag{6.2.4}$$

with

$$x_1 = 0, 1, \ldots, \min[n, N_1],$$
$$x_2 = 0, 1, \ldots, \min[n, N_2],$$
$$\max[0, n - N_3] \leq x_1 + x_2 \leq n.$$

Guldberg (1934) seems to have been the first one to develop this model as a bivariate generalization of the hypergeometric distribution.

6.2.3 Comparison of the models

Much confusion has arisen in the literature as to what constitutes a bivariate hypergeometric distribution, or its more general analog the multivariate hypergeometric distribution. Perhaps, it would be helpful to examine the limiting forms of the distributions in (6.2.3) and (6.2.4) to identify their true nature.

Let us consider the limiting form of (6.2.3) as N_{11}, N_{12}, N_{21} and N_{22} all tend to infinity with

$$\frac{N_{11}}{N} = p_{11}, \quad \frac{N_{12}}{N} = p_{12}, \quad \frac{N_{21}}{N} = p_{21}, \quad \frac{N_{22}}{N} = p_{22}$$

and $p_{11} + p_{12} + p_{21} + p_{22} = 1$.

Before taking this limit of (6.2.3), we see that (6.2.1) tends to

$$\frac{n!}{x_{11}!\, x_{12}!\, x_{21}!\, x_{22}!}\, p_{11}^{x_{11}}\, p_{12}^{x_{12}}\, p_{21}^{x_{21}}\, p_{22}^{x_{22}}, \tag{6.2.5}$$

for each fixed $(x_{11}, x_{12}, x_{21}, x_{22})$ with $x_{11} + x_{12} + x_{21} + x_{22} = n$.

Hence, the limit of (6.2.3) is

$$\sum_{x} \frac{n!\, p_{11}^{x}\, p_{12}^{r-x}\, p_{21}^{u-x}\, p_{22}^{n-r-u+x}}{x!\,(r-x)!\,(u-x)!\,(n-r-u+x)!}, \tag{6.2.6}$$

which corresponds to (3.3.3), the probability function of the Type I bivariate binomial distribution. In this respect one can refer to (6.2.3) as a bivariate hypergeometric distribution.

The limiting form of (6.2.4) as N_1, N_2 and N_3 tend to infinity with

$$\frac{N_1}{N} = p_1, \quad \frac{N_2}{N} = p_2, \quad \frac{N_3}{N} = p_3$$

and $p_1 + p_2 + p_3 = 1$ can be found directly from (6.2.4) to be

$$\frac{n!}{x_1!\, x_2!\, (n - x_1 - x_2)!}\, p_1^{x_1}\, p_2^{x_2}\, p_3^{n-x_1-x_2}, \tag{6.2.7}$$

with

$$x_1 = 0, 1, \ldots, n, \quad x_2 = 0, 1, \ldots, n, \quad x_1 + x_2 \leq n.$$

It is obvious that (6.2.7) is the probability function of a trinomial distribution (sometimes fallaciously) referred to as a bivariate binomial.

It is important at this point to clarify the nomenclature to be used in this book. We will follow most of the literature and refer to Guldberg's form (6.2.4) as the bivariate hypergeometric. Strictly speaking, based on the limiting form of this distribution a better term would be the *trihypergeometric* -- as a precursor to the trinomial. By the same argument, Wicksell's form (6.2.3) of the distribution should be called the *bivariate hypergeometric* as its limiting form is given by the bivariate binomial (3.3.3), which is itself a precursor to the bivariate Poisson distribution. However, we shall follow the published literature [*viz.*, Janardan (1973, 1975, 1976) and Janardan and Patil (1972)] and reserve the term bivariate hypergeometric for (6.2.4). This is perhaps a questionable term, carried over from sampling theory for finite populations.

6.3 Marginal and conditional distributions

In this section marginal and conditional distributions are developed for both forms of the bivariate hypergeometric model. The conditional distributions are used to determine the regression functions and the corresponding correlation coefficient.

6.3.1 Wicksell form

For the pf given in (6.2.3), the marginal distribution of R is readily obtained from the 2 x 2 table for the random variables; hence,

$$P\{R = r\} = \frac{\binom{N_{1+}}{r}\binom{N_{2+}}{n-r}}{\binom{N}{n}}, \quad \max[0, n - N_{2+}] \le r \le \min[n, N_{1+}] \qquad (6.3.1)$$

from the counting principle. Thus each of the variables, R and U, has the hypergeometric distribution with the appropriate parameters.

Using (6.2.3) and (6.3.1) we can find the conditional distribution of U given R = r:

$$P\{U = u \mid R = r\} = \sum_x \frac{\binom{N_{11}}{x}\binom{N_{12}}{r-x}\binom{N_{21}}{u-x}\binom{N_{22}}{n-r-u+x}}{\binom{N_{1+}}{r}\binom{N_{2+}}{n-r}},$$

which can be simplified to yield

$$\sum_x \frac{\binom{r}{x}\binom{N_{1+}-r}{N_{11}-x}\binom{N_{2+}-(n-r)}{N_{21}-(u-x)}\binom{n-r}{u-x}}{\binom{N_{1+}}{N_{11}}\binom{N_{2+}}{N_{21}}}. \qquad (6.3.2)$$

Since (6.3.2) is a proper pf, we have the following identity, which will be useful later:

$$\sum_u \sum_x \binom{r}{x}\binom{N_{1+}-r}{N_{11}-x}\binom{N_{2+}-(n-r)}{N_{21}-(u-x)}\binom{n-r}{u-x} = \binom{N_{1+}}{N_{11}}\binom{N_{2+}}{N_{21}}. \qquad (6.3.3)$$

The conditional distribution given in (6.3.2) is not of a very tractable form; however, the conditional expectation of U given R = r can be obtained directly as

$$E[U \mid R = r] = \sum_{u} u \; P\{U = u \mid R = r\}.$$

Substituting from (6.3.2) for the probability on the right hand side yields

$$\frac{\displaystyle\sum_{u,x} u \binom{r}{x}\binom{N_{1+}-r}{N_{11}-x}\binom{N_{2+}-(n-r)}{N_{21}-(u-x)}\binom{n-r}{u-x}}{\displaystyle\binom{N_{1+}}{N_{11}}\binom{N_{2+}}{N_{21}}},$$

which can be rearranged to give

$$\sum_{u,x} (u - x) \, P\{U = u \mid R = r\} + \sum_{u,x} x \, P\{U = u \mid R = r\}. \tag{6.3.4}$$

The first term in (6.3.4) is

$$\frac{\displaystyle(n - r) \sum_{u,x} \binom{r}{x}\binom{N_{1+}-r}{N_{11}-x}\binom{N_{2+}-1-(n-r-1)}{N_{21}-1-(u-x-1)}\binom{n-r-1}{u-x-1}}{\displaystyle\binom{N_{1+}}{N_{11}}\binom{N_{2+}}{N_{21}}},$$

which upon using the identity (6.3.3) becomes

$$(n - r)\frac{\binom{N_{2+}-1}{N_{21}-1}}{\binom{N_{2+}}{N_{21}}} = (n - r)\frac{N_{21}}{N_{2+}}.$$

Similarly, the second term in (6.3.4) can be shown to be equal to $r\,\dfrac{N_{11}}{N_{1+}}$.

Hence, the regression of U on R is

$$E[U \mid R = r] = n\frac{N_{21}}{N_{2+}} + r\left(\frac{N_{11}}{N_{1+}} - \frac{N_{21}}{N_{2+}}\right), \tag{6.3.5}$$

which is linear in r. Equation (6.3.5) can be expressed in the form given in Marshall and Olkin (1985) if we rewrite the second term on the right hand side as

$$r\left(\frac{N_{11}N_{2+} - N_{21}N_{1+}}{N_{1+}N_{2+}}\right) = r\left(\frac{N_{11}N_{22} - N_{12}N_{21}}{N_{1+}N_{2+}}\right). \tag{6.3.6}$$

Thus

$$E[U \mid R = r] = n\frac{N_{21}}{N_{2+}} + r\left(\frac{N_{11}N_{22} - N_{12}N_{21}}{N_{1+}N_{2+}}\right). \tag{6.3.7}$$

By a similar argument we can write

$$E[R \mid U = u] = n\frac{N_{12}}{N_{+2}} + u\left(\frac{N_{11}N_{22} - N_{12}N_{21}}{N_{+1}N_{+2}}\right). \tag{6.3.8}$$

Combining (6.3.7) and (6.3.8), the coefficient of correlation between R and U is given by

$$\rho_{X,Y} = \frac{N_{11}N_{22} - N_{12}N_{21}}{(N_{+1}N_{+2}N_{1+}N_{2+})^{\frac{1}{2}}}. \tag{6.3.9}$$

6.3.2 Bivariate hypergeometric distribution

The marginal distributions corresponding to the joint pf in (6.2.4) can be determined directly from the definition of a marginal pf in terms of the joint pf; hence,

$$P\{X_1 = x_1\} = \sum_{x_2} \frac{\binom{N_1}{x_1}\binom{N_2}{x_2}\binom{N_3}{n-x_1-x_2}}{\binom{N}{n}}$$

$$= \frac{\binom{N_1}{x_1}\binom{N_2+N_3}{n-x_1}}{\binom{N}{n}},$$

$$\max[0, n-(N-N_1)] \le x_1 \le \min[n, N_1] \qquad (6.3.10)$$

and

$$P\{X_2 = x_2\} = \frac{\binom{N_2}{x_2}\binom{N_1+N_3}{n-x_2}}{\binom{N}{n}},$$

$$\max[0, n-(N-N_2)] \le x_2 \le \min[n, N_2]. \qquad (6.3.11)$$

By definition, the conditional distributions are given by

$$P\{X_1 = x_1 \mid X_2 = x_2\} = \frac{\binom{N_1}{x_1}\binom{N_3}{n-x_1-x_2}}{\binom{N_1+N_3}{n-x_2}},$$

$$\max[0, n-N_3-x_2] \le x_1 \le \min[n-x_2, N_1] \qquad (6.3.12)$$

and

$$P\{X_2 = x_2 \mid X_1 = x_1\} = \frac{\binom{N_2}{x_2}\binom{N_3}{n-x_1-x_2}}{\binom{N_2+N_3}{n-x_1}}$$

$$\max[0, n-N_3-x_1] \le x_2 \le \min[n-x_1, N_2], \qquad (6.3.13)$$

both of which are univariate hypergeometric distributions.

As N_1, N_2 and N_3 tend to infinity, the marginal distributions given in (6.3.10) and (6.3.11) tend to binomial distributions:

$$X_1 \approx B(n, p_1), \qquad p_1 = \frac{N_1}{N}$$

$$X_2 \approx B(n, p_2), \qquad p_2 = \frac{N_2}{N}.$$

Also, the conditional distributions (6.3.12) and (6.3.13) tend to binomial distributions with

$$\{X_1 \mid X_2 = x_2\} \approx B(n - x_2, \frac{N_1}{N_1 + N_3})$$

$$\{X_2 \mid X_1 = x_1\} \approx B(n - x_1, \frac{N_2}{N_2 + N_3}).$$

Recalling the expectation for a hypergeometric random variable, we have

$$E[X_1 \mid X_2 = x_2] = (n - x_2) \frac{N_1}{N_1 + N_3}$$

$$= \frac{n N_1}{N_1 + N_3} - \frac{N_1}{N_1 + N_3} x_2 \qquad (6.3.14)$$

and

$$E[X_2 \mid X_1 = x_1] = (n - x_1) \frac{N_2}{N_2 + N_3}$$

$$= \frac{n N_2}{N_2 + N_3} - \frac{N_2}{N_2 + N_3} x_1, \qquad (6.3.15)$$

which are linear in x_2 and x_1, respectively.

It is interesting to note that the coefficient of correlation in this case is

$$\rho_{X,Y} = -\left\{\frac{N_1 N_2}{(N_1 + N_3)(N_2 + N_3)}\right\}^{\frac{1}{2}},$$

which is the coefficient of correlation between X_1 and X_2 in the trinomial distribution with $p_1 = \frac{N_1}{N}$ and $p_2 = \frac{N_2}{N}$.

6.4 Probability generating functions

The pgf of the univariate hypergeometric distribution in (6.1.1) can be expressed in terms of Gauss' hypergeometric function in one variable as

$$\Pi_X(t) = \frac{(N - n)!\,(N - N_1)!}{N!\,(N - N_1 - n)!}\, {}_2F_1(-n, -N_1; N - N_1 - n + 1; t)$$

with

$$ {}_2F_1(\alpha,\ \beta;\gamma;t)\ =\ \sum_{x=0}^{\infty}\ \frac{(\alpha)_x\,(\beta)_x}{(\gamma)_x}\,\frac{t^x}{x!}\,. $$

The hypergeometric function of two variables is given by the series

$$ F(\alpha;\beta_1,\beta_2;\gamma;t_1,t_2)=\sum_{x_1=0}^{\infty}\ \sum_{x=x_1}^{\infty}\ \frac{(\alpha)_x\,(\beta_1)_{x_1}\,(\beta_2)_{x-x_1}}{(\gamma)_x}\ \frac{t_1^{x_1}}{x_1!}\,\frac{t_2^{x-x_1}}{(x-x_1!)}\,. \tag{6.4.1} $$

[See Erdelyi *et al.* (1953, Vol. 1, p. 224, eq. 16).]

It is possible to write the probability generating function of the bivariate hypergeometric distribution in terms of this function.

Consider

$$ \zeta(t_1,t_2;u)=(1+t_1\,u)^{N_1}\,(1+t_2\,u)^{N_2} \tag{6.4.2} $$

expanded as a series

$$ \zeta(t_1,t_2;u)=\prod_{i=1}^{2}\ \{\sum_{j=0}^{N_i}\ \binom{N_i}{j}(t_i\,u)^j\}\,. $$

Then the probability function given in (6.2.4)

$$ f(x_1,x_2)=\frac{\binom{N_1}{x_1}\binom{N_2}{x_2}\binom{N_3}{n-x_1-x_2}}{\binom{N}{n}} $$

is related to $\zeta(t_1, t_2; u)$ as the coefficient of $t_1^{x_1}\,t_2^{x_2}$ in

$$ \Pi(t_1,t_2)\ =\ \frac{(N-n)!}{N!}\,\frac{\partial^n}{\partial u^n}\,\zeta(t_1,t_2;u)\Big|_{u\,=\,0} \tag{6.4.3} $$

[Tsao (1965)].

Janardan and Patil (1972) represent this pgf in terms of the hypergeometric function in (6.4.1). Relating the parameters given in the pf in (6.2.4) with those in Janardan and Patil (1972, equation 2.6), the pgf is given by

$$\Pi(t_1, t_2) = \frac{N_3!\,(N-n)!}{(N_3-n)!\,N!} F(-n; -N_1, -N_2; N_3-n+1; t_1, t_2). \quad (6.4.4)$$

It should be recalled that

$$(-n)! = \frac{(-1)^{n-1}}{(n-1)!}$$

and

$$\frac{(-n)!}{(-n-m)!} = \frac{(-1)^m (n+m-1)!}{(n-1)!}$$

with n and m as integers.

From (6.4.4) the pgf's of X_1 and X_2 can be found to be

$$\Pi_{X_1}(t) = \frac{N_3!\,(N-n)!}{(N_3-n)!\,N!} F(-n; -N_1, -N_2; N_3-n+1; t, 1)$$

$$= \frac{N_3!\,(N-n)!}{(N_3-n)!\,N!} \left\{\frac{\Gamma(N_2+N_3+1)\,\Gamma(N_3-n+1)}{\Gamma(N_3+1)\Gamma(N_2+N_3-n+1)}\right\} {}_2F_1(-n, -N_1; N_2+N_3-n+1; t)$$

from Erdelyi *et al.* (1953, Vol. 1, p. 239, eq. 10). Thus

$$\Pi_{X_1}(t) = \frac{(N-n)!}{N!} \frac{(N-N_1)!}{(N-N_1-n)!} {}_2F_1(-n, -N_1; N-N_1-n+1; t) \quad (6.4.5)$$

and

$$\Pi_{X_2}(t) = \frac{(N-n)!}{N!} \frac{(N-N_2)!}{(N\ N_2-n)!} {}_2F_1(-n, -N_2; N-N_2-n+1; t), \quad (6.4.6)$$

which are the pgf's of univariate hypergeometric distributions corresponding to the pf's given in (6.3.10) and (6.3.11), respectively.

Now setting $t_1 = t_2 = t$ in (6.4.4), the distribution of the sum $X_1 + X_2$ is seen to have pgf

$$\Pi_{X_1+X_2}(t) = \frac{N_3!\,(N\ n)!}{(N_3-n)!\,N!}\,F(-n; -N_1, -N_2; N_3 - n + 1; t, 1)$$

$$= \frac{(N-n)!}{N!}\frac{(N-N_1-N_2)!}{(N-N_1-N_2-n)!}\,{}_2F_1(-n, -N_1 - N_2; N - N_1 - N_2 - n + 1; t), \tag{6.4.7}$$

which is the pgf of a univariate hypergeometric distribution.

6.5 Models for the bivariate hypergeometric distribution

Several stochastic models can give rise to the pf given in (6.2.4). We will paraphrase the discussion of Janardan (1973) relating to the models for the multivariate hypergeometric distribution and discuss the bivariate cases:

(1) *Categorical data.* Suppose that N individuals are placed into three cells with the ith cell having N_i individuals. Let a random sample of n individuals be taken from this population *without replacement.* Then the random variables X_1, X_2, X_3, which count the number of individuals of each type in the sample, have the pf given in (6.2.4).

(2) *Conditional distribution of independent binomials.* Let the random variables Y_i be independent $B(N_i, p)$ for i = 1, 2, 3. Then the conditional distribution of Y_1, Y_2, Y_3 given $Y_1 + Y_2 + Y_3 = Y = y$ is bivariate hypergeometric.

To prove this result, we see that

$$P\{Y_1 = y_1, Y_2 = y_2, Y_3 = y_3\} = \prod_{i=1}^{3} \binom{N_i}{y_i} p^{y_i} (1 - p)^{N_i - y_i}$$

and

$$P\{Y = y\} = \binom{N}{y} p^y (1 - p)^{N-y}.$$

Then the conditional distribution of (Y_1, Y_2, Y_3) given Y = y is

$$P\{Y_1 = y_1, Y_2 = y_2, Y_3 = y_3 \mid Y = y\} =$$

$$\frac{P\{Y_1 = y_1, Y_2 = y_2, Y_3 = y - y_1 - y_2\}}{P\{Y = y\}}.$$

Hence, the joint distribution of the random variables Y_1, Y_2 is

$$P\{Y_1 = y_1, Y_2 = y_2\} = \frac{\binom{N_1}{y_1}\binom{N_2}{y_2}\binom{N_3}{y-y_1-y_2}}{\binom{N}{y}}, \quad y_1 + y_2 \leq y,$$

which is the same form as (6.2.4).

(3) *Haemocytometer count.* Suppose that a haemocytometer is divided into three squares, each having m stalls or spaces. Consider n red blood cells being distributed at random into the haemocytometer with each stall accommodating only one cell. Then the probability that x_i stalls are occupied in the ith square with $n = x_1 + x_2 + x_3$ is

$$P\{X_1 = x_1, X_2 = x_2\} = \frac{\binom{m}{x_1}\binom{m}{x_2}\binom{m}{n-x_1-x_2}}{\binom{3m}{n}}.$$

(4) *Conditional distribution in the trinomial.* Let $\underline{X}$ and $\underline{Y}$ be independent random variables with the pf's

$$P\{X_1 = x_1, X_2 = x_2\}$$

$$= \frac{n!}{x_1!\, x_2!\, (n - x_1 - x_2)!} p_1^{x_1} p_2^{x_2} (1 - p_1 - p_2)^{n - x_1 - x_2}$$

and

$$P\{Y_1 = y_1, Y_2 = y_2\}$$

$$= \frac{(N - n)!}{y_1!\, y_2!\, (N - n - y_1 - y_2)!} p_1^{y_1} p_2^{y_2} (1 - p_1 - p_2)^{N - n - y_1 - y_2}.$$

Now consider the conditional distribution of $\underline{X}$ given $\underline{Z} = \underline{X} + \underline{Y} = \underline{z}$; that is,

$$P\{\underline{X} = \underline{x} \mid \underline{Z} = \underline{z}\} = \frac{P\{\underline{X} = \underline{x}\} P\{\underline{Y} = \underline{z} - \underline{x}\}}{P\{\underline{Z} = \underline{z}\}}.$$

Since

$$P\{\{\underline{Z} = \underline{z}\} = \frac{N!}{z_1!\, z_2!\, (N - z_1 - z_2)!} p_1^{z_1} p_2^{z_2} (1 - p_1 - p_2)^{N - z_1 - z_2},$$

we have

$$P\{\underline{X} = \underline{x} \mid \underline{Z} = \underline{z}\} = \frac{\binom{n}{x_1,\, x_2} \binom{N-n}{z_1-x_1,\, z_2-x_2}}{\binom{N}{z_1,\, z_2}}, \tag{6.5.1}$$

where

$$\binom{n}{r,\, s} = \frac{n!}{r!\, s!\, (n-r-s)!}.$$

The right hand side of (6.5.1) can be rearranged to yield

$$P\{\underline{X} = \underline{x} \mid \underline{Z} = \underline{z}\} = \frac{\binom{z_1}{x_1} \binom{z_2}{x_2} \binom{N-z_1-z_2}{n-x_1-x_2}}{\binom{N}{n}},$$

the probability function of the bivariate hypergeometric distribution.

6.6 Related models based on sampling without replacement

Several other distributions arise in practice when dealing with sampling without replacement: bivariate inverse hypergeometric, bivariate negative hypergeometric, bivariate inverse negative hypergeometric, bivariate Polya and bivariate inverse Polya. The form of each of these five distributions and the chance mechanisms giving rise to them will be discussed. Then a unified model due to Janardan and Patil (1972) will be presented. By an appropriate choice of the parameters, this model gives rise to the bivariate hypergeometric as well as each of the five distributions mentioned above.

6.6.1 Probability distributions

A. *Bivariate inverse hypergeometric*

(1) *Waiting times.* In section 5.2 we saw that the bivariate negative binomial and the bivariate geometric distribution arise through *inverse sampling with replacement* or by counting the waiting times in Bernoulli trials. Now consider sampling *without replacement* from a finite population with three types of elements of sizes N_1, N_2 and N_3. Sampling is continued until k items of the third type are drawn. The random variables of interest are

$$X_1 = \text{number of Type I elements in the sample}$$
$$X_2 = \text{number of Type II elements in the sample.}$$

The random variables (X_1, X_2) are said to have the bivariate inverse hypergeometric distribution with the joint pf given by

$$P\{X_1 = x_1, X_2 = x_2\} = \frac{\binom{N_1}{x_1}\binom{N_2}{x_2}\binom{N_3}{k-1}}{\binom{N}{k+x_1+x_2-1}} \frac{N_3 - k + 1}{N - k + x_1 + x_2 + 1}, \tag{6.6.1}$$

where $x_i = 0, 1, \ldots, N_i$ for $i = 1, 2$ and the parameters $k = 1, 2, \ldots, N_3$ with $N = N_1 + N_2 + N_3$. See Steyn (1951) for details of the multivariate analog of (6.6.1).

(2) *Conditional distribution.* The bivariate inverse hypergeometric distribution given in (6.6.1) can also be derived as the conditional distribution of independent negative trinomials given their sum.

Let $\underline{X}$ and $\underline{Y}$ be independent negative trinomial random variables with pf's given by

$$P\{X_1 = x_1, X_2 = x_2\} = \frac{(k+x_1+x_2-1)!}{x_1!\, x_2!\, (k-1)!} p_1^{x_1} p_2^{x_2} p_3^{k}$$

$$P\{Y_1 = y_1, Y_2 = y_2\} = \frac{(N_3-k+y_1+y_2)!}{y_1!\, y_2!\, (N_3-k)!}\, p_1^{y_1}\, p_2^{y_2}\, p_3^{N_3-k+1},$$

where $p_3 = 1 - p_1 - p_2$.

Now consider the sum $\underline{X} + \underline{Y} = \underline{Z}$. Since the sum of two negative trinomials is itself a negative trinomial with the index parameter equal to the sum of the indices, the pf of $\underline{Z}$ is

$$P\{Z_1 = z_1, Z_2 = z_2\} = \frac{(N_3+z_1+z_2)!}{z_1!\, z_2!\, N_3!}\, p_1^{z_1}\, p_2^{z_2}\, p_3^{N_3+1}.$$

Therefore, the conditional probability function

$$P\{X_1 = x_1, X_2 = x_2 | Z_1 = z_1, Z_2 = z_2\}$$

$$= \frac{P\{X_1 = x_1, X_2 = x_2\}\, P\{Y_1 = z_1 - x_1, Y_2 = z_2 - x_2\}}{P\{Z_1 = z_1, Z_2 = z_2\}}$$

$$= \frac{(k+x-1)!\, (N-x-k)!\, N_1!\, N_2!\, N_3!}{x_1!\, x_2!\, (k-1)!\, (N_1-x_1)!\, (N_2-x_2)!\, (N_3-k)!\, N!}, \qquad (6.6.2)$$

where $z_1 = N_1$, $z_2 = N_2$, $N = N_1 + N_2 + N_3$ and $x = x_1 + x_2$. The conditional probability given in (6.6.2) is equivalent to the pf given in (6.6.1).

(3) *Compounding.* In a series of papers, Mosimann (1962, 1963, 1970) has studied the distribution of pollen counts. His derivation, based on compounding, gives rise to a multivariate analog of (6.6.1).

Here the bivariate distribution given in (6.6.1) is derived by compounding independent binomial random variables with a beta distribution. Let X_1, X_2 be independent random variables with $X_i \sim B(N_i, p)$, $i = 1, 2$ and p be a random variable with the beta distribution

$$f(p) = \frac{\Gamma(N_3 + 1)}{\Gamma(k)\, \Gamma(N_3 - k + 1)}\, p^{k-1}\, (1 - p)^{N_3-k}, \qquad 0 < p < 1.$$

Then the joint distribution of (X_1, X_2) is given by

$$P\{X_1 = x_1, X_2 = x_2\} = \int_0^1 P\{X_1 = x_1, X_2 = x_2 | p\} f(p)\, dp$$

$$= \frac{\Gamma(N_3 + 1)}{\Gamma(k)\,\Gamma(N_3 - k + 1)} \binom{N_1}{x_1}\binom{N_2}{x_2} \int_0^1 p^{k+x-1} (1-p)^{N-k-x}\, dp$$

$$= \frac{\Gamma(N_3 + 1)}{\Gamma(k)\,\Gamma(N_3 - k + 1)} \binom{N_1}{x_1}\binom{N_2}{x_2} \frac{\Gamma(k+x)\,\Gamma(N-k-x+1)}{\Gamma(N+1)}. \qquad (6.6.3)$$

Since all the quantities involved in the the gamma functions in (6.6.3) are integers, these functions can be replaced by the corresponding factorials and simplified; hence, (6.6.3) is equivalent to (6.6.1).

B. Bivariate negative hypergeometric distribution.

The random variables (X_1, X_2) are said to have the bivariate negative hypergeometric distribution if their joint pf is given by

$$P\{X_1 = x_1, X_2 = x_2\} = \frac{n!\,\Gamma(N)}{\Gamma(n+N)} \prod_{i=1}^{3} \frac{\Gamma(N_i+x_i)}{\Gamma(N_i)\, x_i!}$$

$$= \frac{\binom{N_1+x_1-1}{x_1}\binom{N_2+x_2-1}{x_2}\binom{N_3+x_3-1}{x_3}}{\binom{N+n-1}{n}}, \qquad (6.6.4)$$

where $x_i = 0, 1, 2, \ldots,\ i = 1,2$ with $N = N_1 + N_2 + N_3$; $n = x_1 + x_2 + x_3$; $N_i > 0$, $N < \infty$.

Janardan (1973) has given an excellent review of the literature in connection with the multivariate negative hypergeometric. Of particular interest are the contributions of Ishii and Hayakawa (1960), Mosimann (1962,1963), Sarndal (1964, 1965) and Hoadley (1969). We will consider three stochastic models under which the bivariate pf given in (6.6.4) arises.

(1) *Compounding.* Let the random variables (X_1, X_2) given

(p_1, p_2) have the trinomial distribution with pf

$$f(x_1, x_2 \mid p_1, p_2) = \frac{n!}{x_1!\, x_2!\, (n-x_1-x_2)!}\, p_1^{x_1}\, p_2^{x_2}\, (1 - p_1 - p_2)^{n-x_1-x_2}$$

where (p_1, p_2) have the joint pdf

$$g(p_1, p_2) = \frac{\Gamma(N)}{\Gamma(N_1)\,\Gamma(N_2)\,\Gamma(N_3)}\, p_1^{N_1-1}\, p_2^{N_2-1}\, (1 - p_1 - p_2)^{N_3-1},$$

$$0 < p_1 < 1,\ 0 < p_2 < 1,\ p_1 + p_2 < 1.$$

Since the marginal distribution of (X_1, X_2) is

$$h(x_1, x_2) = \int \int f(x_1, x_2 \mid p_1, p_2)\ g(p_1, p_2)\, dp_1\, dp_2,$$

we have

$$h(x_1, x_2) = C \int_0^1 \int_0^{1-p_1} p_1^{N_1+x_1-1}\, p_2^{N_2+x_2-1}\, (1 - p_1 - p_2)^{n+N_3-x_1-x_2-1}\, dp_2\, dp_1, \tag{6.6.5}$$

where

$$C = \frac{n!}{x_1!\, x_2!\, (n-x_1-x_2)!}\ \frac{\Gamma(N)}{\Gamma(N_1)\,\Gamma(N_2)\,\Gamma(N_3)}.$$

Integrating (6.6.5) by substituting $p_2 = (1 - p_1)t$ in the first integral, it can be shown to yield

$$\int_0^1 p_1^{N_1+x_1-1}\, (1 - p_1)^{n+N_2+N_3-x_1-1}\, dp_1 \int_0^1 t^{N_2+x_2-1}\, (1 - t)^{n+N_3-x_1-x_2-1}\, dt,$$

which equals

$$\frac{\Gamma(N_1+x_1)\,\Gamma(n+N_2+N_3-x_1)}{\Gamma(n+N)}\ \frac{\Gamma(N_2+x_2)\,\Gamma(n+N_3-x_1-x_2)}{\Gamma(n+N_2+N_3-x_1)}.$$

Using C and simplifying, the pf in (6.6.4) is obtained.

The above approach and its multivariate analog are due to Mosimann (1962).

(2) *Conditional argument.* Let X_1, X_2, X_3 be independent random variables having the negative binomial pf

$$f(x_i) = \frac{\Gamma(N_i+x_i)}{\Gamma(N_i)\, x_i!}\, p^{N_i}(1-p)^{x_i}, \qquad x_i = 0, 1, \ldots$$

Consider the conditional distribution of (X_1, X_2) given $x_1 + x_2 + x_3 = n$. It can be seen that $X_1 + X_2 + X_3$ has a negative binomial distribution with the pf

$$f(x) = \frac{\Gamma(N+x)}{\Gamma(N)\, x!}\, p^N (1-p)^x.$$

Therefore,

$$P\{X_1 = x_1, X_2 = x_2 | X = n\} =$$

$$\frac{\Gamma(N_1+x_1)\,\Gamma(N_2+x_2)\,\Gamma(n+N_3-x_1-x_2)\,\Gamma(N)\, x!}{\Gamma(N_1)\,\Gamma(N_2)\,\Gamma(N_3)\, x_1!\, x_2!\, (n-x_1-x_2)!\,\Gamma(N+x)};$$

that is, the conditional pf of (X_1, X_2) given $X = n$ is equal to the distribution in (6.6.4).

(3) *Sarndal's model.* The multivariate negative hypergeometric distribution has also been derived by Sarndal (1965). He uses an interesting technique based on a two stage sampling scheme. The bivariate analog is considered here.

A box contains N balls with N_i being of the ith color for i = 1, 2, 3. Although the total number of ball N is *known*, the actual numbers of each of the colors, N_1, N_2 and N_3, are *unknown*. Balls are drawn without replacement on two occasions. On the *first* occasion m balls with the observed frequencies Y_i of the ith color (i = 1, 2, 3). Then $m = Y_1 + Y_2 + Y_3$. On the *second* occasion n balls are drawn without replacement from the remaining balls. Let X_i be the observed frequency of the ith color at this second draw with $n = X_1 + X_2 + X_3$. Then

$$P\{Y_1 = y_1, Y_2 = y_2 | N_1 = n_1, N_2 = n_2\} = \frac{\binom{n_1}{y_1}\binom{n_2}{y_2}\binom{n_3}{y_3}}{\binom{N}{m}}, \tag{6.6.6}$$

with $n_1 + n_2 + n_3 = N$ and $y_1 + y_2 + y_3 = m$. Given n_1, n_2, y_1, y_2, the pf of (X_1, X_2) is

$$P\{X_1 = x_1, X_2 = x_2 | y_1, y_2, n_1, n_2\} = \frac{\binom{n_1-y_1}{x_1}\binom{n_2-y_2}{x_2}\binom{n_3-y_3}{x_3}}{\binom{N-m}{n}}, \tag{6.6.7}$$

with $x_1 + x_2 + x_3 = n$. Therefore,

$$P\{n_1, n_2, y_1, y_2, x_1, x_2\} = P\{n_1, n_2\}P\{y_1, y_2|n_1, n_2\}\, P\{x_1, x_2|y_1, y_2, n_1, n_2\},$$

where $P\{y_1, y_2|n_1, n_2\}$ and $P\{x_1, x_2|y_1, y_2, n_1, n_2\}$ are given by (6.6.6) and (6.6.7), respectively. Rewriting (6.6.6) and (6.6.7) and simplifying the expressions,

$$P\{n_1, n_2, y_1, y_2, x_1, x_2\}$$

$$= P\{n_1, n_2\} \frac{\binom{n_1}{x_1+y_1}\binom{n_2}{x_2+y_2}\binom{n_3}{x_3+y_3}}{\binom{N}{m+n}} \frac{\binom{x_1+y_1}{x_1}\binom{x_2+y_2}{x_2}\binom{x_3+y_3}{x_3}}{\binom{m+n}{m}} \tag{6.6.8}$$

In order to obtain $P\{n_1, n_2\}$, Sarndal (1965) considers a prior distribution for the unknown parameters N_1, N_2 and N_3 based on the Bose-Einstein statistics [Feller (1957, p. 38)]. Under this assumption, the probability function is given by

$$P\{n_1, n_2\} = \frac{1}{\binom{N+2}{2}}. \tag{6.6.9}$$

The marginal pf of $(x_1, x_2\ y_1, y_2)$ is obtained by using (6.6.9) in (6.6.8) and then summing over all n_1, n_2, n_3 satisfying $x_i + y_i \leq n_i < N; n_1 + n_2 \leq N$ for i = 1, 2, 3; hence,

$$P\{x_1, x_2, y_1, y_2\} = \frac{\binom{x_1+y_1}{x_1}\binom{x_2+y_2}{x_2}\binom{x_3+y_3}{x_3}}{\binom{m+n+2}{m,n}}.$$

From this the marginal distribution of (Y_1,Y_2) is

$$P\{y_1, y_2\} = \frac{1}{\binom{m+2}{2}}.$$

Therefore, the conditional probability

$$P\{x_1, x_2 | y_1, y_2\} = \frac{\binom{x_1+y_1}{x_1}\binom{x_2+y_2}{x_2}\binom{x_3+y_3}{x_3}}{\binom{m+n+2}{n}}. \qquad (6.6.10)$$

We see that (6.6.10) is of the same form as (6.6.4) with $y_1 = N_1 - 1$, $y_2 = N_2 - 1$, $y_3 = N_3 - 1$, $m = y_1 + y_2 + y_3 = N - 3$ and $n = x_1 + x_2 + x_3$.

C. *Bivariate negative inverse hypergeometric*

(1) *Compounding.* Consider inverse sampling with replacement from a population with three types of items. Suppose that we continue sampling until k items of the third type are obtained. Then we have seen that the joint distribution of the number of items of Type I and the number of Type II will be negative trinomial. If, however, the proportion of the types of items in the population are assumed to be *random*, the joint distribution of (X_1, X_2) is the negative inverse hypergeometric with pf

$$P\{X_1 = x_1, X_2 = x_2\} = \frac{\Gamma(k+x)\,\Gamma(N)\,\Gamma(k+N_3)\,\Gamma(N_1+x_1)\,\Gamma(N_2+x_2)}{\Gamma(k+x+N)\,\Gamma(k)\,\Gamma(N_1)\,\Gamma(N_2)\,\Gamma(N_3)\,x_1!\,x_2!}, \qquad (6.6.11)$$

where $x_i = 0, 1, 2, \ldots, i = 1, 2$; $N_1 + N_2 + N_3 = N$; $x = x_1 + x_2$ with N_i and k as positive real constants.

This development is similar to that in section B(1) above. Here, however, the conditional pf of (X_1, X_2) is a negative trinomial distribution rather than a trinomial:

$$f(x_1, x_2 | p_1, p_2) = \frac{\Gamma(k+x_1+x_2)}{\Gamma(k)\, x_1!\, x_2!} p_1^{x_1} p_2^{x_2} (1 - p_1 - p_2)^k,$$

$$x_1, x_2 = 0, 1, 2, \ldots.$$

As in the previous development, the distribution of the proportions is taken to be a bivariate beta with pdf

$$g(p_1, p_2) = \frac{\Gamma(N)}{\Gamma(N_1)\,\Gamma(N_2)\,\Gamma(N_3)} p_1^{N_1-1} p_2^{N_2-1} (1 - p_1 - p_2)^{N_3-1}$$

$$0 < p_1 < 1,\ 0 < p_2 < 1,\ p_1 + p_2 < 1.$$

Then the unconditional distribution of (X_1, X_2) is

$$h(x_1, x_2) = C \int_0^1 \int_0^{1-p_1} p_1^{N_1+x_1-1} p_2^{N_2+x_2-1} (1 - p_1 - p_2)^{k+N_3-x_1-x_2-1}\, dp_2\, dp_1,$$

where

$$C = \frac{\Gamma(k+x_1+x_2)}{\Gamma(k)\, x_1!\, x_2!} \frac{\Gamma(N)}{\Gamma(N_1)\,\Gamma(N_2)\,\Gamma(N_3)}.$$

Putting $p_2 = (1 - p_1)t$ and integrating with respect to each of the variables, we have $P\{X_1 = x_1, X_2 = x_2\}$ given by (6.6.11). Mosimann (1963) has given the multivariate analog of this result.

D. *Bivariate Polya distribution*

(1) *Urn model.* As in the univariate case, the bivariate Polya distribution can be derived using an urn model. An urn contains N balls, N_1 of color 1, N_2 of color 2 and N_3 of color 3. At each draw, the ball drawn is replaced along with c balls of the color drawn. This is repeated n times. The random variables of interest are X_i, the number of balls of

color i in the sample, for i = 1, 2. Consequently in the sample there are n - x_1 - x_2 balls of the third color. Using counting arguments similar to those in the univariate case, the joint distribution of (X_1, X_2) can be seen to be

$$P\{X_1 = x_1 , X_2 = x_2\} = \frac{\binom{n}{x_1, x_2} \Gamma(\nu)\, \Gamma(\nu_1+x_1)\, \Gamma(\nu_2+x_2)\, \Gamma(\nu_3+x_3)}{\Gamma(\nu_1)\, \Gamma(\nu_2)\, \Gamma(\nu_3)\, \Gamma(\nu+n)}, \tag{6.6.12}$$

where $x_i = 0, 1, 2, \ldots, n$; $N_i = 1, 2, \ldots, N$ for i = 1,2; $N = N_1 + N_2 + N_3$, $x = x_1 + x_2$ and $\nu_i = N_i/c$ for $c \neq 0$, i = 1, 2, 3. The distribution in (6.6.12) is called a bivariate Polya distribution. In practice the constant c can be taken to be either positive or negative [Steyn (1951); Janardan and Patil (1970)].

(2) *Conditional.* Let X_1, X_2, X_3 be independent random variables with pf's given by

$$f(x_i) = \binom{\nu_i+x_i-1}{x_i} p^{x_i} (1-p)^{\nu_i}$$

for i = 1, 2, 3. Since X_1, X_2, X_3 are independent,

$$f(x_1, x_2, x_3) = \prod_{i=1}^{3} \binom{\nu_i+x_i-1}{x_i} p^{x_i} (1-p)^{\nu_i}.$$

For $n = x_1 + x_2 + x_3$

$$g(n) = \binom{\nu+n-1}{n} p^n (1-p)^{\nu}$$

with $\nu = \nu_1 + \nu_2 + \nu_3$.

It can easily be seen that

$$P\{X_1 = x_1 , X_2 = x_2 | X = n\} = \frac{\binom{\nu_1+x_1-1}{x_1}\binom{\nu_2+x_2-1}{x_2}\binom{\nu_3+n-x_1-x_2-1}{n-x_1-x_2}}{\binom{\nu+n-1}{n}}$$

which can be shown to be equal to (6.6.12). Janardan and Patil (1970) refer to this method of generating the bivariate Polya as a case of true (positive) contagion.

It is interesting to note that the bivariate Polya distribution in (6.6.12) reduces to the bivariate hypergeometric if $c = -1$ and to the bivariate negative hypergeometric if $c = +1$.

E. *Bivariate inverse Polya distribution*

(1) *Urn model.* This distribution, like the bivariate Polya, can be derived by considering an urn model. Again consider an urn with balls of three colors with N_i balls of each color. Suppose that balls are drawn at random from the urn until k balls of, say, color 3 are obtained. As in the preceding situation, we add c balls of the color drawn at each stage. Then the random variables X_1 and X_2, counting the number of balls of color 1 and of color 2 in the sample, are jointly distributed with the inverse Polya distribution

$$P\{X_1 = x_1, X_2 = x_2\} = \frac{\Gamma(k+x)}{x_1!\, x_2!\, \Gamma(k)} \prod_{i=1}^{3} \frac{\Gamma(\nu_i+x_i)\,\Gamma(\nu)}{\Gamma(\nu_i)\,\Gamma(\nu+k+x)}, \qquad (6.6.13)$$

where $x_i = 0, 1, 2, \ldots$; $N_i = 1, 2, \ldots, N$ for $i = 1,2$; $N = N_1 + N_2 + N_3$, $x = x_1 + x_2$ and $\nu_i = N_i/c$ for $c \neq 0$, $i = 1, 2, 3$.

(2) *Positive contagion.* Let X_1, X_2 be independent random variables with pf

$$f(x_i|p) = \binom{\nu_i+x_i-1}{x_i} p^{\nu_i} (1-p)^{x_i}, \qquad i = 1, 2$$

with

$$g(p) = \frac{\Gamma(\nu_3+k)}{\Gamma(\nu_3)\,\Gamma(k)} p^{\nu_3-1} (1-p)^{k-1}, \qquad 0 < p < 1.$$

Then the unconditional pf of (X_1, X_2) is

$$h(x_1, x_2) = C \int_0^1 p^{\nu-1} (1-p)^{k+x-1}\, dp,$$

where

$$C = \frac{\Gamma(v_3+k)}{\Gamma(v_3)\,\Gamma(k)} \binom{v_1+x_1-1}{x_1} \binom{v_2+x_2-1}{x_2}.$$

Integration leads to (6.6.13).

Note that the bivariate inverse Polya distribution reduces to the inverse hypergeometric if c = -1 and to the negative inverse hypergeometric if c = +1.

6.6.2 Unified model

Janardan and Patil (1972) show that the bivariate hypergeometric type distributions discussed in section 6.6.1 can be given a unified representation. Let

$$f(x_1, x_2) = \frac{\binom{a_1}{x_1}\binom{a_2}{x_2}\binom{a_3}{x_3}}{\binom{a}{n}}, \qquad (6.6.14)$$

where $x_i = 0, 1, 2, \ldots$ for $i = 1, 2$; $n = x_1 + x_2 + x_3$ and $a = a_1 + a_2 + a_3$.

Under this representation all the preceding distributions can be generated by appropriate choice of the parameters. This relationship is given in the following table.

Table 6.1 Structure of Unified Distribution

Distribution	**Parameter Structure**
Bivariate hypergeometric (6.2.4)	$a_1 = N_1$, $a_2 = N_2$, $a_3 = N_3$; $a = N$; $n = n$
Bivariate inverse hypergeometric (6.6.1)	$a_1 = N_1$, $a_2 = N_2$, $a = -N_3-1$; $a_3 = -N-1$; $n = -k$
Bivariate negative hypergeometric (6.6.4)	$a_1 = -N_1$, $a_2 = -N_2$, $a_3 = -N_3$; $a = -N$; $n = n$

Bivariate negative inverse hypergeometric (6.6.11)	$a_1 = -N_1,\ a_2 = -N_2,\ a = N_3-1;$ $a_3 = n-1;\ N = -k$
Bivariate Polya (6.6.12)	$a_1 = -\nu_1,\ a_2 = -\nu_2,\ a_3 = -\nu_3;$ $a = -\nu,\ n = n$
Bivariate inverse Polya (6.6.13)	$a_1 = -\nu_1,\ a_2 = -\nu_2;\ a = \nu-1;$ $n = -k$

Janardan and Patil (1972) credit Steyn (1951) with a more general form of the hypergeometric families of distributions having the probability function of the form

$$f(x_1, x_2; \alpha, \beta) = \frac{(\gamma - \beta_1 - \beta_2 - 1)^{(\alpha)}}{(\gamma - 1)^{(\alpha)}} \frac{(\alpha + x - 1)^{(x)}}{(\gamma + x - 1)^{(x)}} \prod_{i=1}^{2} \binom{\beta_i + x_i - 1}{x_i} \tag{6.6.15}$$

with $x_i = 0, 1, 2, \ldots$, $i = 1, 2$; α, β, γ are real parameters and $x = x_1 + x_2$. For an appropriate choice of these parameters the pf in (6.6.12) reduces to that in (6.6.11).

Using the form of the pf in (6.6.14), it is possible to find the pgf of this family of distributions.

Theorem 6.6.1

The pgf of (6.6.14) is given by

$$\Pi(t_1, t_2) = \sum_{j=0} C_j H^{(j)}[t_1, t_2; 0], \tag{6.6.16}$$

where

$$C_j = \frac{\binom{a_3}{n-j}}{j! \binom{a}{n}} \quad \text{and} \quad H[t_1, t_2; u] = \prod_{i=1}^{2} (1 + u t_i)^{a_i}$$

with u having the same sign as a_i. Also $H^{(j)}[t_1, t_2; 0]$ is the jth partial derivative of H with respect to u, evaluated at u = 0.

Proof:

Let us expand H as

$$H[t_1, t_2; u] = \prod_{i=1}^{2} \left[\sum_{x_i=0}^{a_i} \binom{a_i}{x_i} (u\, t_i)^{x_i} \right]$$

$$= \sum_{v} \left[\sum_{x_1+x_2=v} \left\{ \prod_{i=1}^{2} \binom{a_i}{x_i} t_i^{x_i} \right\} u^{v} \right].$$

Differentiating with respect to u, j times, and setting u = 0, we have

$$H^{(j)}[t_1, t_2; 0] = j! \sum_{x_1+x_2=j} \left\{ \prod_{i=1}^{2} \binom{a_i}{x_i} t_i^{x_i} \right\}. \tag{6.6.17}$$

By definition

$$\Pi(t_1, t_2) = \sum_{x_1} \sum_{x_2} t_1^{x_1} t_2^{x_2} \frac{\binom{a_1}{x_1}\binom{a_2}{x_2}\binom{a_3}{x_3}}{\binom{a}{n}}. \tag{6.6.18}$$

Rearranging the summation and substituting from (6.6.17), we obtain (6.6.16) from (6.6.18). □

Janardan and Patil refer to the multivariate version of the distribution given in (6.6.15) as the unified multivariate hypergeometric (UMH) distribution. We shall designate (6.6.15) as the UBVH distribution.

Th pgf's of the various specialized bivariate hypergeometric type distributions can be given in terms of the (bivariate) hypergeometric series introduced in equation (6.4.1). Now with $x_1 + x_2 = x$, (6.4.1) can be rewritten as

$$F(\alpha; \beta_1, \beta_2; \gamma; t_1, t_2) = \sum_{x_2=0}^{\infty} \sum_{x_1=0}^{\infty} \frac{(\alpha)_x (\beta_1)_{x_1} (\beta_2)_{x_2}}{(\gamma)_x} \frac{t_1^{x_1}}{x_1!} \frac{t_2^{x_2}}{x_2!}. \tag{6.6.19}$$

In the following table the forms for the pgf's corresponding to the unified distribution and its six special cases are given.

Table 6.2 Probability Generating Functions

Distribution	Parameter Structure
Bivariate unified hypergeometric (6.6.14)	$\frac{a_3^{(n)}}{a^{(n)}} F(-n; -a_1, -a_2; a_3-n+1; t_1, t_2)$
Bivariate hypergeometric (6.2.4)	$\frac{N_3^{(n)}}{N^{(n)}} F(-n; -N_1, -N_2; N_3-n+1; t_1, t_2)$
Bivariate inverse hypergeometric (6.6.1)	$\frac{N_3^{(k)}}{N^{(k)}} F(k; -N_1, -N_2; -N+k; t_1, t_2)$
Bivariate negative hypergeometric (6.6.4)	$\frac{(N_3+n-1)^{(n)}}{(N+n-1)^{(n)}} F(-n; N_1, N_2; -N_3-n+1; t_1, t_2)$
Bivariate negative inverse hypergeometric (6.6.11)	$\frac{(N_3+k-1)^{(k)}}{(N+k-1)^{(k)}} F(k; N_1, N_2; N+k; t_1, t_2)$
Bivariate Polya (6.6.12)	$\frac{(\nu_3+n-1)^{(n)}}{(\nu+n-1)^{(n)}} F(-n; \nu_1, \nu_2; -\nu_3 -n+1; t_1, t_2)$
Bivariate inverse Polya (6.6.13)	$\frac{(\nu_3+k-1)^{(k)}}{(\nu+k-1)^{(k)}} F(k; \nu_1, \nu_2; \nu+k; t_1, t_2)$

6.7 Estimation in the hypergeometric distributions

The problem of estimation of the parameters in the hypergeometric family of distributions has received considerable attention. Unfortunately there seems to be a great deal of difficulty in establishing many of the results available for the exponential family of distributions. Of particular interest is the construction of the unbiased minimum variance (UMV) and maximum likelihood (ML) estimators. In connection with the UMV estimators, deep results have been established by Hartley and Rao (1968).

They point out some of the existing 'fallacies' in understanding the problem and clarify these difficulties. For the wider class of distributions, Janardan has attempted to construct UMV estimators with partial success. The ML estimation is complicated due to the finite structure of the population and the integer restriction on the parameters. However, these difficulties have been resolved to a great extent. While ML estimates can be found in most situations by iterative techniques, no closed form solutions appear feasible.

In addition to the classical estimation procedures suggested above, several authors have examined the problem of Bayesian estimation in this case. Hald (1960) applied this procedure in the univariate case using a variety of priors. The multivariate analogs are given by Hoadley (1969). Other contributors in this area for the multivariate hypergeometric distribution are Sarndal (1964) and Janardan (1976).

6.7.1 Completeness and sufficiency

The multivariate hypergeometric distribution has been considered by Hartley and Rao (1968). In that case, they have shown by induction that for the parameter vector $(N_1, N_2, ..., N_T)$, the sample $(X_1,X_2, ...,X_T)$ is complete and sufficient. For the multivariate inverse hypergeometric family Janardan (1976) has established the completeness. In both cases the proof depends upon an induction technique due to B. K. Kale [see Hartley and Rao (1968)]. Unfortunately, other members of the hypergeometric family do not seem to share a similar trait needed for using this technique. Janardan (1975, 1976) gives a discussion of the inference problems in this case.

6.7.2 Maximum likelihood estimation

(1) *Bivariate hypergeometric.* The maximum likelihood estimation for the multivariate case has been presented by Hartley and Rao (1968). Here we will consider the bivariate case in which the parameters N_1, N_2, N_3 are to be estimated under the restriction that $\sum_{i=1}^{3} N_i = N$ and that N_1,

N_2, N_3 are integers. The likelihood function of N_1, N_2, N_3 for the sample x_1, x_2, x_3 with $x_1 + x_2 + x_3 = n$ is

$$L(N_1, N_2, N_3) = \frac{\binom{N_1}{x_1}\binom{N_2}{x_2}\binom{N_3}{x_3}}{\binom{N}{n}}. \qquad (6.7.1)$$

It is preferable to maximize (6.7.1) after eliminating N_3 and X_3; that is,

$$L(N_1, N_2) = \frac{\binom{N_1}{x_1}\binom{N_2}{x_2}\binom{N-N_1-N_2}{n-x_1-x_2}}{\binom{N}{n}}. \qquad (6.7.2)$$

In this situation two cases arise: (i) $\frac{N}{n} = r$ is an integer

$$\hat{N}_1 = \frac{N}{n}x_1, \quad \hat{N}_2 = \frac{N}{n}x_2, \quad \hat{N}_3 = \frac{N}{n}x_3. \qquad (6.7.3)$$

(ii) $\frac{N}{n} = r$ is not an integer

Here the ML estimators $\hat{N}_1$, $\hat{N}_2$, $\hat{N}_3$ will be found from (6.7.3), except that they have to be 'integerized' by rounding up or down. The actual process involves iteration and the details are provided in Hartley and Rao (1968).

(2) *Bivariate negative inverse hypergeometric.* The likelihood function in this case is the pf for X_1, X_2, X_3 given by

$$L(N_1, N_2, N_3) = \frac{\Gamma(k+x)\,\Gamma(N)\,\Gamma(k+N_3)\,\Gamma(N_1+x_1)\,\Gamma(N_2+x_2)}{\Gamma(k+x+N)\,\Gamma(k)\,\Gamma(N_1)\,\Gamma(N_2)\,\Gamma(N_3)\,x_1!\,x_2!} \qquad (6.7.4)$$

with $x = x_1 + x_2$ and $N = N_1 + N_2 + N_3$.

In the multidimensional case, Janardan (1975) was able to write the likelihood equations using the expansion for

$$\frac{d}{dv}\log\Gamma(u+v) = \Psi(u+v)$$

which for v = 0 is

$$\Psi(u) = 0.57722 - \sum_{j=0}^{\infty} \left(\frac{1}{u+j} - \frac{1}{j+1}\right)$$

and

$$\Psi^{(m)}(u) = (-1)^{m+1}\, m! \sum_{j=0}^{\infty} \frac{1}{(u+j)^{m+1}}$$

[see Davis (1963)].

In the bivariate case the likelihood equations become

$$\sum_{j=0}^{k+x-1} \frac{1}{N+j} - \sum_{j=0}^{k-1} \frac{1}{N_3+j} = 0 \tag{6.7.5}$$

and for i = 1, 2

$$\sum_{j=0}^{k+x-1} \frac{1}{N+j} - \sum_{j=0}^{x_i-1} \frac{1}{N_i+j} = 0. \tag{6.7.6}$$

The equations in (6.7.5) and (6.7.6) have to be solved by iterative methods starting with the moment estimators as the initial values. The asymptotic variance matrix can be obtained using the information matrix. Details are available in Janardan (1976).

7

BIVARIATE LOGARITHMIC SERIES DISTRIBUTION

7.1. Construction of the distribution

The univariate logarithmic series distribution was introduced by Fisher *et al.* (1943) as a limit of the zero-truncated negative binomial distribution. For this purpose, let us consider the random variable X with the pgf

$$\Pi(t) = q^n(1 - pt)^{-n}$$

with $p + q = 1$. The pgf of the zero-truncated distribution is given by

$$\Pi_T(t) = \frac{\Pi(t) - \Pi(0)}{1 - \Pi(0)}.$$

Taking limits in $\Pi_T(t)$ as n tends to zero, the pgf becomes

$$\Pi^*(t) = \lim_{n \to 0} \frac{(1 - pt)^{-n} - 1}{(1 - p)^{-n} - 1}$$

$$= \frac{\log (1 - pt)}{\log (1 - p)},$$

yielding the pf

$$f(x) = \frac{p^x}{- x \log (1 - p)}. \qquad x = 1, 2, 3, \ldots \qquad (7.1.1)$$

The random variable X is said to have the logarithmic series distribution (LSD) in this case.

In Chapter 5 a general form of the bivariate negative binomial distribution has been introduced with the pgf

$$\Pi(t_1, t_2) = (\tau_0 - \tau_1 t_1 - \tau_2 t_2 - \tau_3 t_1 t_2)^{-n}$$
$$= \theta^n (1 - \theta_1 t_1 - \theta_2 t_2 - \theta_3 t_1 t_2)^{-n} \qquad (7.1.2)$$

with $\theta = 1 - \theta_1 - \theta_2 - \theta_3$. The parameters θ_1, θ_2, θ_3 and θ are all restricted to lie between zero and one. Truncating the distribution at the origin, the pgf is

$$\Pi_T(t_1, t_2) = \frac{(1 - \theta_1 t_1 - \theta_2 t_2 - \theta_3 t_1 t_2)^{-n} - 1}{\theta^{-n} - 1}. \qquad (7.1.3)$$

Letting $n \to 0$ the joint pgf of the random variables (X, Y) becomes

$$\Pi(t_1, t_2) = \frac{-\log(1 - \theta_1 t_1 - \theta_2 t_2 - \theta_3 t_1 t_2)}{-\log \theta}. \qquad (7.1.4)$$

The generating function in equation (7.1.4) was developed by Subrahmaniam (1966) and the random variables are said to have the bivariate logarithmic series distribution. In the special case when θ_3 is zero, the distribution has the pgf

$$\Pi(t_1, t_2) = \frac{-\log(1 - \theta_1 t_1 - \theta_2 t_2)}{-\log(1 - \theta_1 - \theta_2)}. \qquad (7.1.5)$$

The multivariate analog of (7.1.5) was studied by Patil and Bildikar (1967). The more general bivariate form (7.1.4) with $\theta_3 \neq 0$ has been studied at some length by Kocherlakota and Kocherlakota (1990).

7.2. Properties of the distribution

In what follows we use the notation $\delta = -\log \theta$, $T_1 = \theta_2 + \theta_3 t_1$, $T_2 = \theta_1 + \theta_3 t_2$ and $T = 1 - \theta_1 t_1 - \theta_2 t_2 - \theta_3 t_1 t_2$.

Differentiating $\Pi(t_1, t_2)$ with respect to t_1, we have

$$\frac{\partial^r \Pi(t_1, t_2)}{\partial t_1^r} = \frac{(r-1)!\, T_2^r}{\delta\, T^r}, \qquad r \geq 1. \tag{7.2.1}$$

Also,

$$\frac{\partial^i T_2^r}{\partial t_2^i} = \frac{r!\, \theta_3^i\, T_2^{r-i}}{(r-i)!}, \qquad r \geq i \tag{7.2.2}$$

and

$$\frac{\partial^j T^{-r}}{\partial t_2^j} = \frac{\Gamma(r+j)\, T_1^j}{\Gamma(r) T^{r+j}}, \qquad j \geq 1. \tag{7.2.3}$$

Differentiating (7.2.1) s times with respect to t_2 and applying Leibnitz's theorem, we have

$$\Pi^{(r,s)}(t_1, t_2) = \frac{(r-1)!}{\delta} \sum_{i=0}^{s} \binom{s}{i} \frac{\partial^i T_2^r}{\partial t_2^i} \frac{\partial^{s-i} T^{-r}}{\partial t_2^{s-i}}. \tag{7.2.4}$$

Substituting from (7.2.1), (7.2.2) and (7.2.3) into (7.2.4) yields

$$\Pi^{(r,s)}(t_1, t_2) = \frac{(r-1)!}{\delta} \sum_{i=0}^{\min(r,s)} \binom{s}{i} \frac{r!\, \Gamma(r+s-i)}{(r-i)!\, \Gamma(r)} \cdot T_1^{s-i}\, T_2^{r-i}\, T^{-(r+s-i)}\, \theta_3^i, \tag{7.2.5}$$

which upon simplification gives

$$\Pi^{(r,s)}(t_1, t_2) = \frac{r!\, s!}{\delta} \sum_{i=0}^{\min(r,s)} \frac{\Gamma(r+s-i)\, T_1^{s-i}\, T_2^{r-i} \theta_3^i}{(r-i)!\, (s-i)!\, i!\, T^{r+s-i}}. \tag{7.2.6}$$

An alternative representation for (7.2.6) is

$$\Pi^{(r,s)}(t_1, t_2) = \frac{r!\, s! T_1^s\, T_2^r}{\delta T^{r+s}} \sum_{i=0}^{\min(r,s)} \frac{\Gamma(r+s-i)\, (\tau\, \theta_3)^i}{(r-i)!\, (s-i)!\, i!}, \tag{7.2.7}$$

where

$$\tau = \frac{T}{T_1 T_2} = \frac{1 - \theta_1 t_1 - \theta_2 t_2 - \theta_3 t_1 t_2}{(\theta_2 + \theta_3 t_1)(\theta_1 + \theta_3 t_2)}.$$

It should be noted that in these results (r, s) is not equal to (0, 0). Two important consequences of the above representation of the partial derivative of the pgf are :

Probability function

Using the relationship

$$f(r, s) = \frac{\Pi^{(r,s)}(0, 0)}{r!\, s!},$$

the joint pf of (X, Y) is seen to be

$$f(r, s) = \frac{\theta_1^r\, \theta_2^s}{\delta} \sum_{i=0}^{\min(r,s)} \frac{\Gamma(r+s-i)}{(r-i)!\,(s-i)!\,i!} \left\{\frac{\theta_3}{\theta_1\theta_2}\right\}^i, \tag{7.2.8}$$

where $(r, s) \in I x I$, $(r, s) \neq (0, 0)$ and $I = (0, 1, 2, \ldots)$.

In the special case when the parameter θ_3 is zero, we have

$$f(r, s) = \frac{\Gamma(r+s)}{r!\, s!\, \delta^*} \theta_1^r\, \theta_2^s \tag{7.2.9}$$

with $\delta^* = -\log(1 - \theta_1 - \theta_2)$ and (r, s) taking on values as above.

Factorial moments

The factorial moments can be found from the relationship of the moments to the derivative $\Pi^{(r,\, s)}(t_1, t_2)$:

$$\mu_{[r,s]} = \Pi^{(r,\, s)}(1, 1).$$

Or

$$\mu_{[r,s]} = \frac{r!\, s!}{\delta} (\theta_1 + \theta_3)^r (\theta_2 + \theta_3)^s\, \theta^{-(r+s)} \sum_{i=0}^{\min(r,s)} \frac{\Gamma(r+s-i)\,(\gamma\,\theta)^i}{(r-i)!\,(s-i)!\,i!}, \tag{7.2.10}$$

where $\gamma = \theta_3 / \{(\theta_1 + \theta_3)(\theta_2 + \theta_3)\}$.

In the special case of $\theta_3 = 0$, we have

$$\mu_{[r,s]} = \frac{\Gamma(r+s)}{\delta^* \; (1 - \theta_1 - \theta_2)^{r+s}} \theta_1^r \, \theta_2^s.$$

Marginal distributions

The marginal distributions are obtained from the pgf $\Pi(t_1, t_2)$ given in equation (7.1.4). Thus the random variable X has the pgf

$$\begin{aligned} \Pi_X(t) &= \Pi(t, 1) \\ &= - \log [1 - \theta_2 - (\theta_1 + \theta_3)\, t] / \delta, \end{aligned} \tag{7.2.11}$$

where $\delta = - \log \theta$. A comparison of the pgf in (7.2.11) with that of the univariate logarithmic series distribution shows that this is not a (univariate) logarithmic series distribution. Expanding (7.2.11) in powers of t we have

$$\Pi_X(t) = \frac{- \log [1 - \theta_2]}{\delta} + \sum_{r=1}^{\infty} \left\{\frac{t\,(\theta_1 + \theta_3)}{(1 - \theta_2)}\right\}^r \frac{1}{r\,\delta}, \tag{7.2.12}$$

which is valid since

$$\frac{(\theta_1 + \theta_3)}{(1 - \theta_2)} < 1.$$

Hence the probability function of X is

$$\left.\begin{aligned} f_X(0) &= \frac{- \log [1 - \theta_2]}{\delta} \\ f_X(r) &= \left\{\frac{(\theta_1 + \theta_3)}{(1 - \theta_2)}\right\}^r \frac{1}{r\,\delta}, \qquad r \ge 1 \end{aligned}\right\}. \tag{7.2.13}$$

This is a *modified logarithmic series distribution* introduced by Patil and Bildikar (1967). In contrast to the usual logarithmic series distribution, this distribution has a probability mass at the origin.

Similarly, the random variable Y has the probability function

$$\left.\begin{aligned} f_Y(0) &= \frac{-\log[1-\theta_1]}{\delta} \\ f_Y(s) &= \left\{\frac{(\theta_2+\theta_3)}{(1-\theta_1)}\right\}^s \frac{1}{s\,\delta}, \qquad s \geq 1 \end{aligned}\right\}. \qquad (7.2.14)$$

Conditional distributions

The conditional distributions can be found either by using Theorem 1.3.1 or directly from the definition of the conditional probability function in the discrete case. Both techniques are illustrated below:

(i) By definition

$$P\{X = x|Y = y\} = \frac{f(x, y)}{f_Y(y)}.$$

Now for the bivariate logarithmic series distribution f(0, 0) is zero. Thus, for y = 0, we have the conditional probability function

$$P\{X = x|y = 0\} = \frac{\theta^x}{x\{-\log(1-\theta_1)\}}, \qquad x = 1, 2, \ldots \qquad (7.2.15)$$

and, if $y \neq 0$, the conditional probability function is

$$P\{X = x|Y = y\} = y\left\{\frac{(1-\theta_1)}{\theta_2+\theta_3}\right\}^y \sum_{i=0}^{\min(x,y)} \frac{\Gamma(x+y-i)}{(x-i)!\,(y-i)!\,i!}\,\theta_1^{x-i}\,\theta_2^{y-i}\,\theta_3^{i}$$

$$x = 0, 1, 2, \ldots \qquad (7.2.16)$$

It is quite clear that the pf in (7.2.15) is that of the univariate logarithmic series distribution. However, the pf in (7.2.16) is not of a standard form. To get a better idea of its nature we use the second method for finding the conditional pf.

(ii) The conditional pgf of X, given Y = y, is the ratio

$$\Pi_X(t|y) = \frac{\Pi^{(0,y)}(t, 0)}{\Pi^{(0,y)}(1, 0)}.$$

To evaluate this ratio, we recall that two cases arise, depending on whether y is zero or otherwise.

If y = 0, then

$$\Pi_X(t|y=0) = \frac{\Pi(t, 0)}{\Pi(1, 0)}$$

$$= \frac{-\log(1-\theta_1 t)}{-\log(1-\theta_1)}, \qquad (7.2.17)$$

which yields the pf

$$P\{X = x \mid y = 0\} = \frac{\theta_1^x}{x\{-\log(1-\theta_1)\}}, \qquad x = 1, 2, \ldots \qquad (7.2.18)$$

showing that in this case the random variable X has the univariate logarithmic series distribution.

For $y \neq 0$, the conditional pgf can be seen to be

$$\Pi_X(t|y \neq 0) = \left\{\frac{\theta_2 + \theta_3 t}{\theta_2 + \theta_3}\right\}^y \left\{\frac{1-\theta_1 t}{1-\theta_1}\right\}^{-y}, \qquad (7.2.19)$$

which is the pgf of the convolution of the random variables

$$X_1 \sim B(y, \theta_3/[\theta_2+\theta_3]) \quad \text{and} \quad X_2 \sim NB(y, 1-\theta_1).$$

When $\theta_3 = 0$, the conditional distributions take special forms. These are:

(i) For y = 0, X has the univariate logarithmic series distribution with the pgf given by (7.2.17).

(ii) For $y \neq 0$, the conditional distribution is $NB(y, 1-\theta_1)$.

Similar results can be obtained for the conditional distribution of Y given X. In this case the roles of θ_1 and θ_2 are interchanged.

Regressions and correlation

The regression of X on Y can be found directly from the conditional distribution derived in the previous section. It is obvious that this will depend upon the value of the conditioning variable Y. Thus

$$E(X| y = 0) = \frac{\theta_1}{(1-\theta_1)\{-\log(1-\theta_1)\}} \qquad (7.2.20)$$

and, for $y \neq 0$

$$E(X|\, y \neq 0) = E(X_1) + E(X_2)$$
$$= y\Big[\frac{\theta_3}{\theta_2 + \theta_3} + \frac{\theta_1}{1 - \theta_1}\Big]. \qquad (7.2.21)$$

These equations show that the regression of X on Y is not linear. From a similar argument the regression of Y on X is found to be

$$E(Y|\, x = 0) = \frac{\theta_2}{(1 - \theta_2)\{-\log(1 - \theta_2)\}} \qquad (7.2.22)$$

and for $x \neq 0$

$$E(Y|\, x \neq 0) = x\Big[\frac{\theta_3}{\theta_1 + \theta_3} + \frac{\theta_2}{1 - \theta_2}\Big]. \qquad (7.2.23)$$

These equations show that the regression in both cases is non-linear. Thus the correlation cannot be found from the regression function. It can, however, be determined from the joint moments. Using the expression for the joint factorial moments given in (7.2.10), we have

$$\left.\begin{aligned}
\mu_{[1,0]} &= E(X) = \frac{(\theta_1 + \theta_3)}{\delta\,\theta} \\
\mu_{[0,1]} &= E(Y) = \frac{(\theta_2 + \theta_3)}{\delta\,\theta} \\
\mu_{[2,0]} &= E[X(X-1)] = \frac{(\theta_1 + \theta_3)^2}{\delta\,\theta^2} \\
\mu_{[0,2]} &= E[Y(Y-1)] = \frac{(\theta_2 + \theta_3)^2}{\delta\,\theta^2} \\
\mu_{[1,1]} &= E(XY) = \frac{(\theta_1 + \theta_3)(\theta_1 + \theta_2)}{\delta\,\theta^2}[1 + \gamma\,\theta]
\end{aligned}\right\}. \qquad (7.2.24)$$

Hence

$$\text{Var}(X) = \Big[\frac{(\theta_1 + \theta_3)}{\delta\theta}\Big]^2 \Big[\frac{\delta\,(1 - \theta_2)}{(\theta_1 + \theta_3)} - 1\Big], \qquad (7.2.25)$$

$$\mathrm{Var}\,(Y) = \left[\frac{(\theta_2 + \theta_3)}{\delta\theta}\right]^2 \left[\frac{\delta\,(1 - \theta_1)}{(\theta_2 + \theta_3)} - 1\right] \qquad (7.2.26)$$

and

$$\mathrm{Cov}\,(X, Y) = \frac{(\theta_1 + \theta_3)\,(\theta_2 + \theta_3)}{(\delta\theta)^2}\left[\delta\,(1 + \gamma\theta) - 1\right]. \qquad (7.2.27)$$

Using the equations (7.2.24) - (7.2.27) we can determine the coefficient of correlation

$$\rho_{X,Y} = \frac{\delta\,(1 + \gamma\,\theta) - 1}{\left[\frac{\delta\,(1 - \theta_1)}{(\theta_2 + \theta_3)} - 1\right]^{\frac{1}{2}}\left[\frac{\delta\,(1 - \theta_2)}{(\theta_1 + \theta_3)} - 1\right]^{\frac{1}{2}}}. \qquad (7.2.28)$$

Setting $\theta_3 = 0$ in (7.2.28) we get

$$\rho_{X,Y} = \frac{\delta^* - 1}{\left[\frac{\delta^*(1 - \theta_1)}{\theta_2} - 1\right]^{\frac{1}{2}}\left[\frac{\delta^*(1 - \theta_2)}{\theta_1} - 1\right]^{\frac{1}{2}}}, \qquad (7.2.29)$$

where $\delta^* = -\log\,(1 - \theta_1 - \theta_2)$.

As pointed out by Patil and Bildikar (1967), $\rho_{X,Y} = 0$ if and only if $\delta^* = 1$. This condition is equivalent to $\theta_1 + \theta_2 = 1 - e^{-1}$. They also point out that the random variables X and Y cannot be perfectly correlated in this case.

7.3 Sum and difference

Kemp (1981c) has examined the distribution of the random variables X and Y given their sum and their difference for the case when the parameter $\theta_3 = 0$. The general model of Subrahmaniam (1966) is a little more difficult to handle. Both the cases will be presented in this section. It is relevant to recall the main results in section 1.4 on the conditional distributions of the random variables given their sum and the difference. In this connection let the joint pgf of (X, Y) be denoted by $\Pi(t_1, t_2)$ and let $Z = X + Y$, $W = Y - X$. Then the joint probability generating functions of X with Z and with W are

$$\Pi_{X,Z}(t_1, t_2) = \Pi(t_1 t_2, t_2) \text{ and } \Pi_{X,W}(t_1, t_2) = \Pi(t_1/t_2, t_2). \qquad (7.3.1)$$

7.3.1 $\theta_3 = 0$

The joint pgf of X and Y is

$$\Pi(t_1, t_2) = \frac{-\log(1 - \theta_1 t_1 - \theta_2 t_2)}{\delta^*}.$$

Therefore, the joint pgf of X and Z is

$$\Pi_{X,Z}(t_1, t_2) = \frac{-\log(1 - \theta_1 t_1 t_2 - \theta_2 t_2)}{\delta^*}, \qquad (7.3.2)$$

while that of X and W is

$$\Pi_{X,W}(t_1, t_2) = \frac{-\log(1 - \theta_1 t_1/t_2 - \theta_2 t_2)}{\delta^*}. \qquad (7.3.3)$$

Marginal distributions

The marginal of Z is clearly a univariate logarithmic series distribution of the same type as the random variable X. The probability function is therefore

$$h(z) = \frac{(\theta_1 + \theta_2)^z}{z[-\log(1 - \theta_1 - \theta_2)]} \qquad z = 1, 2, \ldots$$

The marginal of W is a little more complicated. The pgf is

$$\Pi_W(t) = \frac{-\log(1 - \theta_1/t - \theta_2 t)}{\delta^*}.$$

To expand the pgf, it is necessary to write the numerator as

$$\log(1 - \theta_1/t - \theta_2 t) = -\log[(1 - \lambda\theta_1/t)(1 - \lambda\theta_2 t)/\lambda],$$

where

$$\lambda = \frac{[1 - (1 - 4\,\theta_1\theta_2)^{1/2}]}{2\,\theta_1\theta_2}.$$

Hence

$$\Pi_W(t) = \frac{-\log(1 - \lambda\theta_1/t) - \log(1 - \lambda\theta_2 t) + \log\lambda}{-\log(1 - \theta_1 - \theta_2)}.$$

Upon expanding the terms on the right hand side we see that the probability function of W is

$$P\{W = 0\} = \frac{\log\lambda}{-\log(1 - \theta_1 - \theta_2)} \tag{7.3.4a}$$

while for j = 1, 2, 3, ...

$$\left.\begin{aligned} P\{W = j\} &= \frac{(\theta_2\lambda)^j}{j\{-\log(1 - \theta_1 - \theta_2)\}} \\ P\{W = -j\} &= \frac{(\theta_1\lambda)^j}{j\{-\log(1 - \theta_1 - \theta_2)\}} \end{aligned}\right\}. \tag{7.3.4b}$$

This yields the

$$\sum_{j=1}^{\infty} P\{W = j\} = \frac{-\log(1 - \lambda\theta_1)}{-\log(1 - \theta_1 - \theta_2)}$$

and hence, given W > 0, the conditional distribution of W is

$$P\{W = j \mid W > 0\} = \frac{(\theta_2\lambda)^j}{j\{-\log(1 - \theta_2\lambda)\}}, \qquad j \geq 1.$$

Thus, given W > 0, conditionally W has a univariate logarithmic series distribution.

Similarly

$$P\{W = -j \mid W < 0\} = \frac{(\theta_1\lambda)^j}{j\{-\log(1 - \theta_1\lambda)\}}, \qquad j \geq 1$$

which is also a univariate logarithmic series distribution for the random variable - W, with $-W > 0$.

Conditional distributions

From Theorem 1.3.1, since $\Pi(t_1, t_2)$ is of the homogeneous type, the conditional distribution of X given the sum Z = z has the pgf

$$\Pi_X(t|z) = \left[\frac{\theta_2 + \theta_1 t}{\theta_1 + \theta_2}\right]^z,$$

which is that of the binomial distribution $B[z; \theta_1/(\theta_1 + \theta_2)]$.

On the other hand the conditional distribution of X given the difference W = j has to be determined in two situations:

(a) j = 0. The pgf of interest is

$$\Pi_X(t|W{=}0) = \frac{\text{coefficient of } t_2^0 \text{ in } \log(1 - \theta_1 t/t_2 - \theta_2 t_2)}{\text{coefficient of } t_2^0 \text{ in } \log(1 - \theta_1/t_2 - \theta_2 t_2)}.$$

Proceeding as before, the numerator can be written as

$$\log(1 - \theta_1 t/t_2 - \theta_2 t_2) = \log(1 - \theta_1 t\eta/t_2) + \log(1 - \theta_2 t_2 \eta) - \log\eta,$$

where

$$\eta = \frac{[1 - (1 - 4\,\theta_1\theta_2 t)^{\frac{1}{2}}]}{2\,\theta_1\theta_2 t}.$$

The denominator can be written as

$$\log(1 - \theta_1/t_2 - \theta_2 t_2) = \log(1 - \theta_1\lambda/t_2) + \log(1 - \theta_2 t_2 \lambda) - \log\lambda,$$

where

$$\lambda = \frac{[1 - (1 - 4\,\theta_1\theta_2)^{\frac{1}{2}}]}{2\,\theta_1\theta_2}.$$

Hence

$$\Pi_X(t|w=0) = \frac{\log \eta}{\log \lambda},$$

which can be written as

$$\Pi_X(t|w=0) \quad = \frac{t\; {}_3F_2[3/2, 1, 1; 2, 2; 4\,\theta_1\theta_2 t]}{{}_3F_2[3/2, 1, 1; 2, 2; 4\,\theta_1\theta_2]}. \tag{7.3.5}$$

For details we refer to Erdelyi *et al.* (1953, Vol. 1). Kemp and Kemp (1969) have designated the pgf in (7.3.5) as that of the *lost games* distribution.

(b) $j \neq 0$. Here the pgf is given by

$$\Pi_X(t|w=j) = \frac{\text{coefficient of } t_2^j \text{ in } \log(1 - \theta_1 t/t_2 - \theta_2 t_2)}{\text{coefficient of } t_2^j \text{ in } \log(1 - \theta_1/t_2 - \theta_2 t_2)}$$

$$= \frac{\text{coefficient of } t_2^j \text{ in } \log(1 - \theta_2 t_2 \eta)}{\text{coefficient of } t_2^j \text{ in } \log(1 - \theta_2 t_2 \lambda)}$$

$$= \left[\frac{\eta}{\lambda}\right]^j,$$

which can be seen to be

$$= \frac{{}_2F_1[j/2, (j+1)/2; j+1; 4\theta_1\theta_2 t]}{{}_2F_1[j/2, (j+1)/2; j+1; 4\theta_1\theta_2]}. \tag{7.3.6}$$

Equation (7.3.6) is the pgf of a *lost games* distribution of Kemp and Kemp (1968). The above development in the case of $\theta_3 = 0$ is given by Kemp (1981c).

7.3.2 $\theta_3 \neq 0$

In this case the joint pgf of (X, Y) is given by

$$\Pi(t_1, t_2) = \frac{-\log(1 - \theta_1 t_1 - \theta_2 t_2 - \theta_3 t_1 t_2)}{-\log(1 - \theta_1 - \theta_2 - \theta_3)}.$$

The joint pgf of X and Z is from (7.3.1)

$$\Pi_{X,Z}(t_1,t_2) = \frac{-\log(1 - \theta_1 t_1 t_2 - \theta_2 t_2 - \theta_3 t_1 t_2^2)}{-\log(1 - \theta_1 - \theta_2 - \theta_3)}$$

while that of X and W is

$$\Pi_{X,W}(t_1,t_2) = \frac{-\log(1 - \theta_1 t_1/t_2 - \theta_2 t_2 - \theta_3 t_1)}{-\log(1 - \theta_1 - \theta_2 - \theta_3)}.$$

Marginal distribution of Z

The pgf of the sum Z is seen to be

$$\Pi_Z(t) = \frac{-\log(1 - \theta_1 t - \theta_2 t - \theta_3 t^2)}{\delta}.$$

Factoring the quadratic in the numerator, the pgf becomes

$$\Pi_Z(t) = \frac{-\log(1 - \alpha t) - \log(1 - \beta t)}{\delta} \tag{7.3.7}$$

where

$$\alpha = \frac{1}{2}[\theta^* + (\theta^{*2} + 4\theta_3)^{\frac{1}{2}}], \qquad \beta = \frac{1}{2}[\theta^* - (\theta^{*2} + 4\theta_3)^{\frac{1}{2}}]$$

with $\theta^* = \theta_1 + \theta_2$. Equation (7.3.7) can be written as

$$\Pi_Z(t) = \frac{\delta_1}{\delta}\left[\frac{-\log(1 - \alpha t)}{-\log(1 - \alpha)}\right] + \frac{\delta_2}{\delta}\left[\frac{-\log(1 - \beta t)}{-\log(1 - \beta)}\right], \tag{7.3.8}$$

where $\delta_1 = -\log(1 - \alpha)$ and $\delta_2 = -\log(1 - \beta)$. It can be seen that $\delta = \delta_1 + \delta_2$. Therefore $\Pi_Z(t)$ is a proper pgf.

When we examine the two terms on the right hand side of equation (7.3.8) we see that the first term, $\frac{-\log(1 - \alpha t)}{-\log(1 - \alpha)}$, is the pgf of a logarithmic series distribution. Unfortunately, as $\beta < 0$, the second term cannot be viewed as a proper pgf since the odd powers of t have negative coefficients. However, we can write

$$\frac{-\log(1-\beta t)}{\delta_2} = \frac{1}{\delta_2}\sum_{r=1}^{\infty}\frac{\beta^{2r}t^{2r}}{2r} + \frac{1}{\delta_2}\sum_{r=1}^{\infty}\frac{\beta^{2r-1}t^{2r-1}}{2r-1}$$

or

$$\frac{-\log(1-\beta t)}{\delta_2} = \frac{-\log(1-\beta^2 t^2)}{2\delta_2} + \frac{\beta t}{\delta_2}\ {}_2F_1[1, \tfrac{1}{2}; \tfrac{3}{2}; \beta^2 t^2].$$

(7.3.9)

Upon rearranging the terms on the right hand side of (7.3.9) it can be written as

$$\frac{-\log(1-\beta^2)}{2\,\delta_2}\ \Pi_1^*(t) + \frac{\beta}{\delta_2}\ {}_2F_1[1, \tfrac{1}{2}; \tfrac{3}{2}; \beta^2]\ \Pi_2^*(t),$$

where

$$\Pi_1^*(t) = \frac{-\log(1-\beta^2 t^2)}{-\log(1-\beta^2)}, \tag{7.3.10}$$

$$\Pi_2^*(t) = \frac{t\ {}_2F_1[1, \tfrac{1}{2}; \tfrac{3}{2}; \beta^2 t^2]}{{}_2F_1[1, \tfrac{1}{2}; \tfrac{3}{2}; \beta^2]}. \tag{7.3.11}$$

The weight functions of these pgf's can be seen to add to one since, from Erdelyi *et al.* (1953, Vol. 1, p. 102),

$$\log\frac{1+z}{1-z} = 2z\ {}_2F_1[1, \tfrac{1}{2}; \tfrac{3}{2}; z^2) \tag{7.3.12}$$

and hence

$$\frac{-\log(1-\beta^2)}{2\,\delta_2} + \frac{\beta}{\delta_2}\ {}_2F_1[1, \tfrac{1}{2}; \tfrac{3}{2}; \beta^2] = \frac{-\log(1-\beta)}{\delta_2} = 1.$$

The pgf in (7.3.10) is that of a logarithmic series distribution on the even integers. The pgf (7.3.11), which is defined on the odd integers, gives rise to a new discrete distribution. Thus

$$\Pi_Z(t) = \frac{\delta_1}{\delta}\Pi_1(t) + \frac{[-\log(1-\beta^2)]}{2\,\delta}\Pi_1^*(t) + \frac{\beta}{\delta}\,{}_2F_1[1, \tfrac{1}{2}; \tfrac{3}{2}; \beta^2]\,\Pi_2^*(t)$$

(7.3.13)

It should be noted that $\Pi_Z(t)$ is the mixture of three discrete distributions. The first two have positive coefficients while the third has a negative coefficient. Such mixtures have been discussed in Titterington *et al.* (1985, p. 50). The above representation was suggested by Kemp and Kemp in a personal communication.

Using the identity in (7.3.12) it is possible to write equation (7.3.13) as

$$\Pi_Z(t) = \frac{\delta_1}{\delta}\Pi_1(t) + \frac{[-\log(1-\beta^2)]}{2\,\delta}\Pi_1^*(t) + \frac{1}{2\delta}\log\left[\frac{1+\beta}{1-\beta}\right]\Pi_2^*(t)$$

Marginal distribution of W

The pgf of the difference W = Y - X is

$$\begin{aligned}\Pi_W(t) &= \frac{-\log\left[1-\theta_1/t-\theta_2\,t-\theta_3\right]}{\delta}\\ &= \frac{-\log(1-\theta_3)}{\delta} - \frac{\log(1-\phi_1/t-\phi_2\,t)}{\delta},\end{aligned}$$

where $\phi_i = \theta_i/[1-\theta_3]$, for i = 1, 2.

Proceeding as before, the second term can be written as

$$-\frac{1}{\delta}\left[\log(1-\phi_1\,\lambda/t) + \log(1-\phi_2\lambda\,t) - \log\lambda\right],$$

where

$$\lambda = \frac{1}{2\,\phi_1\,\phi_2}\left[1-(1-4\,\phi_1\,\phi_2)^{\frac{1}{2}}\right].$$

Therefore,

$$\Pi_W(t) = \frac{1}{\delta}\left[\log\left\{\frac{\lambda}{1-\theta_3}\right\} - \log(1-\phi_1\,\lambda/t) - \log(1-\phi_2\lambda t)\right].$$

(7.3.14)

Expanding $\Pi_W(t)$ in powers of t, we have

$$\Pi_W(t) = \frac{1}{\delta}\left[- \log \left\{\frac{(1 - \theta_3)}{\lambda}\right\}\right] + \frac{1}{\delta}\sum_{i=1} \frac{1}{i}\left[\frac{\lambda\phi_1}{t}\right]^i + \frac{1}{\delta}\sum_{i=1} \frac{1}{i}\left[\lambda\phi_2 t\right]^i,$$

which shows that

$$P\{W = 0\} = \frac{- \log \{(1 - \theta_3)/\lambda\}}{- \log (1 - \theta_1 - \theta_2 - \theta_3)} \tag{7.3.15a}$$

while for i = 1, 2, 3, ... we have

$$\left.\begin{aligned} P\{W = i\} &= \frac{(\lambda\phi_2)^i}{i\,\delta} \\ P\{W = - i\} &= \frac{(\lambda\phi_1)^i}{i\,\delta} \end{aligned}\right\}. \tag{7.3.15b}$$

Conditional distributions

The conditional distribution of X given Z = z can be determined from Theorem 1.3.1. Now

$$\Pi^{(0,z)}(t_1 t_2, t_2) = \frac{z!\,(\theta_2 + \theta_1 t_1 + \theta_3 t_1^2)^z}{\delta(1 - \theta_1 t_1 t_2 - \theta_2 t_2 - \theta_3 t_1^2 t_2)^z}.$$

Hence

$$\Pi_X(t|z) = \left[\frac{\theta_2 + \theta_1 t + \theta_3 t^2}{\theta_1 + \theta_2 + \theta_3}\right]^z$$

$$= \left\{\frac{t}{\phi_1} + \frac{(\phi_1 - 1)}{\phi_1}\right\}^z \left\{\frac{t}{\phi_2} + \frac{(\phi_2 - 1)}{\phi_2}\right\}^z, \tag{7.3.16}$$

where

$$\phi_1 = 1 + \frac{\theta_1}{2\,\theta_3} + \frac{[\theta_1^2 - 4\,\theta_2\theta_3]^{\frac{1}{2}}}{2\,\theta_3}, \quad \phi_2 = 1 + \frac{\theta_1}{2\,\theta_3} - \frac{[\theta_1^2 - 4\,\theta_2\theta_3]^{\frac{1}{2}}}{2\,\theta_3}.$$

That is, the conditional distribution of X, given Z = z, is the convolution of B(z, $1/\phi_1$) and B(z, $1/\phi_2$) provided $\theta_1^2 - 4\,\theta_2\theta_3 \geq 0$.

Similarly,

$$\Pi_Y(t|z) = \left\{\frac{t}{\zeta_1} + \frac{\zeta_1 - 1}{\zeta_1}\right\}^z \left\{\frac{t}{\zeta_2} + \frac{\zeta_2 - 1}{\zeta_2}\right\}^z, \qquad (7.3.17)$$

where

$$\zeta_1 = 1 + \frac{\theta_2}{2\,\theta_3} + \frac{[\theta_2^2 - 4\,\theta_1\theta_3]^{\frac{1}{2}}}{2\,\theta_3}, \quad \zeta_2 = 1 + \frac{\theta_2}{2\,\theta_3} - \frac{[\theta_2^2 - 4\,\theta_1\theta_3]^{\frac{1}{2}}}{2\,\theta_3},$$

giving the convolution of B(z, $1/\zeta_1$) and B(z, $1/\zeta_2$). The restriction in this case is that $\theta_2^2 - 4\,\theta_1\theta_3 \geq 0$.

7.4 Estimation

The problem of estimation of the parameters is examined under three conditions on the parameter θ_3. These give rise to different levels of complexity in the estimation process. The procedures suggested in the literature depend upon the nature of this parameter. To a large extent the method of maximum likelihood will be studied although other estimation methods will also be presented.

It will be useful to recall that the joint pf of (X, Y) can be written as

$$f(r, s) = \frac{\theta_1^r\,\theta_2^s}{\delta} \sum_{i=0}^{\min(r,s)} \frac{\Gamma(r + s - i)}{(r - i)!\,(s - i)!\,i!} \left\{\frac{\theta_3}{\theta_1\theta_2}\right\}^i, \qquad (7.4.1)$$

$(r, s) \in I \times I$ with $(r, s) \neq (0, 0)$.

7.4.1 $\theta_3 = 0$

The problem of estimation in the bivariate distribution when $\theta_3 = 0$ and the analogous multivariate version of the distribution has been dealt with by Patil and Bildikar (1967). As mentioned earlier in this case the joint pf of (X, Y) is

$$f(r, s) = \frac{\Gamma(r+s)}{r!\, s!} \frac{\theta_1^r \theta_2^s}{-\log(1 - \theta_1 - \theta_2)}, \tag{7.4.2}$$

$(r, s) \in I \times I$, with $(r, s) \neq (0, 0)$. The restrictions on the parameters are

$0 < \theta_1 < 1, \quad 0 < \theta_2 < 1, \quad 0 < \theta_1 + \theta_2 < 1.$

Let (x_i, y_i), $i = 1, 2, \ldots, n$ be n pairs of independent observations obtained from this distribution. Equivalently, it is possible to consider the observed frequency of the pair (r, s) to be given by $\{n_{rs}\}$ with $\sum_{r,s} n_{rs} = n$. As usual, let $\bar{x}$ and $\bar{y}$ represent, respectively, the x and y sample means.

Method of maximum likelihood

Since the likelihood function is

$$L \alpha \frac{\theta_1^{n\bar{x}} \theta_2^{n\bar{y}}}{\{-\log(1 - \theta_1 - \theta_2)\}^n}, \tag{7.4.3}$$

the likelihood equations for θ_1, θ_2 are

$$\left.\begin{aligned} \bar{x} &= \frac{\hat{\theta}_1}{\{(1 - \hat{\theta}_1 - \hat{\theta}_2)\}\{-\log(1 - \hat{\theta}_1 - \hat{\theta}_2)\}} \\ \bar{y} &= \frac{\hat{\theta}_2}{\{(1 - \hat{\theta}_1 - \hat{\theta}_2)\}\{-\log(1 - \hat{\theta}_1 - \hat{\theta}_2)\}} \end{aligned}\right\}, \tag{7.4.4}$$

which yield the equation

$$T = \frac{\hat{\Phi}}{(1 - \hat{\Phi})\{-\log(1 - \hat{\Phi})\}}, \tag{7.4.5}$$

where $T = \bar{x} + \bar{y}$ and $\Phi = \theta_1 + \theta_2$.

In (7.4.5) $\hat{\Phi}$ is the maximum likelihood estimator of Φ, the parameter of a univariate logarithmic series distribution. This distribution has been tabulated extensively by Patil and Wani (1965). Thus the maximum likelihood estimators of θ_i, i = 1, 2, are

$$\hat{\theta}_1 = \frac{\bar{x}}{T}\hat{\Phi} \quad , \quad \hat{\theta}_2 = \frac{\bar{y}}{T}\hat{\Phi} \tag{7.4.6}$$

with the variance matrix $\Sigma_{ML} = \frac{1}{n}\Gamma^{-1}$ and Γ given by

$$\frac{1}{\theta^{*2}\delta^{*2}} \begin{bmatrix} \frac{\delta^*(1-\theta_2)}{\theta_1} - 1 & \delta^* - 1 \\ \cdots & \frac{\delta^*(1-\theta_1)}{\theta_2} - 1 \end{bmatrix}, \tag{7.4.7}$$

where $\theta^* = (1 - \theta_1 - \theta_2)$ and $\delta^* = -\log\theta^*$.

It is also possible to solve the equations in (7.4.7) iteratively, using the modified moment estimators given below as the initial values. However, the use of the Patil-Wani tables, if available, is recommended.

Minimum variance unbiased estimation

Writing S_x^n as Stirling's number of the first kind with arguments n and x, it can be shown that

$$\left\{-\log(1-\theta_1-\theta_2)\right\}^n = \sum_{u_1}\sum_{u_2} \frac{n!\,|S_u^n|}{u_1!\,u_2!}\,\theta_1^{u_1}\,\theta_2^{u_2}, \tag{7.4.8}$$

where $u = u_1 + u_2 \geq n$.

Patil and Bildikar (1967) have given the minimum variance unbiased estimator of θ_i for i =1, 2 as

$$\left.\begin{aligned} \tilde{\theta}_i &= z_i \frac{|S_{z-1}^n|}{|S_z^n|} && \text{for } z > n \\ &= 0 && \text{for } z = n \end{aligned}\right\}. \tag{7.4.9}$$

where z_1 and z_2 are the totals of the x and y sample values, respectively, and z is the sum of these two quantities.

The minimum variance unbiased estimator of θ_i can also be determined from that of the parameter of a univariate logarithmic series distribution. This is given by

$$\tilde{\theta}_i = \frac{z_i}{z}\tilde{\Phi}, \qquad i = 1, 2 \tag{7.4.10}$$

where the minimum variance unbiased estimator of $\Phi = \theta_1 + \theta_2$ is

$$\tilde{\Phi} = z\frac{|S^n_{z-1}|}{|S^n_z|}. \tag{7.4.11}$$

It should be noted that $\tilde{\Phi}$ given by the equation (7.4.11) is the minimum variance unbiased estimator of the parameter of a univariate logarithmic series distribution, which has been tabled extensively by Patil and Bildikar (1966).

These results point to the close resemblance between the maximum likelihood estimators and the minimum variance unbiased estimators.

A modified method of moments estimator

The method of moments yields the same equations for θ_i as the maximum likelihood procedure. To avoid an iterative solution, it is possible to obtain modified estimators by using the first two moments in each of the marginal distributions. Let us recall from (7.2.24) with $\theta_3 = 0$

$$\tau_1 = \frac{\mu_{[2,0]}}{\mu_{[1,0]}} = \frac{\theta_1}{1 - \theta_1 - \theta_2} \quad \text{and} \quad \tau_2 = \frac{\mu_{[0,2]}}{\mu_{[0,1]}} = \frac{\theta_2}{1 - \theta_1 - \theta_2}.$$

Then taking T_1 and T_2 as the corresponding sample ratios,

$$T_1 = \frac{m_{[2,0]}}{\bar{x}} = \frac{m'_{2,0}}{\bar{x}} - 1 \quad \text{and} \quad T_2 = \frac{m_{[0,2]}}{\bar{y}} = \frac{m'_{0,2}}{\bar{y}} - 1$$

and solving for $\approx\theta_1$, $\approx\theta_2$ from

$$T_1 = \frac{\tilde{\tilde{\theta}}_1}{1 - \tilde{\tilde{\theta}}_1 - \tilde{\tilde{\theta}}_2} \quad \text{and } T_2 = \frac{\tilde{\tilde{\theta}}_2}{1 - \tilde{\tilde{\theta}}_1 - \tilde{\tilde{\theta}}_2},$$

we get

$$\left.\begin{aligned} \tilde{\tilde{\theta}}_1 &= \frac{T_1}{1 + T_1 + T_2} \\ \tilde{\tilde{\theta}}_2 &= \frac{T_2}{1 + T_1 + T_2} \end{aligned}\right\}. \tag{7.4.12}$$

The variance matrix of $\tilde{\tilde{\theta}}_1$ and $\tilde{\tilde{\theta}}_2$ can be found as usual.

7.4.2 The ratio $\dfrac{\theta_3}{\theta_1 \theta_2} = \rho$ is known

Under this assumption the pf of (X, Y) becomes

$$f(r, s) = \frac{\theta_1^r \theta_2^s}{\{-\log(1 - \theta_1 - \theta_2 - \rho\,\theta_1\theta_2)\}} \sum_{i=0}^{\min(r,s)} \frac{\Gamma(r+s-i)}{(r-i)!\,(s-i)!\,i!} \rho^i. \tag{7.4.13}$$

Since ρ is assumed known, the only unknown parameters in (7.4.13) are θ_1 and θ_2. For the sample (x_i, y_i), $i = 1, 2, ..., n$, the statistics

$$Z = \sum_{i=1}^{n} x_i \ , \quad W = \sum_{i=1}^{n} y_i$$

have been shown by Charalambides (1984) to be complete sufficient statistics for (θ_1, θ_2). To construct the minimum variance unbiased estimators, the Rao-Blackwell theorem can be used to determine the functions of Z and W which are, respectively, unbiased for θ_1 and θ_2.

Now the joint pf of Z and W is

$$p^*(z, w) = \frac{n!\, Q(z, w, n; \rho)}{\{-\log(1 - \theta_1 - \theta_2 - \rho\,\theta_1\theta_2)\}^n} \frac{\theta_1^z}{z!} \frac{\theta_2^w}{w!}$$

$$z, w = 0, 1, 2, ..., \quad z + w \geq n, \tag{7.4.14}$$

where

$$Q(z, w, n; \rho) = \sum_{i=0}^{\min(z,w)} \frac{z!\, w!\, \rho^i}{(z-i)!\,(w-i)!\,i!} |S^n_{z+w-i}|. \qquad (7.4.15)$$

In (7.4.15) S^n_m is the Stirling number of the first kind as defined previously.

It follows that for this case the uniformly minimum variance unbiased estimators of θ_1, θ_2 are

$$\left.\begin{aligned} \hat{\theta}_1 &= z\,\frac{Q(z-1,w,n;\rho)}{Q(z,w,n;\rho)}, \quad z \geq 1,\ w \geq 0,\ z+w-1 \geq n \\ &= 0 \qquad \text{otherwise} \end{aligned}\right\} \qquad (7.4.16)$$

and

$$\left.\begin{aligned} \hat{\theta}_2 &= w\,\frac{Q(z,w-1,n;\rho)}{Q(z,w,n;\rho)}, \quad z \geq 0,\ w \geq 1,\ z+w-1 \geq n \\ &= 0 \qquad \text{otherwise} \end{aligned}\right\} \qquad (7.4.17)$$

Charalambides has also considered the minimum variance unbiased estimation of the probability function in this case and when the parameter $\theta_3 = 0$.

7.4.3 $\theta_3 \neq 0$

The presence of the parameter θ_3 complicates the estimation procedures in this case. Fortunately the method of maximum likelihood with the Newton-Raphson iterative technique or the method of scoring can be used to solve the equations involved quite rapidly. In addition, the method of moments modified as in the case of $\theta_3 = 0$ can be applied to the problem.

Method of moments

We shall be using the first two marginal raw moments of the random variables X and Y and the raw mixed moment to solve from the relationships in equation (7.2.24) for the parameters θ_1, θ_2 and θ_3.

Let $T_1^* = m'_{2,0} / \bar{x}$ and $T_2^* = m'_{0,2} / \bar{y}$. Substituting for the moments in terms of the parametric functions and solving for the parameters, we have the moment estimators $\tilde{\theta}_1, \tilde{\theta}_2, \tilde{\theta}_3$ as the solutions to the equations

$$T_1^* = \frac{1 - \tilde{\theta}_2}{\tilde{\theta}}, \quad T_2^* = \frac{1 - \tilde{\theta}_1}{\tilde{\theta}},$$

$$m'_{1,1} = \frac{(\tilde{\theta}_1+\tilde{\theta}_3)(\tilde{\theta}_2+\tilde{\theta}_3)}{(-\log \tilde{\theta})\,\tilde{\theta}^2}\left\{1+\frac{\tilde{\theta}_3\tilde{\theta}}{(\tilde{\theta}_1+\tilde{\theta}_3)(\tilde{\theta}_2+\tilde{\theta}_3)}\right\}, \tag{7.4.18}$$

where $\tilde{\theta} = 1 - \tilde{\theta}_1 - \tilde{\theta}_2 - \tilde{\theta}_3$.

Solving for $\tilde{\theta}_1, \tilde{\theta}_2$ in terms of $\tilde{\theta}_3$ from the first two equations yields

$$\left.\begin{aligned} \tilde{\theta}_1 &= 1 - \frac{T_2^*}{T_1^*+T_2^* - 1}\{\tilde{\theta}_3 + 1\} \\ \tilde{\theta}_2 &= 1 - \frac{T_1^*}{T_1^* + T_2^* - 1}\{\tilde{\theta}_3 + 1\} \end{aligned}\right\}. \tag{7.4.19}$$

From (7.4.19) we get

$$m'_{1,1} = \frac{\tilde{\theta}_3}{(-\log \tilde{\theta})\,\tilde{\theta}} + (-\log \tilde{\theta})\,\bar{x}\,\bar{y},$$

where $\tilde{\theta} = \dfrac{1 + \tilde{\theta}_3}{(T_1^* + T_2^* - 1)}$. This can be solved iteratively for $\tilde{\theta}_3$ and hence $\tilde{\theta}_1, \tilde{\theta}_2$ can be obtained from equation (7.4.19) .

Method of maximum likelihood

The logarithm of the likelihood function for θ_1, θ_2, θ_3 is

$$\log L = \sum_{r,s} n_{rs}\,[r \log \theta_1 + s \log \theta_2 - \log \delta + \log C(r, s)], \tag{7.4.20}$$

where

$$C(r, s) = \sum_{i=0}^{\min(r,s)} \frac{\Gamma(r + s - i)}{(r - i)!\,(s - i)!\,i!} \left\{\frac{\theta_3}{\theta_1\theta_2}\right\}^i . \tag{7.4.21}$$

Rao (1973, p. 360) suggests the method of scoring for an efficient scheme of solving the likelihood equations. Differentiating (7.4.20) successively with respect to θ_1, θ_2, θ_3, we have

$$\left.\begin{aligned}
\frac{\partial \log L}{\partial \theta_1} &= \sum_{r,s} n_{rs}\left[\frac{r}{\theta_1} - \frac{U(r, s)}{\theta_1} - \frac{1}{\theta\,(-\log \theta)}\right] \\
\frac{\partial \log L}{\partial \theta_2} &= \sum_{r,s} n_{rs}\left[\frac{s}{\theta_2} - \frac{U(r, s)}{\theta_2} - \frac{1}{\theta\,(-\log \theta)}\right] \\
\frac{\partial \log L}{\partial \theta_3} &= \sum_{r,s} n_{rs}\left[-\frac{1}{\theta\,(-\log \theta)} + \frac{U(r, s)}{\theta_3}\right]
\end{aligned}\right\} \tag{7.4.22}$$

where

$$U(r, s) = \frac{D(r, s)}{C(r, s)}$$

and

$$D(r, s) = \sum_{i=1}^{\min(r,s)} \frac{i\,\Gamma(r + s - i)}{(r - i)!\,(s - i)!\,i!} \left\{\frac{\theta_3}{\theta_1\theta_2}\right\}^i \qquad r \geq 1,\ s \geq 1.$$

The method of scoring is discussed in section 2.1.

Example 7.4.1

The following results are for the two bodies of data, A and B, given in section 7.9. The maximum likelihood estimates and their variances have been determined using the method of scoring. The iterative steps started with the values of the estimates obtained for θ_1 and θ_2 under the assumption that θ_3 is known.

Data set A

$\theta_1 = 0.333, \theta_2 = 0.333, \theta_3 = 0.111$

$\hat{\theta}_1 = 0.343, \hat{\theta}_2 = 0.344, \hat{\theta}_3 = 0.101$

$$\Sigma_{ML} = \begin{bmatrix} .1756 & .0626 & -.1378 \\ \cdots & .1758 & -.1377 \\ \cdots & \cdots & .2157 \end{bmatrix}$$

Data set B

$\theta_1 = 0.425, \theta_2 = 0.225, \theta_3 = 0.150$

$\hat{\theta}_1 = 0.343, \hat{\theta}_2 = 0.344, \hat{\theta}_3 = 0.146$

$$\Sigma_{ML} = \begin{bmatrix} .1502 & .0514 & -.1193 \\ \cdots & .1541 & -.1387 \\ \cdots & \cdots & .2177 \end{bmatrix}.$$

7.5 Tests of hypotheses

Two hypotheses are of interest in connection with the distribution when $\theta_3 \neq 0$. The first one refers to the goodness-of-fit of the distribution and the second one is in reference to parameter θ_3. As discussed in section 2.2, there are several tests available to assess the goodness-of-fit of a distribution. We shall be concerned here with the application of Pearson's χ^2 and the test based on the empirical probability generating function developed in Kocherlakota and Kocherlakota (1986). The methods are illustrated with the help of the data sets A and B generated in section 7.7.

The hypothesis for the parameter θ_3 being equal to zero is tested using the two methods presented in section 2.2. These include Rao's efficient score test and the likelihood ratio test.

7.5.1 Goodness-of-fit

The hypothesis under consideration is that the data conform to the distribution of (X, Y) when the parameter θ_3 is not equal to zero. In this case the probability function is given by

$$f(r, s) = \frac{\theta_1^r \theta_2^s}{\delta} \sum_{i=0}^{\min(r,s)} \frac{\Gamma(r+s-i)}{(r-i)!\,(s-i)!\,i!} \left\{\frac{\theta_3}{\theta_1\theta_2}\right\}^i. \qquad (7.5.1)$$

The test is performed using the procedures mentioned above.

Pearson's χ^2 test

Let the observed frequency in the cell (r, s) be n_{rs}, then the test is based on the statistic

$$\underline{X}^2 = \sum_{r,s} \frac{[n_{rs} - n\hat{f}(r, s)]^2}{n\hat{f}(r, s)},$$

where the summations extend over the cells that have been appropriately combined to keep the expected cell frequencies larger than one. In the bivariate case this can give rise to some difficulties since the pooling of cells can be done in a variety of ways. We have attempted to retain the structure of the 'pooled' data as close to the original distribution as possible.

Example 7.5.1

Applying the test to the data sets A and B generated in section 7.7 using the maximum likelihood estimates found in the previous section, we have the following results:

	$\underline{X}^2$	k	v	Prob-value
A	34.48	30	26	0.1234
B	24.55	35	31	0.7876

k = effective number of cells v = degrees of freedom

We see that the prob-value is large in both cases leading us to accept the hypothesis of the data being from the bivariate logarithmic series distribution of the form given in (7.5.1).

Tests of fit based on the probability generating function

To perform the test the partial derivatives of the pgf with respect to the parameters are needed (see equation 2.2.5.) Writing A for the numerator of the pgf in (7.1.4) and using the other notation introduced earlier we have

$$\left.\begin{aligned}\frac{\partial\Pi(t_1, t_2)}{\partial\theta_1} &= -\frac{A}{\delta^2\theta} + \frac{t_1}{\delta(1-\theta_1 t_1 - \theta_2 t_2 - \theta_3 t_1 t_2)}\\ \frac{\partial\Pi(t_1, t_2)}{\partial\theta_2} &= -\frac{A}{\delta^2\theta} + \frac{t_2}{\delta(1-\theta_1 t_1 - \theta_2 t_2 - \theta_3 t_1 t_2)}\\ \frac{\partial\Pi(t_1, t_2)}{\partial\theta_3} &= -\frac{A}{\delta^2\theta} + \frac{t_1\, t_2}{\delta(1-\theta_1 t_1 - \theta_2 t_2 - \theta_3 t_1 t_2)}\end{aligned}\right\}. \quad (7.5.4)$$

Example 7.5.2

In using the technique based on the probability generating function, values for t_1 and t_2 have to be selected. As in Kocherlakota and Kocherlakota (1986) the test statistic is evaluated for values $.1 \le t_1 \le .9$ and $.1 \le t_2 \le .9$. For both sets of data the prob-values associated with Z for all combinations of t_1 and t_2 are between .11 and .35; this supports the null hypothesis that each of the data sets is from bivariate logarithmic series distribution in (7.5.1).

7.5.2 Test of the hypothesis $\theta_3 = 0$

As discussed in section 2.2.2, the test of the hypothesis H_0: $\theta_3 = 0$ against the alternative H_a: $\theta_3 \neq 0$ can be performed using either the Generalized likelihood ratio test or Rao's efficient score procedure. These are both used in the present context.

Generalized likelihood ratio test

This test is based on the statistic

$$2[\log L(\hat{\underset{\sim}{\theta}}; \underset{\sim}{x}) - \log L(\hat{\underset{\sim}{\theta}}^{*}; \underset{\sim}{x})] \quad (7.5.5)$$

where $\hat{\underset{\sim}{\theta}}$ and $\hat{\underset{\sim}{\theta}}^{*}$ stand for the unrestricted maximum likelihood estimate of the parameter and that under null hypothesis, respectively. In this instance the test statistic is asymptotically distributed as χ^2 on one degree of freedom.

Example 7.5.3

Using the data sets A and B, maximum likelihood estimates under

H_0 can be obtained from equation (7.4.6). For data set A the estimates are (0.4116, 0.4129) and for data set B they are (0.5260, 0.3323). The estimates under H_a are given in Example 7.4.1. The calculation of the test statistics in (7.5.5) is summarized as:

	$\log L(\hat{\underline{\theta}}^*; \underline{x})$	$\log L(\hat{\underline{\theta}}; \underline{x})$	Test statistic
A	-2742.89	-2721.16	43.46
B	-2888.00	-2845.48	85.68

Since the critical value for the test with $\alpha = .05$ is 3.84, the null hypothesis is rejected for both data sets A and B and we conclude that the model with $\theta_3 \neq 0$ is more appropriate.

Efficient score test

As indicated in equation (2.2.9) the test is based on the statistic

$$S_C = \underline{\phi}^{*'} \Gamma^{*-1} \underline{\phi}^* \tag{7.5.6}$$

where $\underline{\phi}^{*'} = (u_1, u_2, u_3)$ with

$$u_1 = \frac{\sqrt{n}\,\bar{x}}{\hat{\theta}_{10}} - \frac{\sqrt{n}}{\hat{\delta}^* \hat{\theta}^*}, \quad u_2 = \frac{\sqrt{n}\,\bar{y}}{\hat{\theta}_{20}} - \frac{\sqrt{n}}{\hat{\delta}^* \hat{\theta}^*}$$

$$u_3 = \frac{1}{\sqrt{n}\,\hat{\theta}_{10}\hat{\theta}_{20}} \sum_{r,s} \frac{r\,s}{r+s-1} n_{rs} - \frac{\sqrt{n}}{\hat{\delta}^* \hat{\theta}^*}.$$

The (r, s)th element of the matrix Γ^*

$$\gamma^*_{r,s} = \sum_{i,j} f_0(i, j)\, P_r\, P_s$$

with

$$P_1 = \frac{i}{\theta_1} - \frac{1}{\delta^* \theta^*}, \quad P_2 = \frac{j}{\theta_1} - \frac{1}{\delta^* \theta^*} \quad \text{and} \quad P_3 = -\frac{1}{\delta^* \theta^*}.$$

For the determination of the test statistic S_C, Γ^* is evaluated using the maximum likelihood estimates.

Example 7.5.4

Using the maximum likelihood estimates obtained earlier, we can now evaluate the statistic S_c for each of the two data sets. For A we have

$$S_c = (.2472, .2267, 13.2971)\begin{bmatrix} .1971 & .0710 & -.1696 \\ \dots & .1968 & -.1693 \\ \dots & \dots & .2542 \end{bmatrix}\begin{pmatrix} .2472 \\ .2267 \\ 13.2971 \end{pmatrix}$$

$$= 42.8250$$

and for B

$$S_c = (-.4019, -.3925, 16.6556)\begin{bmatrix} .1597 & .0637 & -.1490 \\ \dots & .2042 & -.1953 \\ \dots & \dots & .2748 \end{bmatrix}\begin{pmatrix} -.4019 \\ -.3925 \\ 16.6556 \end{pmatrix}$$

$$= 80.8737.$$

In both cases S_c is distributed as a χ^2 on one degree of freedom; hence for $\alpha = .05$ the critical value is 3.84 and H_0: $\theta_3 = 0$ is again rejected.

7.6 Other models for the bivariate LSD

It has been seen that the bivariate logarithmic series distribution, as does its univariate counterpart, arises from the negative binomial distribution when it is truncated at the origin and the index parameter tends to zero. This is the model adopted by Subrahmaniam (1966) and Patil and Bildikar (1967). As has been pointed out by Kemp (1981c), any model leading to the bivariate negative binomial distribution would, as a consequence, lead to and be a model for the bivariate logarithmic series distribution. Several situations other than the Fisher-limit model give rise to the logarithmic series distribution in the univariate case. Some such models for the case $\theta_3 = 0$ have been discussed by Kemp (1981c). In the following models we relax the condition of $\theta_3 = 0$.

7.6.1 Mixture models

The negative binomial distribution has its origins in the compounding of the Poisson distribution by the gamma distribution with the parameter k. Since the logarithmic series distribution arises from the negative binomial distribution by truncating the distribution at the origin and taking the limit as $k \to 0$, we have two possibilities. We examine two such situations below.

Truncation after compounding

Let the pdf of the mixing distribution be

$$f(\lambda) = \frac{\exp\{-\frac{\lambda}{\alpha}\}}{\lambda\, E_1[\frac{\varepsilon}{\alpha}]}, \qquad \varepsilon > 0, \qquad \varepsilon < \lambda < \infty, \tag{7.6.1}$$

where $E_1(\varepsilon) = \int_{\varepsilon}^{\infty} x^{-1} e^{-x}\, dx$ is the exponential integral [Abramowitz and Stegun (1968, p.231)].

The compounded distribution has the pgf

$$\Pi(t_1, t_2) = \frac{1}{E_1[\frac{\varepsilon}{\alpha}]} \int_{\varepsilon}^{\infty} \lambda^{-1} e^{-\frac{\lambda}{\alpha}}\, \Pi(t_1, t_2 \mid \lambda)\, d\lambda,$$

where

$$\Pi(t_1, t_2 \mid \lambda) = \exp \lambda \left\{\theta_1 (t_1 - 1) + \theta_2 (t_2 - 1) + \theta_3 (t_1 t_2 - 1)\right\}. \tag{7.6.2}$$

Thus

$$\Pi(t_1, t_2) = \frac{1}{E_1[\frac{\varepsilon}{\alpha}]} E_1\left[\frac{\varepsilon}{\alpha} - \varepsilon \left\{\theta_1 (t_1 - 1) + \theta_2 (t_2 - 1) + \theta_3 (t_1 t_2 - 1)\right\}\right]. \tag{7.6.3}$$

We recall that if we truncate the distribution at (0, 0), the resulting distribution has the pgf

$$\Pi_T(t_1, t_2) = \frac{\Pi(t_1, t_2) - \Pi(0, 0)}{1 - \Pi(0, 0)}$$

Substituting from equation (7.6.3), the numerator can be seen to be

$$E_1\left[\frac{\varepsilon}{\alpha} - \varepsilon\{\theta_1(t_1 - 1) + \theta_2(t_2 - 1) + \theta_3(t_1 t_2 - 1)\}\right] - E_1\left[\frac{\varepsilon}{\alpha} + \varepsilon(\theta_1 + \theta_2 + \theta_3)\right]$$

(7.6.4)

while the denominator is

$$E_1\left[\frac{\varepsilon}{\alpha}\right] - E_1\left[\frac{\varepsilon}{\alpha} + \varepsilon(\theta_1 + \theta_2 + \theta_3)\right]. \qquad (7.6.5)$$

Consider the expansion

$$E_1(x) = -\gamma - \log x - \sum_{n=1}^{\infty} \frac{(-x)^n}{n!\,n},$$

[equation (5.1.11) of Abramowitz and Stegun (1968, p. 229)], where γ is Euler's constant. This gives for the numerator the expansion

$$-\log\left\{1 - \frac{\alpha\left[\theta_1 t_1 + \theta_2 t_2 + \theta_3 t_1 t_2\right]}{1 + \alpha\left[\theta_1 + \theta_2 + \theta_3\right]}\right\} + T(\varepsilon), \qquad (7.6.6)$$

while the denominator can be seen to become

$$-\log\left\{\frac{1}{1 + \alpha\left[\theta_1 + \theta_2 + \theta_3\right]}\right\} + T^*(\varepsilon). \qquad (7.6.7)$$

In (7.6.6) and (7.6.7), $T(\varepsilon)$ and $T^*(\varepsilon)$ tend to zero as $\varepsilon \to 0$. Thus as we let ε tend to zero in the pgf $\Pi_T(t_1, t_2)$ we have

$$\Pi(t_1, t_2) = \frac{-\log\{1 - \beta_1 t_1 - \beta_2 t_2 - \beta_3 t_1 t_2\}}{-\log\{1 - \beta_1 - \beta_2 - \beta_3\}},$$

(7.6.8)

the pgf of a bivariate logarithmic series distribution.

Truncation before compounding

Let (X, Y) have the truncated bivariate Poisson distribution. Then from (7.6.2) the pgf of (X, Y) given λ is, after simplification,

$$\Pi_T(t_1, t_2 \mid \lambda) = \frac{\exp\left[\lambda\{\theta_1 t_1 + \theta_2 t_2 + \theta_3 (t_1 t_2)\}\right] - 1}{\exp\left[\lambda\{\theta_1 + \theta_2 + \theta_3\}\right] - 1}. \tag{7.6.9}$$

Let the pdf of λ be

$$f(\lambda) = \frac{\theta\left[\{\exp(\lambda\theta) - 1\} \exp\{\frac{(-\lambda\theta)}{\alpha}\}\right]}{\left[-(\lambda\theta) \log(1 - \alpha)\right]}, \qquad \lambda > 0,\ 0 < \alpha < 1 \tag{7.6.10}$$

with $\theta = \theta_1 + \theta_2 + \theta_3$. Compounding (7.6.9) with (7.6.10), we have

$$\Pi(t_1, t_2) = \int_0^\infty \Pi_T(t_1, t_2 \mid \lambda)\, f(\lambda)\, d\lambda.$$

To integrate, we substitute $\lambda\theta = \tau$ and obtain

$$\Pi(t_1, t_2) =$$

$$-\frac{1}{\log(1 - \alpha)} \int_0^\infty \frac{\exp\left\{-\tau\left[\frac{1}{\alpha} - \frac{\theta_1 t_1 + \theta_2 t_2 + \theta_3 t_1 t_2}{\theta}\right]\right\} - \exp(-\frac{\tau}{\alpha})}{\tau} d\tau$$

From Gradshteyn and Ryzhik [1980, p. 341, eq. (3.476)], we see that this yields

$$\Pi(t_1, t_2) = -\frac{1}{\log(1 - \alpha)} \log\left\{\frac{\theta}{\theta - \alpha(\theta_1 t_1 + \theta_2 t_2 + \theta_3 t_1 t_2)}\right\}$$

with appropriate restrictions on t_1, t_2, t_3. Or

$$\Pi(t_1, t_2) = \frac{-\log\left[1 - \frac{\alpha}{\theta}(\theta_1 t_1 + \theta_2 t_2 + \theta_3 t_1 t_2)\right]}{-\log(1 - \alpha)}, \tag{7.6.11}$$

which is once again the pgf of a bivariate logarithmic series distribution.

7.6.2 Model for a modified LSD

The bivariate negative binomial distribution based on the Edwards and Gurland (1961) and Subrahmaniam (1966) model has a pgf of the form

$$\left[1 - \left\{\theta_1 (t_1 - 1) + \theta_2 (t_2 - 1) + \theta_3 (t_1 t_2 - 1)\right\}\right]^{-k}.$$

The probability at (0, 0) is readily seen to be $\left[1 + \theta_1 + \theta_2 + \theta_3\right]^{-k}$.

Kemp (1981c) considers the Fisher-limit of the bivariate negative binomial distribution when only a fraction of the (0, 0)-cell probability is truncated. In the case of the present model let us say that this fraction is β^{-k}. Then $\Pi_T(t_1, t_2)$ is

$$\frac{\left[1 - \left\{\theta_1(t_1 - 1)+\theta_2 (t_2 - 1)+\theta_3 (t_1 t_2 - 1)\right\}\right]^{-k} - \left[\beta\left\{1+\theta_1+\theta_2+\theta_3\right\}\right]^{-k}}{1 - \left[\beta\left\{1+\theta_1+\theta_2+\theta_3\right\}\right]^{-k}},$$

which, upon taking the limit as $k \to 0$, yields

$$\Pi (t_1, t_2) = \frac{\log \left[\frac{1}{\beta} - \frac{\theta_1 t_1 + \theta_2 t_2 + \theta_3 t_1 t_2}{\beta (1 + \theta_1 + \theta_2 + \theta_3)}\right]}{\log \left[\frac{1}{\beta (1 + \theta_1 + \theta_2 + \theta_3)}\right]}. \qquad (7.6.12)$$

This is the pgf of a modified bivariate logarithmic series distribution with the probability at (0, 0) given by

$$\frac{\log \beta}{\log \left[\beta\left\{1 + \theta_1 + \theta_2 + \theta_3\right\}\right]}.$$

Several other models for constructing the bivariate logarithmic series distribution and the modified bivariate logarithmic series distribution with $\theta_3 = 0$ are discussed by Kemp (1981c). These can be readily generalized to the present situation.

7.7 Computer generation of the distribution

The computer generation of the general bivariate log series distribution when $\theta_3 \neq 0$, uses two basic properties of the distribution:

(1) Marginally, each of the random variables X and Y has the modified logarithmic series distribution with the probability functions given by equations (7.2.13) and (7.2.14), respectively.

(2) The conditional distribution of X given Y = s is

(i) the univariate logarithmic series distribution (7.2.17), for s = 0

(ii) the convolution of $Y_1 \sim B(s, \theta_3/[\theta_2+\theta_3])$ and $Y_2 \sim NB(s, 1-\theta_1)$, for $s \neq 0$.

Several techniques for simulating the univariate log series distribution have been discussed by Kemp (1981a). In the present case Kemp's Algorithm LS has been used for generating random samples from the univariate log series distribution. This algorithm has been modified here to generate observations from the modified log series distribution (7.2.13).

The generating procedure can be summarized as: Generate a realization s of Y from the modified log series distribution. If the returned value s = 0, X is generated from a univariate log series distribution. If the returned value $s \neq 0$, X is generated as a convolution: Generate y_1 from $B(s, \theta_3/[\theta_2+\theta_3])$ using IMSL subroutine GGBN and y_2 from $NB(s, 1-\theta_1)$ using IMSL subroutine GGBNR. The realized value of X is taken to be $y_1 + y_2$. Note that the index parameters of the distribution of Y_1 and of Y_2 are each a function of s since we are considering a conditional distribution. This procedure is repeated n times to give a random sample from the bivariate log series distribution.

Using the techniques discussed above, two sets of data, A and B, were generated. For data set A, $\theta_1 = \theta_2$ giving rise to a symmetric distribution. For B, $\theta_1 \neq \theta_2$ gives rise to an asymmetric distribution.

Data Set A: $\theta_1 = .333$, $\theta_2 = .333$, $\theta_3 = .111$ and $N = 1000$

x \ y	0	1	2	3	4	5	6	7	8	9	10
0		221	29	14	1	1	0	0	0	0	0
1	226	144	50	15	5	4	1	0	1	0	0
2	35	41	28	13	6	4	0	2	0	0	0
3	9	18	17	17	7	9	2	1	0	1	0
4	5	4	6	6	2	2	1	1	0	0	0
5	0	1	6	6	8	2	2	0	2	0	1
6	0	1	0	4	0	3	2	0	1	0	0
7	0	0	0	1	4	2	1	0	1	0	0
8	0	0	0	0	0	0	0	0	0	0	0
9	0	0	0	0	2	0	0	0	0	1	0

Data Set B:* $\theta_1 = .425$, $\theta_2 = .225$, $\theta_3 = .150$ and $N = 1000$

x \ y	0	1	2	3	4	5	6	7	8	9	10
0		133	13	2	0	0	0	0	0	0	0
1	265	155	26	6	5	1	0	0	0	0	0
2	54	54	34	11	1	3	0	1	0	0	0
3	23	29	20	13	7	3	0	0	0	0	0
4	6	16	14	10	3	4	0	0	1	0	0
5	5	5	3	6	3	1	0	0	0	0	0
6	0	1	4	8	4	2	3	3	0	0	0
7	0	0	3	3	4	1	1	0	0	1	0
8	0	0	2	1	2	4	1	1	0	0	0
9	0	0	2	0	1	2	0	0	0	0	0
10	0	0	0	0	3	1	0	1	1	0	0

*$n_{12,4} = 1$, $n_{15,11} = 1$, $n_{17,9} = 1$ and $n_{56,18} = 1$

8

COMPOUNDED BIVARIATE POISSON DISTRIBUTIONS

8.1 Introduction

As we saw in section 1.5, a wide class of distributions can be constructed by *compounding* or *generalizing.* The bivariate negative binomial distribution discussed in Chapter 5 is an example of a distribution which can be derived by compounding a bivariate Poisson distribution with a gamma distribution. In this chapter these definitions and results will be extended to the bivariate case.

8.2 Bivariate Poisson: General results

In this chapter we study the class of distributions closely related to those introduced in section 4.4. The following theorem casts it in terms of the general notation introduced in section 1.5.

Theorem 8.2.1

Given the parameter τ, let the random variables (X, Y) have the bivariate Poisson distribution with the pgf

$$\Pi(t_1, t_2|\tau) = \exp[\tau\{\lambda_1(t_1 - 1) + \lambda_2(t_2 - 1) + \lambda_3(t_1 t_2 - 1)\}]. \qquad (8.2.1)$$

Let τ be a random variable with the moment generating function M(t), then $(X_1, X_2) \wedge \tau$ has the pgf $M\{\lambda_1(t_1 - 1) + \lambda_2(t_2 - 1) + \lambda_3(t_1 t_2 - 1)\}$.

Proof:

Given in section 4.4. □

Independence

Under what conditions are the random variables X_1 and X_2 independent? The following result provides a partial answer to this problem.

Theorem 8.2.2

Conditional on the random variable τ, let (X_1, X_2) be jointly distributed with the pgf

$$\Pi(t_1, t_2|\tau) = \exp[\tau\{\lambda_1(t_1 - 1) + \lambda_2(t_2 - 1)]. \tag{8.2.2}$$

A necessary and sufficient condition for the random variables (X_1, X_2) to be unconditionally independent is that the random variable τ has the mgf $M(t) = \exp[ct]$ for some constant c.

Proof [Kocherlakota (1988)}:

If the random variable τ has the degenerate distribution with the probability at the point c, it follows that the unconditional pgf of the random variables (X_1, X_2) is

$$\Pi(t_1, t_2) = \exp[c\{\lambda_1(t_1 - 1) + \lambda_2(t_2 - 1)\}], \tag{8.2.3}$$

from which it follows that they are independent.

To prove the converse, recall that the marginal distribution of X_1 and X_2 can be found from the joint pgf as

$$\Pi_{X_1}(t) = M[\lambda_1(t - 1)] \quad \text{and} \quad \Pi_{X_2}(t) = M[\lambda_2(t - 1)]. \tag{8.2.4}$$

For the random variables X_1 and X_2 to be independent we require that

$$\Pi(t_1, t_2) = \Pi_{X_1}(t)\, \Pi_{X_2}(t)$$

or

$$M[\lambda_1(t_1 - 1) + \lambda_2(t_2 - 1)] = M[\lambda_1(t - 1)]\, M[\lambda_2(t - 1)]; \tag{8.2.5}$$

that is, M(.) has to satisfy the functional equation

$$M(z_1 + z_2) = M(z_1)\, M(z_2),$$

which admits the solution $M(t) = \exp(ct)$ for some constant c. □

Convolutions

In the univariate case, Teicher (1960) has proved that the convolution of two compound Poisson distributions is again a compound Poisson distribution when the mixing cdf is the convolution of the two given mixing cdf's. The bivariate extension is presented in Theorem 8.2.3.

Theorem 8.2.3

Let $\underline{X}_i$ be independent bivariate Poisson random variables with the pgf's, for i= 1, 2,

$$\Pi_i(t_1, t_2|\tau_i) = \exp[\tau_i\{\lambda_1(t_1 - 1) + \lambda_2(t_2 - 1) + \lambda_3(t_1 t_2 - 1)\}]. \qquad (8.2.6)$$

Then the convolution of the compound random variables $\underline{X}_i \wedge \tau_i$, i = 1, 2, is the compound of the bivariate Poisson random variable with the pgf

$$\Pi(t_1, t_2|\tau) = \exp[\tau\{\lambda_1(t_1 - 1) + \lambda_2(t_2 - 1) + \lambda_3(t_1 t_2 - 1)\}] \qquad (8.2.7)$$

and the random variable τ which is the convolution of τ_1 and τ_2.

Proof:

Let $\underline{X}_1$ and $\underline{X}_2$ be independent bivariate Poisson random variables with the pgf's as stated in the theorem. Let τ_i be independent for i = 1, 2 with the mgf's $M_i(t)$. Then their convolution has the mgf $M_1(t)\, M_2(t)$. Also the compounds $\underline{X}_i \wedge \tau_i = \underline{Z}_i$ for i=1, 2, have the pgf $M_i(u)$, where

$$u = \lambda_1(t_1 - 1) + \lambda_2(t_2 - 1) + \lambda_3(t_1 t_2 - 1). \qquad (8.2.8)$$

Therefore, the pgf of the convolution of $\underline{Z}_1$ and $\underline{Z}_2$ is

$$\Pi(t_1, t_2) = M_1(u)M_2(u). \qquad (8.2.9)$$

Now if we consider (W_1, W_2) having the bivariate Poisson distribution with pgf

$$\Pi(t_1, t_2|\tau) = \exp[\tau\{\lambda_1(t_1 - 1) + \lambda_2(t_2 - 1) + \lambda_3(t_1 t_2 - 1)\}]$$

and the mgf of τ as $M_1(t)M_2(t)$, then, by theorem 8.2.1, the compound

$$(W_1, W_2) \wedge \tau$$

has the pgf

$$\Pi(t_1, t_2) = M_1(u)M_2(u), \tag{8.2.10}$$

which is the same as (8.2.9). By the uniqueness of the pgf's it follows that the convolution of $\underline{Z}_1$ and $\underline{Z}_2$ is the compound of (W_1, W_2) with the random variable τ. However, from the mgf of τ it is clear that τ is the convolution of τ_1 and τ_2. Hence the theorem. □

8.2.1 Some properties of the compound Poisson

As has been seen in the previous section if $\underline{X}$ has the bivariate Poisson distribution (8.2.1) and τ a distribution of which the mgf is M(t) then the pgf of $\underline{X} \wedge \tau$ is M(u) where u is given in (8.2.8). In Kocherlakota (1988) it is proved that the

$$\frac{\partial^{r+s}}{\partial t_1^r \partial t_2^s} \Pi(t_1 t_2) = \sum_{k=0}^{\min(r,s)} \binom{r}{k} \binom{s}{k} k!\; M^{(t-k)}(u)\; T_1^{r-k}\, T_2^{s-k} \lambda_3^k, \tag{8.2.11}$$

where $M^{(j)}$ is the jth derivative of M, $t = r+s$, $T_1 = \lambda_1 + \lambda_3 t_2$, $T_2 = \lambda_2 + \lambda_3 t_1$ and u is defined above.

Probability function

From this result it is possible to write down the pf of the compound distribution as

$$f(r, s) = \frac{\lambda_1^r \lambda_2^s}{r!\; s!} \sum_{k=0}^{\min(r,s)} \binom{r}{k} \binom{s}{k} k!\; M^{(t-k)}(\gamma)\, \delta^k, \tag{8.2.12}$$

where $\gamma = -(\lambda_1 + \lambda_2 + \lambda_3)$ and $\delta = \lambda_3/\lambda_1\lambda_2$. If $\lambda_3 = 0$, the pf becomes

$$f(r, s) = \frac{\lambda_1^r \lambda_2^s}{r!\; s!} M^{(t)}(-\lambda_1 - \lambda_2).$$

Factorial moments

Putting $t_1 = t_2 = 1$ in (8.2.11),

$$\mu_{[r,s]} = (\lambda_1 + \lambda_3)^r (\lambda_2 + \lambda_3)^s \sum_{k=0}^{\min(r,s)} \frac{r!\, s!\, \delta^{*k}\, \mu'_{r+s-k}}{(r-k)!\,(s-k)!\,k!}, \qquad (8.2.13)$$

where $\delta^* = \lambda_3/\{(\lambda_1 + \lambda_3)(\lambda_1 + \lambda_2)\}$ and μ'_k is the k-th moment around zero in the distribution with the mgf M(t).

If $\lambda_3 = 0$, the factorial moment reduces to $\mu_{[r,s]} = \lambda_1^r \lambda_2^s \mu'_{r+s}$.

Recurrence relations

Kocherlakota (1988) obtains the recurrence relations for the pgf. Let $\Pi_{r,s}$ denote the (r, s)th partial derivative of the pgf with respect to t_1, t_2. Then, for $r \geq 1$, $s \geq 1$,

$$\left.\begin{aligned} \Pi_{r,s} &= \frac{T_2}{T_1}\Pi_{r+1,s-1} + (r-s+1)\frac{\lambda_3}{T_1}\Pi_{r,s-1} \qquad r \geq s \\ \Pi_{r,s} &= \frac{T_1}{T_2}\Pi_{r-1,s+1} + (s-r+1)\frac{\lambda_3}{T_2}\Pi_{r-1,s} \qquad r \leq s \end{aligned}\right\}.$$

(8.2.14)

From these relations it is possible to give the recurrence relations for the pf:

$$\left.\begin{aligned} f(r, s) &= \frac{(r+1)\lambda_2}{s\lambda_1} f(r+1, s-1) + (r-s+1)\frac{\lambda_3}{s\lambda_1} f(r, s-1) \qquad r \geq s \\ f(r, s) &= \frac{(s+1)\lambda_1}{r\lambda_2} f(r-1, s+1) + (s-r+1)\frac{\lambda_3}{r\lambda_2} f(r-1, s) \qquad r \leq s \end{aligned}\right\}.$$

(8.2.15)

These equations are helpful in the rapid calculation of the probabilities of the random variables for $r \geq 1$, $s \geq 1$. It is however necessary to determine the probabilities in the first row and first column. These will require the setting up of recurrence relations specific to the distribution under consideration. The following schematic representation is useful in developing an algorithm for the calculation. Our experience with the use of these equations is that they are extremely rapid and yield very precise results.

$\underline{r}$	$\underline{s}$ 0	1	2	3	4
0	f(0, 0)	f(0, 1)	f(0, 2)	f(0, 3)	f(0, 4)
1	f(1, 0) →	f(1, 1) →	f(1, 2) →	f(1, 3) →	f(1, 4)
2	↑ f(2,0) →	↑ f(2, 1) →	↑ f(2, 2) →	↑ f(2, 3) →	f(2, 4)
3	↑ f(3, 0) →	↑ f(3, 1) →	↑ f(3, 2) →	↑ f(3, 3) →	f(3, 4)
4	↑ f(4, 0)	↑ f(4, 1)	↑ f(4, 2)	↑ f(4, 3)	f(4, 4)

Similarly the recurrence relations for the factorial moments are

$$\left.\begin{aligned} \mu_{[r,s]} &= \frac{\lambda_2+\lambda_3}{\lambda_1+\lambda_3}\mu_{[r+1,s-1]} + (r-s+1)\frac{\lambda_3}{\lambda_1+\lambda_3}\mu_{[r,s-1]} \qquad r \geq s \\ \mu_{[r,s]} &= \frac{\lambda_1+\lambda_3}{\lambda_2+\lambda_3}\mu_{[r-1,s+1]} + (s-r+1)\frac{\lambda_3}{\lambda_2+\lambda_3}\mu_{[r-1,s]} \qquad r \leq s \end{aligned}\right\}. \tag{8.2.16}$$

Conditional distributions

From theorem 1.3.1 the conditional pgf of X_1 given $X_2 = x_2$ is

$$\Pi_{X_1}(t|x_2) = \frac{M^{(x_2)}[-(\lambda_2+\lambda_3) + \lambda_1(t-1)]}{M^{(x_2)}[-(\lambda_2+\lambda_3)]}\left[\frac{\lambda_2}{\lambda_2+\lambda_3} + \frac{\lambda_3 t}{\lambda_2+\lambda_3}\right]^{x_2}. \tag{8.2.17}$$

This shows that the conditional distribution is the convolution of

(i) $Y_1 \sim B[x_2, \lambda_3/(\lambda_2+\lambda_3)]$

and

(ii) Y_2 of which the pgf is the first ratio on the right hand side of (8.2.17).

In the cases considered in the following sections, this ratio takes special forms. However, the following result is of interest in this connection. Expanding the numerator in the ratio as

$$M^{(x)}[-(\lambda_2+\lambda_3)+\lambda_1(t-1)] = \sum_{r=0}^{\infty} \frac{M^{[x+r]}\{-(\lambda_2+\lambda_3)\}}{r!} [\lambda_1(t-1)]^r,$$

the pgf of Y_2 can be written as

$$\sum_{r=0}^{\infty} \frac{M^{[x+r]}\{-(\lambda_2+\lambda_3)\}}{r!\, M^{[x]}\{-(\lambda_2+\lambda_3)\}} [\lambda_1(t-1)]^r. \tag{8.2.18}$$

From (8.2.18) we see that if $M(t) = \exp(ct)$, c a constant, the pgf of Y_2 becomes $\exp\{\theta[t-1]\}$; that is, Y_2 has the Poisson distribution.

On the other hand, if the pgf of Y_2 is $\exp\{\theta[t-1]\}$, then

$$\sum_{r=0}^{\infty} \frac{M^{[x+r]}\{-(\lambda_2+\lambda_3)\}}{r!\, M^{[x]}\{-(\lambda_2+\lambda_3)\}} [\lambda_1(t-1)]^r = \sum_{r=0}^{\infty} \frac{[\theta(t-1)]^r}{r!}.$$

Equating the coefficients of $(t-1)^r$, we see that

$$\frac{M^{[x+r]}(t)}{M^{[x]}(t)} = \text{constant}, \qquad r = 0, 1, 2, \dots,$$

which implies that $M(t) = \exp(ct)$. This shows that a necessary and sufficient condition for distribution of Y_2 to be Poisson is that the compounding distribution be degenerate. In this case the conditional distribution of X_1 given $X_2 = x_2$ is the convolution of $B[x_2, \lambda_3/(\lambda_2+\lambda_3)]$ and $P[\theta]$.

8.3 Bivariate Neyman type A

The univariate version of the distribution was obtained by Neyman (1939). Briefly, the procedure consists of considering the random variable X having the Poisson distribution with the parameter $\theta\phi$ while ϕ itself is a random variable with the Poisson distribution with the parameter λ. Then the unconditional distribution of X is given by the compounding $P(\theta\phi) \underset{\phi}{\wedge} P(\lambda)$. The pgf of this distribution is readily seen to be $\exp[\lambda\{\exp\theta(t-1)-1\}]$. In this case the random variable X is said to have the Neyman Type A distribution. We will denote this by writing NTA(λ, θ). It should be noted that the order in which the parameters appear in the parentheses is of consequence. Another way of constructing this distribution is through generalizing: $P(\lambda) \vee P(\theta)$. From Theorem 8.1.1, we know that these two procedures lead to equivalent distributions.

The bivariate case extension of this development is due to Holgate (1966). He defines three types of Neyman Type A distributions. We will, however, consider only the type I version of the type A distribution as this falls in the general format of the previous section. Let the random variables (X_1, X_2) be the number of two types of individuals arising in clusters in a population. Their numbers per cluster are assumed to have the bivariate Poisson distribution with the parameters $\lambda_1\tau$, $\lambda_2\tau$ and $\lambda_3\tau$. The number of clusters are themselves assumed to have the Poisson distribution with the parameter λ. We are interested in the unconditional distribution of the random variables (X_1, X_2) in the population.

Applying the results in the preceding section, we see that since

$$\Pi(t_1, t_2|\tau) = \exp[\tau\{\lambda_1(t_1 - 1) + \lambda_2(t_2 - 1) + \lambda_3(t_1 t_2 - 1)\}]$$

and

$$M(t) = \exp[\lambda\{\exp(t) - 1\}],$$

the joint pgf of (X_1, X_2) is, from theorem 8.2.1,

$$\Pi(t_1, t_2) = \exp[\lambda\{\exp[\lambda_1(t_1 - 1) + \lambda_2(t_2 - 1) + \lambda_3(t_1 t_2 - 1)] - 1\}]. \tag{8.3.1}$$

8.3.1 Properties of the distribution

Marginal distributions

The marginal distribution of X_1 is given by the pgf

$$\Pi_{X_1}(t) = \exp[\lambda\{\exp[(\lambda_1 + \lambda_3)(t-1)] - 1\}],$$

which is obtained by setting $t_2 = 1$ in (8.3.1). This shows that the marginal of X_1 is NTA(λ, $\lambda_1 + \lambda_3$). Similarly, the marginal of X_2 is NTA(λ, $\lambda_2 + \lambda_3$).

Probability function

The probability function can be found using equation (8.2.12). Now

$$M^{(r)}(t) = [\exp\{\lambda(e^t - 1)\}]\, \eta_r,$$

where η_r = the rth raw moment of P[λexp(t)]. Applying this to the present case, we have

$$f(r, s) = \frac{\lambda_1^r \lambda_2^s}{r!\ s!} \exp[\lambda\,(e^{\gamma} - 1)] \sum_{k=0}^{\min(r,s)} \binom{r}{k} \binom{s}{k} k!\, \delta^k \eta_{r+s-k}, \qquad (8.3.2)$$

where $\delta = \lambda_3/[\lambda_1\lambda_2]$ and

$$\eta_r = \text{the rth raw moment of } P[\lambda\exp(\gamma)]$$

with $\gamma = -(\lambda_1 + \lambda_2 + \lambda_3)$. It should be noted that Gillings (1974) also obtained the pf in (8.3.2) but by using a different technique.

The probability function in (8.3.2) is complicated and it may be time consuming to evaluate. The recurrence relations in equation (8.2.14) can be used for determining f(r, s) for $r \geq 1$, $s \geq 1$. Thus from the pgf we have, with $\eta_1 = \lambda \exp(\gamma)$,

$$f(0, 0) = M(\gamma),\ \ f(1, 0) = \eta_1\, \lambda_1\, f(0, 0),\ \ f(0,1) = \eta_1\, \lambda_2\, f(0, 0).$$

From the representation

$$\eta_r = \sum_{k=0}^{r-1} \binom{r-1}{k} \eta_{r-1-k}\eta_1, \qquad r \geq 1$$

and

$$f(r, 0) = \frac{\lambda_1^r}{r!} M(\gamma)\eta_r,$$

the recurrence relation for $r \geq 1$ is obtained as

$$f(r, 0) = \frac{\eta_1}{r} \sum_{k=0}^{r-1} \frac{\lambda_1^{k+1}}{k!} f(r-k-1, 0).$$

Similarly, for $s \geq 1$,

$$f(0, s) = \frac{\eta_1}{s} \sum_{k=0}^{s-1} \frac{\lambda_2^{k+1}}{k!} f(0, s-k-1).$$

Factorial moments

The (r, s)th factorial moment is obtained from equation (8.2.13) as

$$\mu_{[r,s]} = (\lambda_1 + \lambda_3)^r (\lambda_2 + \lambda_3)^s \sum_{k=0}^{\min(r,s)} \frac{r!\, s!\, \tau^{*k} \mu'_{r+s-k}}{(r-k)!\,(s-k)!\,k!} \qquad (8.3.3)$$

where

$$\tau^* = \frac{\lambda_3}{[\{\lambda_1 + \lambda_3\}\{\lambda_2 + \lambda_3\}]}$$

and

$$\mu'_r = \text{rth raw moment of } P(\lambda).$$

In particular we have

$$E(X_1) = \{\lambda_1 + \lambda_3\}\lambda \ , \ E(X_2) = \{\lambda_2 + \lambda_3\}\lambda,$$

$$\text{Var}(X_1) = \lambda\{\lambda_1 + \lambda_3\}\{1 + \lambda_1 + \lambda_3\} \ , \ \text{Var}(X_2) = \lambda\{\lambda_2 + \lambda_3\}\{1 + \lambda_2 + \lambda_3\}$$

and

$$\text{Cov}(X_1, X_2) = \lambda[\{\lambda_1 + \lambda_3\}\{\lambda_2 + \lambda_3\} + \lambda_3].$$

From these we have

$$\rho_{X_1,X_2} = \frac{\{\lambda_1+\lambda_3\}\{\lambda_2+\lambda_3\}+\lambda_3}{[\{\lambda_1+\lambda_3\}\{1+\lambda_1+\lambda_3\}\{\lambda_2+\lambda_3\}\{1+\lambda_2+\lambda_3\}]^{\frac{1}{2}}}.$$

Conditional distributions

Using the result in equation (8.2.17) with

$$M^{(r)}(t) = M(t)\,\eta_r(\lambda e^t), \tag{8.3.4}$$

the conditional distribution of X_1 given $X_2 = x_2$ can be written as the convolution of: (i) $Y \sim B[\,x_2, \lambda_3/(\lambda_2+\lambda_3)\,]$ and (ii) W with pgf given by the first ratio on the right hand side of (8.2.17). Applying (8.3.4) the pgf of W is

$$\exp\{\lambda\,[\exp-(\lambda_2+\lambda_3)][\exp\lambda_1(t-1) - 1]\} \cdot \left\{\frac{\eta_{x_2}[\lambda\exp\{-(\lambda_2+\lambda_3) + \lambda_1(t-1)\}]}{\eta_{x_2}[\lambda\exp\{-(\lambda_2+\lambda_3)\}]}\right\}. \tag{8.3.5}$$

From (8.3.5) we see that W can further be expressed as the convolution of two random variables, W_1 and W_2:

$$\Pi_{W_1}(t) = \exp\{\lambda\,[\exp-(\lambda_2+\lambda_3)][\exp\lambda_1(t-1) - 1]\} \tag{8.3.6}$$

and

$$\Pi_{W_2}(t) = \frac{\eta_{x_2}[\lambda\exp\{-(\lambda_2+\lambda_3) + \lambda_1(t-1)\}]}{\eta_{x_2}[\lambda\exp\{-(\lambda_2+\lambda_3)\}]}. \tag{8.3.7}$$

Using (8.3.6) we see that

$$W_1 \sim NTA[\lambda\exp\{-(\lambda_2+\lambda_3)\}, \lambda_1]$$

or

$$W_1 \sim P(\lambda_1\phi) \underset{\phi}{\wedge} P(\lambda\exp[\{-(\lambda_2+\lambda_3)\}].$$

The next problem is to identify the distribution of W_2 with pgf given by (8.3.7). Haight (1967, p. 6) gives the expansion

$$\eta_r(\mu) = \sum_{m=1}^{r} \mu^m S_{m;r}, \tag{8.3.8}$$

where

$$S_{m;r} = \frac{1}{m!}\sum_{k=0}^{m} (-1)^{m-k}\binom{m}{k}k^r$$

are Stirling's numbers. Therefore, (8.3.7) can be written as

$$\sum_{m=1}^{x_2} \omega_m(x_2) \exp[m\, \lambda_1(t-1)], \qquad (8.3.9)$$

where

$$\omega_m(x_2) = \frac{\lambda^m \exp[-m(\lambda_2 + \lambda_3)] S_{m;x_2}}{\sum_{m=1}^{x_2} \lambda^m \exp[-m(\lambda_2 + \lambda_3)] S_{m;x_2}}. \qquad (8.3.10)$$

Hence,

$W_2 \sim$ mixture of x_2 Poisson random variables with the m-th component having the parameter $m\lambda_1$ and weight $\omega_m(x_2)$ as given in equation (8.3.10) for $m = 1, 2, \ldots, x_2$. (8.3.11)

In summary, the conditional distribution of X_1 given $X_2 = x_2$ is the convolution of

(i) $Y \sim B[x_2, p]$ with $p = \lambda_3/(\lambda_2+\lambda_3)$

(ii) $W_1 \sim NTA[\lambda \exp[-(\lambda_2+\lambda_3), \lambda_1)]$

(iii) $W_2 \sim$ mixture of x_2 Poisson random variables as given in (8.3.11).

Using the above representation, we can determine the conditional moments

$$E[X_1|X_2 = x_2] = px_2 + \lambda\lambda_1 \exp[-(\lambda_2+\lambda_3)] + \lambda_1 \sum_{i=1}^{x_2} i\omega_i(x_2),$$

$$\text{Var}[X_1|X_2 = x_2] = x_2 p(1-p) + \lambda\lambda_1[1+\lambda_1]\exp[-(\lambda_2+\lambda_3)] + \lambda_1 \sum_{i=1}^{x_2} i\, \omega_i(x_2) + \lambda_1^2 \sum_{i=1}^{x_2} i^2 \omega_i(x_2) - [\lambda_1 \sum_{i=1}^{x_2} i\omega_i(x_2)]^2.$$

This development for the conditional distribution is adapted from Kocherlakota (1988).

8.3.2 Estimation

For this distribution estimation using the method of moments, including the variance matrix, has been developed by Holgate (1966). Later Gillings (1974) considered the problem using the maximum likelihood and minimum χ^2 procedures. He has also compared the estimators by means of an efficiency study In this section we will summarize estimation using method of moments and maximum likelihood.

Method of moments

Since, in the population moments, the parameters λ_1, λ_2, λ_3 and λ are inextricably joined together, they cannot *all* be estimated separately. We will consider two cases depending upon the assumption we are willing to make.

λ known

From the marginal means and the covariance we write the following estimating equations for the parameters λ_1, λ_2 and λ_3:

$$\tilde{\lambda}_1 + \tilde{\lambda}_3 = \frac{\bar{x}_1}{\lambda}, \quad \tilde{\lambda}_2 + \tilde{\lambda}_3 = \frac{\bar{x}_2}{\lambda}$$

and

$$(\tilde{\lambda}_1 + \tilde{\lambda}_3)(\tilde{\lambda}_2 + \tilde{\lambda}_3) + \tilde{\lambda}_3 = \frac{m_{1,1}}{\lambda}.$$

Hence the estimates are

$$\tilde{\lambda}_1 = \frac{\bar{x}_1}{\lambda}\{1 + \frac{\bar{x}_2}{\lambda}\} - \frac{m_{1,1}}{\lambda}, \quad \tilde{\lambda}_2 = \frac{\bar{x}_2}{\lambda}\{1 + \frac{\bar{x}_1}{\lambda}\} \frac{m_{1,1}}{\lambda}$$

and

$$\tilde{\lambda}_3 = \frac{m_{1,1}}{\lambda} - \frac{\bar{x}_1\bar{x}_2}{\lambda^2}.$$

λ unknown

In this case it is useful to introduce a reparameterization in order to eliminate the indeterminacy. For this purpose we write $\alpha = \lambda_1 + \lambda_3$ and $\beta = \lambda_2 + \lambda_3$. Holgate (1966) suggests the use of the estimates under the assumption that the marginal variances are almost equal. Thus we have

$$\tilde{\lambda} = \frac{m_{2,0} + m_{0,2} - \bar{x}_1 - \bar{x}_2}{\bar{x}_1^2 + \bar{x}_2^2},$$

$$\tilde{\alpha} = \frac{\bar{x}_1}{\tilde{\lambda}}, \quad \tilde{\beta} = \frac{\bar{x}_2}{\tilde{\lambda}} \quad \text{and} \quad \tilde{\lambda}_3 = \frac{m_{1,1}}{\tilde{\lambda} - \tilde{\alpha}\,\tilde{\beta}}.$$

Maximum likelihood

Representing the observed frequency in the cell (r, s) by n_{rs}, the log of the likelihood is $\log L = \sum_{r,s} n_{rs} \log f(r, s)$, where f(r, s) is given by (8.3.2). The following difference equations can be used to determine the partial derivatives $\frac{\partial \log L}{\partial \theta_i}$ with θ_i standing for λ, λ_1, λ_2 and λ_3. Denote the partial derivatives by

$$d_0(r, s) = \frac{\partial f(r, s)}{\partial \lambda}$$

and

$$d_i(r, s) = \frac{\partial f(r, s)}{\partial \lambda_i} \qquad i = 1, 2, 3.$$

Then for r = 0, s = 0,

$$d_0(0, 0) = (\exp\gamma - 1)\, f(0, 0)$$

$$d_1(0, 0) = d_2(0, 0) = d_3(0, 0) = -\eta_1\, f(0, 0)$$

with $\eta_1 = \lambda \exp(\gamma)$, $\eta_2 = \eta_1^2 + \eta_1$ and $\gamma = -(\lambda_1 + \lambda_2 + \lambda_3)$.

r = 1, s = 0,

$$d_0(1, 0) = \lambda_1\, f(0, 0)\, [\frac{\eta_2}{\lambda} - \eta_1],$$

$$d_1(1, 0) = f(0, 0)\, [\eta_1 - \lambda_1\, \eta_2],$$

$$d_2(1, 0) = -\lambda_1 \eta_2 f(0, 0),$$

$$d_3(1, 0) = -\lambda_1 \eta_2 f(0, 0);$$

$r = 0, s = 1,$

$$d_0(0, 1) = \lambda_2 f(0, 0) [\frac{\eta_2}{\lambda} - \eta_1],$$

$$d_1(0, 1) = -\lambda_2 \eta_2 f(0, 0),$$

$$d_2(0,1) = f(0, 0) [\eta_1 - \lambda_2 \eta_2],$$

$$d_3(0,1) = -\lambda_2 \eta_2 f(0, 0).$$

$r \geq 1, s = 0$

$$d_0(r, 0) = \frac{f(r, 0)}{\lambda} + \frac{\eta_1}{r} \sum_{k=0}^{r-1} \frac{d_0(r-k-1, 0) \lambda_1^{k+1}}{k!},$$

$$d_1(r, 0) = -f(r, 0) + \frac{f(r, 0)}{\lambda_1} + \frac{r-1}{r} f(r-1,0) + \frac{\eta_1}{r} \sum_{k=0}^{r-1} \frac{d_1(r-k-1, 0) \lambda_1^{k+1}}{k!},$$

$$d_2(r, 0) = -f(r, 0) + \frac{\eta_1}{r} \sum_{k=0}^{r-1} \frac{d_2(r-k-1, 0) \lambda_1^{k+1}}{k!},$$

$$d_3(r, 0) = -f(r, 0) + \frac{\eta_1}{r} \sum_{k=0}^{r-1} \frac{d_3(r-k-1, 0) \lambda_1^{k+1}}{k!};$$

$r = 0, s \geq 1$

$$d_0(0, s) = \frac{f(0, s)}{\lambda} + \frac{\eta_1}{s} \sum_{k=0}^{s-1} \frac{d_0(0, s-k-1) \lambda_2^{k+1}}{k!}$$

$$d_1(0, s) = -f(0, s) + \frac{\eta_1}{s} \sum_{k=0}^{s-1} \frac{d_1(0, s-k-1) \lambda_2^{k+1}}{k!},$$

$$d_2(0, s) = -f(0, s) + \frac{f(0, s)}{\lambda_2} + \frac{s-1}{s} f(0, s-1) + \frac{\eta_1}{s} \sum_{k=0}^{s-1} \frac{d_2(0, s-k-1) \lambda_2^{k+1}}{k!},$$

$$d_3(0, s) = -f(0, s) + \frac{\eta_1}{s} \sum_{k=0}^{s-1} \frac{d_3(0, s-k-1) \lambda_2^{k+1}}{k!};$$

$r \geq 1, s \geq 1$

$$d_0(r, s) = \frac{(r+1)\,\lambda_2}{s\,\lambda_1} d_0(r+1, s-1) + \frac{(r-s+1)\,\lambda_3}{s\,\lambda_1} d_0(r, s-1),$$

$$d_1(r, s) = \frac{(r+1)\,\lambda_2}{s\,\lambda_1} d_1(r+1, s-1) + \frac{(r-s+1)\,\lambda_3}{s\,\lambda_1} d_1(r, s-1) - \frac{1}{\lambda_1} f(r, s),$$

$$d_2(r, s) = \frac{(r+1)}{s\,\lambda_1} f(r+1, s-1) + \frac{(r+1)\,\lambda_2}{s\,\lambda_1} d_2(r+1, s-1) + \frac{(r-s+1)\,\lambda_3}{s\,\lambda_1} d_2(r, s-1),$$

$$d_3(r, s) = \frac{(r-s+1)}{s\,\lambda_1} f(r, s-1) + \frac{(r+1)\,\lambda_2}{s\,\lambda_1} d_3(r+1, s-1) + \frac{(r-s+1)\,\lambda_3}{s\,\lambda_1} d_3(r, s-1).$$

To obtain the maximum likelihood equations, the partial derivatives are set equal to zero. The equations can be solved iteratively. For the iterative solution, we require the information matrix

$$\Gamma = \left\{ E[-\frac{\partial^2 \log L}{\partial\theta_i\, \partial\theta_j}] \right\},$$

which can also be written as

$$\Gamma = \{ \sum_{r,s} \frac{n}{f(r,s)} [\frac{\partial f(r, s)}{\partial\theta_i}][\frac{\partial f(r, s)}{\partial\theta_j}] \}.$$

The maximum likelihood equations and the information can easily be obtained by using the difference equations for the pf and for the partial derivatives.

Example 8.3.1

The following data refer to the number of plants of the species *Lacistema aggregatum* (x_1) and *Protium guianense* (x_2) in each of 100 systematically laid and contiguous quadrats. They are obtained in the study of secondary rain forest in Trinidad.

x_1 \ x_2	0	1	2	3	4	Total
0	34	8	3	1	0	46
1	12	13	6	1	0	32
2	4	3	1	0	0	8
3	5	3	2	1	0	11
4	2	0	0	0	0	2
5	0	0	0	0	1	1
Total	57	27	12	3	1	100

From the data set we obtain the following results

$$\bar{x}_1 = 0.95, \quad \bar{x}_2 = 0.60, \quad m_{2,0} = 1.4217, \quad m_{0,2} = 0.1111,$$

which yield the moment estimates as

$$\tilde{\alpha} = 0.4051, \tilde{\beta} = 0.2559, \tilde{\lambda} = 2.3450 \text{ and } \tilde{\lambda}_3 = 0.000.$$

The actual estimated value of $\tilde{\lambda}_3$ turns out to be negative; hence, it is set equal to zero. From this

$$\tilde{\lambda}_1 = 0.405 \quad \text{and} \quad \tilde{\lambda}_2 = 0.256.$$

Gillings gives the maximum likelihood estimates as

$$\hat{\lambda} = 2.169, \quad \hat{\lambda}_1 = 0.414, \quad \hat{\lambda}_2 = 0.279 \quad \text{and} \quad \hat{\lambda}_3 = 0.000,$$

with the asymptotic variance matrix of $(\hat{\lambda}, \hat{\lambda}_1, \hat{\lambda}_2, \hat{\lambda}_3)$ given by

$$10^{-5} \begin{bmatrix} 815 & -52.2 & -125 & 8.74 \\ \cdots & 11.8 & 1.23 & 3.68 \\ \cdots & \cdots & 26.8 & -6.57 \\ \cdots & \cdots & \cdots & 4.77 \end{bmatrix}.$$

Using the maximum likelihood estimates, the χ^2 goodness-of-fit statistic is 12.33 on 8 degrees of freedom. The expected values and groupings are given in Gillings (1974).

8.3.3 Computer simulation

Observations from the pf given in (8.3.2) can be simulated using the marginal distribution of X_2 and the conditional distribution of X_1 given x_2. The generating procedure can be summarized as:

(i) Generate a realization s of X_2 from a univariate NTA(λ, $\lambda_2 + \lambda_3$). The univariate NTA is generated as

$$X_2 \sim P[(\lambda_2 + \lambda_3)\phi] \underset{\phi}{\wedge} P(\lambda).$$

(ii) For the realized value s, generate a realization r of $X_1 = Y + W_1 + W_2$, where

$$Y \sim B[\, x_2, \lambda_3/(\lambda_2+\lambda_3)\,]\,,$$

$$W_1 \sim \text{NTA}\,[\lambda \exp[-\,(\lambda_2+\lambda_3), \lambda_1)]$$

and W_2 is distributed as in (8.3.11).

To determine a realized value of W_2, evaluate $\omega_m(x_2)$ for m = 1, 2, ..., x_2. Form a table of the S_m, the cumulative partial sums of $\omega_m(x_2)$. Generate u from a uniform distribution on the interval 0 to 1. Select the ith component such that u lies in the interval (S_{i-1}, S_i). The realized value of W_2 is then generated from a P(i λ_1). Hence r is the sum, $y + w_1 + w_2$.

The resulting pair (r, s) is one observation from (8.3.2). To obtain a random sample of size n, this procedure is repeated n times.

Example 8.3.2

The following table summarizes the results of 1000 observations generated by the pf given in (8.3.2) with $\lambda = 2$, $\lambda_1 = .4$, $\lambda_2 = .3$, $\lambda_3 = .1$.

x_1 \ x_2	0	1	2	3	4	5	Total
0	344	85	24	6	0	1	460
	(332.42)	89.62	25.52	6.05	(1.62)		
1	122	95	47	11	3	1	279
	(119.49)	(97.94)	(41.23)	(12.97)	(3.45)	(1.03)	
2	43	72	28	8	2	0	153
	(45.38)	(54.98)	(32.81)	(13.51)	(4.44)	(1.65)	
3	11	28	13	8	7	0	67
	(14.35)	(23.06)	(18.02)	(9.39)	(3.76)	(1.73)	
4	6	9	8	3	1	0	27
	(4.10)	(8.17)	(7.89)	(5.01)	(2.39)	(1.36)	
5	2	2	1	1	1	0	7
	(1.08)	(2.57)	(2.95)	(2.22)	(2.10)		
6	0	0	1	1	0	0	2
		(2.00)			(1.87)		
7	0	0	0	3	0	0	3
			(1.87)				
8	0	0	0	1	1	0	2
Total	528	291	122	42	15	2	1000

The expected values given in the table above were calculated under the hypothesis of a bivariate Neyman type A distribution with parameters $\lambda = 2$, $\lambda_1 = .4$, $\lambda_2 = .3$, $\lambda_3 = .1$. Using a χ^2 goodness-of-fit test, we obtained a value of 35.17 on 36 degrees of freedom and a prob-value of 0.5079.

8.4 Bivariate Hermite distribution

In the univariate case the generalized Poisson distribution has the pgf $\Pi(t) = \exp[\lambda\{f(t) - 1\}]$ with f(t) itself being the pgf of a discrete random variable. Expanding f(t) in powers of t, we can write

$$\Pi(t) = \exp\{\sum_{j=1}^{\infty} a_j(t^j - 1)\}, \tag{8.4.1}$$

where $\sum_{j=1}^{\infty} a_j = \lambda$. The terms in the expansion can be modified so that $a_j = 0$ for $j \geq 3$. Then (8.4.1) becomes

$$\Pi(t) = \exp\{ a_1(t - 1) + a_2(t^2 - 1) \}$$

$$= \exp\{- (a_1 + a_2) + a_1 t + a_2 t^2\}.$$

The distribution so constructed is called the Hermite distribution. It should be noted that the order in which the parameters appear in the parentheses is of consequence.

Kemp and Kemp (1966) derive the distribution by a procedure involving the compounding of the Poisson distribution with a normal distribution giving rise to the pgf

$$\Pi(t) = \exp[(\mu - \sigma^2)(t - 1) + \tfrac{1}{2}\sigma^2(t^2 - 1)] \quad \mu \geq \sigma^2. \tag{8.4.2}$$

It follows from Gurland's (1957) theorem that generalizing of the normal with the Poisson also gives rise to the Hermite distribution.

The term Hermite is used to indicate the relationship of the probabilities to the Hermite polynomials [Kemp and Kemp (1965)]. It should be mentioned that in the literature the term Poisson-normal has also been used synonymously with the term Hermite. To avoid confusion, we shall use the term Hermite.

Greenwood and Yule (1920) are credited with having introduced this distribution. They did not feel comfortable with a parameter of which the values are essentially nonnegative being allowed to have a normal distribution. Patil (1964) circumvents this difficulty by compounding with respect to the truncated normal distribution over the range $\lambda \geq 0$. This, however, leads to a rather complicated pgf.

8.4.1 Genesis of the bivariate Hermite distribution

There are several ways in which the bivariate Hermite distribution has been shown to arise in practice:

(i) Using the result of theorem 8.2.1, Kocherlakota (1988) obtains for $\underline{X} \underset{\tau}{\wedge} N(\mu, \sigma^2)$, where $\underline{X}$ has the bivariate Poisson distribution with the parameters $\tau\lambda_1$, $\tau\lambda_2$ and $\tau\lambda_3$, the pgf

$$\Pi(t_1, t_2) = \exp[\mu u + \tfrac{1}{2}\sigma^2 u^2] \tag{8.4.3}$$

with

$$u = \lambda_1(t_1 - 1) + \lambda_2(t_2 - 1) + \lambda_3(t_1 t_2 - 1). \tag{8.4.4}$$

(ii) Kemp and Papageorgiou (1982) consider the various forms in which the distribution (8.4.3) can be expressed by a process of reparameterization. These are :

$$\Pi(t_1, t_2) = \exp[\, a_1(t_1 - 1) + a_2(t_1^2 - 1) + a_3(t_2 - 1) + a_4(t_2^2 - 1) + a_5(t_1 t_2 - 1)$$
$$+ a_6(t_1^2 t_2 - 1) + a_7(t_1 t_2^2 - 1) + a_8(t_1^2 t_2^2 - 1)], \tag{8.4.5}$$

and, alternatively,

$$\Pi(t_1, t_2) = \exp[\, b_1 t_1^* + b_2 t_1^{*2} + b_3 t_2^* + b_4 t_2^{*2} + b_5 t_1^* t_2^*$$
$$+ b_6 t_1^{*2} t_2^* + b_7 t_1^* t_2^{*2} + b_8 t_1^{*2} t_2^{*2}] \tag{8.4.6}$$

where $t_1^* = t_1 - 1$ and $t_2^* = t_2 - 1$. In both of these equations the coefficients are functions of the five parameters, λ_1, λ_2, λ_3, μ and σ. In (8.4.5) with $\tau = \mu + \sigma^2\gamma$, these parameters are

$$a_1 = \lambda_1\tau,\ a_2 = \tfrac{1}{2}\sigma^2\lambda_1^2,\ a_3 = \lambda_2\tau,\ a_4 = \tfrac{1}{2}\sigma^2\lambda_2^2,\ a_5 = \lambda_3\tau + \sigma^2\lambda_1\lambda_2,$$
$$a_6 = \sigma^2\lambda_1\lambda_3,\ a_7 = \sigma^2\lambda_2\lambda_3,\ a_8 = \tfrac{1}{2}\sigma^2\lambda_3^2$$

with with $\gamma = -(\lambda_1 + \lambda_2 + \lambda_3)$.

If the parameter λ_3 is set equal to zero in each of these pgf's, then the coefficients a_6, a_7, a_8, b_6, b_7, b_8 are all equal to zero. In each case the form of the pgf $\Pi(t_1, t_2)$ is correspondingly simplified involving only the first five terms.

(iii) In addition to these forms it is also possible to develop the distribution by other methods. For example, Consael (1952) has considered the bivariate Poisson-normal model as:

$$\text{bivariate Poisson } (\lambda_1, \lambda_2, \lambda_3 = 0) \underset{\lambda_1, \lambda_2}{\wedge} \text{ bivariate normal } (\mu_1, \mu_2, \sigma_1, \sigma_2, \rho),$$

which yields the pgf with five parameters

$$\Pi(t_1, t_2) = \exp[\mu_1 (t_1 - 1) + \frac{1}{2} \sigma_1^2 (t_1 - 1)^2 + \mu_2 (t_2 - 1) + \frac{1}{2} \sigma_2^2 (t_2 - 1)^2$$
$$+ \rho\sigma_1\sigma_2 (t_1 - 1)(t_2 - 1)] \qquad (8.4.7)$$

with the restrictions $\mu_i - \sigma_i^2 \geq \rho\, \sigma_1\sigma_2$, $i = 1, 2$ and $\rho \geq 0$.

Sym (1971) and Steyn (1976) have considered the multivariate extensions of this model.

We refer to Kemp and Papageorgiou (1982) for some other examples of the derivation of the bivariate Hermite distribution.

8.4.2 Properties of the distribution

Marginal distributions

From the joint pgf given in (8.4.5) or (8.4.6), we can determine the marginal pgf of X_1 by setting $t_2 = 1$:

$$\Pi_{X_1}(t) = \exp[\mu(\lambda_1 + \lambda_3)(t - 1) + \frac{1}{2}\sigma^2(\lambda_1 + \lambda_3)^2(t - 1)^2]$$

or

$$\Pi_{X_1}(t) = \exp[\alpha_1(t - 1) + \alpha_2(t^2 - 1)],$$

where

$$\alpha_1 = (\lambda_1 + \lambda_3)[\mu - \sigma^2(\lambda_1 + \lambda_3)],$$
$$\alpha_2 = \frac{1}{2}\sigma^2(\lambda_1 + \lambda_3)^2.$$

Similarly,

$$\Pi_{X_2}(t) = \exp[\beta_1(t - 1) + \beta_2(t^2 - 1)],$$

where

$$\beta_1 = (\lambda_2 + \lambda_3)[\mu - \sigma^2(\lambda_2 + \lambda_3)],$$
$$\beta_2 = \frac{1}{2}\sigma^2(\lambda_2 + \lambda_3)^2.$$

Probability function

We note that the moment generating function of the normal distribution $\exp[\mu t + (\sigma t)^2/2]$ can be differentiated r times with respect to t yielding

$$M^{(r)}(t) = M(t)\, P_r(t),$$

where $P_r(t)$ is a polynomial of degree r in t. Thus from equation (8.2.12), we have the probability function given by

$$f(r, s) = \frac{\lambda_1^r \lambda_2^s}{r!\ s!} M(\gamma) \sum_{k=0}^{\min(r,s)} \binom{r}{k} \binom{s}{k} k!\, P_{r+s-k}(\gamma)\, \delta^k, \tag{8.4.8}$$

where $\gamma = -(\lambda_1 + \lambda_2 + \lambda_3)$ and $\delta = \lambda_3/[\lambda_1\lambda_2]$. If $\lambda_3 = 0$, the probability function reduces to

$$f(r, s) = \frac{\lambda_1^r \lambda_2^s}{r!\ s!} M[-(\lambda_1 + \lambda_2)]\, P_{r+s}[-(\lambda_1 + \lambda_2)]. \tag{8.4.9}$$

Recurrence relations

From (8.2.14) and (8.2.15) we have recurrence relations for the pgf's and for the determination of the probabilities. In the case of the bivariate Hermite distribution the recurrence relations take on a special form. Note that, in this case, the rth differential coefficient of M(t) satisfies the recurrence relation

$$M^{(r)}(t) = (r-1)\sigma^2 M^{(r-2)}(t) + (\mu + \sigma^2 t) M^{(r-1)}(t).$$

Applying this, we have

$$M^{(r)}(\gamma) = (r-1)\sigma^2 M^{(r-2)}(\gamma) + (\mu + \sigma^2\gamma)\, M^{(r-1)}(\gamma);$$

hence, the recurrence relations for the probabilities f(r, 0) and f(0, s) are

$$\left.\begin{aligned} r\, f(r, 0) &= (\lambda_1\sigma)^2 f(r-2, 0) + (\mu + \sigma^2\gamma)\, \lambda_1 f(r-1, 0) \qquad r \geq 1 \\ s\, f(0, s) &= (\lambda_2\sigma)^2 f(0, s-2) + (\mu + \sigma^2\gamma)\, \lambda_2 f(0, s-1) \qquad s \geq 1 \end{aligned}\right\} \tag{8.4.10}$$

More generally, we have from (8.2.15)

$$\left.\begin{aligned} f(r, s) &= \frac{(r+1)\lambda_2}{s\lambda_1} f(r+1, s-1) + \frac{(r-s+1)\lambda_3}{s\lambda_1} f(r, s-1) \quad r \geq s \\ f(r, s) &= \frac{(s+1)\lambda_1}{r\lambda_2} f(r-1, s+1) + \frac{(s-r+1)\lambda_3}{r\lambda_2} f(r-1, s) \quad s \geq r \end{aligned}\right\} \tag{8.4.11}$$

In addition, $f(0, 0) = M(\gamma)$, $f(1, 0) = \lambda_1\tau\, M(\gamma)$ and $f(0, 1) = \lambda_2\tau\, M(\gamma)$ with $\tau = \mu + \sigma^2\gamma$. These relationships permit a rapid calculation of the probabilities.

Kemp and Papageorgiou (1982) have a different set of recurrence equations for the forms of the pgf considered by them. They have also derived the inequalities, for any pair r and s,

$$P\{X_1 \leq r, X_2 \leq s\} \geq P\{X_1 \leq r\}\, P\{X_2 \leq s\}$$
$$P\{X_1 \geq r, X_2 \geq s\} \geq P\{X_1 \geq r\}\, P\{X_2 \geq s\}.$$

Moments and cumulants

The factorial cumulant generating function is seen to be

$$G(t_1, t_2) = b_1t_1 + b_2t_1^2 + b_3t_2 + b_4t_2^2 + b_5t_1t_2 + b_6t_1^2t_2 + b_7t_1t_2^2 + b_8t_1^2t_2^2,$$

from which the factorial cumulants $\kappa_{[r,s]}$ are readily determined as the coefficients of $t_1^r\, t_2^s/r!\, s!$:

$$\left.\begin{aligned} \kappa_{[1,0]} &= b_1,\ \kappa_{[2,0]} = 2b_2,\ \kappa_{[0,1]} = b_3,\ \kappa_{[0,2]} = 2b_4 \\ \kappa_{[1,1]} &= b_5,\ \kappa_{[2,1]} = 2b_6,\ \kappa_{[1,2]} = 2b_7,\ \kappa_{[2,2]} = 4b_8 \end{aligned}\right\} \tag{8.4.12}$$

All higher order factorial cumulants are zero.

Independence

In the case of the eight parameter family of the Hermite distribution, it does not appear that the random variables X_i, i = 1, 2, can be independent. However, for the five parameter family there are two cases which we can consider. In the first case when the

$$\text{bivariate Poisson } (\lambda_1,\lambda_2, \lambda_3 = 0) \underset{\lambda_1,\lambda_2}{\wedge} \text{ bivariate normal } (\mu_1, \mu_2, \sigma_1, \sigma_2, \rho),$$

the pgf is given by (8.4.7). Here it is readily seen that the random variables are independent if and only if $\rho = 0$; that is, if and only if the independent Poisson random variables are compounded by independent normal distributions.

On the other hand, if we consider the five parameter family from (8.4.6) which arises when the compounding is

$$P(\lambda_1, \tau)\, P(\lambda_2, \tau) \underset{\tau}{\wedge} g(\tau)$$

with g being $N(\mu, \sigma^2)$, the pgf of (X_1, X_2) is

$$\Pi(t_1, t_2) = \exp[\mu\{\lambda_1(t_1-1)+\lambda_2(t_2-1)\} + \tfrac{1}{2}\sigma^2\{\lambda_1(t_1-1)+\lambda_2(t_2-1)\}^2].$$

The necessary and sufficient condition for X_1 and X_2 to be independent is that σ^2 be equal to zero. This, however, is a necessary and sufficient condition for $g(\tau)$ to be degenerate. The resulting distribution is the double Poisson, which may be considered a special case of the bivariate Hermite distribution.

Conditional distributions and regression

If the pgf is of the form in (8.4.5), then using theorem 1.3.1 the pgf of the conditional distribution of X_1 given $X_2 = x_2$ is obtained from equation (8.2.17) as

$$\Pi_{X_1}(t|x_2) = \frac{M^{(x_2)}[-(\lambda_2+\lambda_3) + \lambda_1(t-1)]}{M^{(x_2)}[-(\lambda_2+\lambda_3)]} \left[\frac{\lambda_2}{\lambda_2+\lambda_3} + \frac{\lambda_3\, t}{\lambda_2+\lambda_3}\right]^{x_2}$$

(8.4.13)

It can be seen that the differential coefficient on the right-hand side can be written as

$$M^{(r)}(t) = M(t)\,\sigma^r\,\overset{*}{H}_x[\sigma t + \frac{\mu}{\sigma}] \qquad (8.4.14)$$

with

$$\overset{*}{H}_x(t) = \sum_{r=0}^{[\frac{x}{2}]} \frac{x!\,t^{x-2r}}{(x-2r)!\,r!\,2^r}, \qquad (8.4.15)$$

the modified Hermite polynomials defined by Kemp and Kemp (1965, equation 13).

Substituting in the first term on the right-hand side of (8.4.13), we have

$$\frac{M[\lambda_1(t-1)-\lambda_2-\lambda_3]}{M[-\lambda_2-\lambda_3]}\,\frac{\overset{*}{H}_{x_2}[\sigma\{\lambda_1(t-1)-\lambda_2-\lambda_3\}+\frac{\mu}{\sigma}]}{\overset{*}{H}_{x_2}[\sigma\{-\lambda_2-\lambda_3\}+\frac{\mu}{\sigma}]}. \qquad (8.4.16)$$

The first term in equation (8.4.16) can be expressed as

$$\exp[\{\mu-\sigma^2(\lambda_2+\lambda_3)\}\lambda_1(t-1) + \tfrac{1}{2}(\sigma\lambda_1)^2(t-1)^2],$$

while the second term is

$$\sum_{r=0}^{[\frac{x_2}{2}]} \omega_r(x_2)\,(Q + Pt)^{x_2-2r},$$

where

$$P = \frac{\sigma^2\lambda_1}{\mu-\sigma^2(\lambda_2+\lambda_3)}, \quad Q = 1 - P$$

and

$$\omega_r(x) = \frac{\frac{x!}{(x-2r)!r!}\alpha^{-r}}{\sum_{s=0}^{[\frac{x}{2}]}\frac{x!}{(x-2s)!s!}\alpha^{-s}} \quad \text{with} \quad \alpha = 2\{\mu/\sigma - \sigma(\lambda_2+\lambda_3)\}^2.$$

In the above development we have assumed that the parameters satisfy the condition $\mu/\sigma > \sigma\,(\lambda_2+\lambda_3)$. Under this assumption the conditional distribution of X_1 given $X_2 = x_2$ is the convolution of

(i) $Y_1 \sim B(x_2, p), \quad p = \lambda_3/(\lambda_2+\lambda_3)$;

(ii) $Y_2 \sim$ Hermite obtained by $P(\lambda_1\tau) \underset{\tau}{\wedge} N(\mu-\sigma^2(\lambda_2+\lambda_3), \sigma^2)$

(iii) Y_3 which is a mixture of $B(x_2 - 2r, P)$, $r = 0, 1, 2, \ldots, [\frac{x_2}{2}]$ with the r-th component having the weight function given above.

These results are from Kocherlakota (1988). Kemp and Papageorgiou (1982) consider the distribution for the case $\lambda_3 = 0$.

The regression can be found as usual from the conditional distribution as

$$E[X_1|x_2] = x_2 p + \lambda_1[\mu-\sigma^2(\lambda_2 + \lambda_3)] + \sum_{r=0}^{[\frac{x_2}{2}]} \omega_r(x_2)(x_2-2r)\ P$$

which is not linear in the regressor x_2.

8.4.3 Estimation

Moment and maximum likelihood estimators are developed for the five parameter families discussed in the earlier section. The methods are illustrated by a set of accident data taken from Cresswell and Froggatt (1963) and reproduced in Kemp and Papageorgiou (1982).

Method of moments

Consider the five parameter family of the distribution given by the pgf's in equations (8.4.5) and (8.4.6) with $a_i = b_i = 0$ for i =6, 7, 8. These yield the moments in terms of the b_i's as given in (8.4.12). Solving for the parameters in terms of the sample moments we have

$$\tilde{b}_1 = \bar{x}_1, \quad \tilde{b}_2 = [m_{2,0} - \bar{x}_1]/2,$$
$$\tilde{b}_3 = \bar{x}_2, \quad \tilde{b}_4 = [m_{0,2} - \bar{x}_2]/2$$

and

$$\tilde{b}_5 = m_{1,1}. \tag{8.4.17}$$

Reparameterizing to the a's, we have the equations

$$\left.\begin{aligned} &\tilde{a}_1 + 2\tilde{a}_2 + \tilde{a}_5 = \bar{x}_1, \quad \tilde{a}_3 + 2\tilde{a}_4 + \tilde{a}_5 = \bar{x}_2 \\ &\tilde{a}_1 + 4\tilde{a}_2 + \tilde{a}_5 = m_{2,0}, \quad \tilde{a}_3 + 4\tilde{a}_4 + \tilde{a}_5 = m_{0,2} \\ &\tilde{a}_5 = m_{1,1} \end{aligned}\right\}$$

yielding

$$\left.\begin{aligned} &\tilde{a}_1 = 2\bar{x}_1 - m_{2,0} - m_{1,1}, \quad \tilde{a}_2 = [m_{2,0} - \bar{x}_1]/2 \\ &\tilde{a}_3 = 2\bar{x}_2 - m_{0,2} - m_{1,1}, \quad \tilde{a}_4 = [m_{0,2} - \bar{x}_2]/2 \\ &\tilde{a}_5 = m_{1,1} \end{aligned}\right\} \tag{8.4.18}$$

The covariance matrices of the estimates can be found using the results on the joint moments of the sample moments given in section 1.2.

Maximum likelihood

Maximum likelihood estimators can be developed for the various parameterizations. Let n_{rs} represent the observed frequency in the (r, s) cell and f(r, s) the corresponding probability function. Using the notation of section 8.3.2, we can find

$$\frac{\partial \log L}{\partial \theta_i} = \sum_{r,s} \frac{n_{rs}}{f(r, s)} \frac{\partial \log f(r, s)}{\partial \theta_i},$$

where θ_i represents the parameters a_i for i = 1, 2, ..., 5.

Kemp and Papageorgiou (1982) give recurrence relations between the partial derivatives of f(r, s) with respect to the various parameters. From the pgf (8.4.5) with the five parameters, we have

$$\frac{\partial \log f(r, s)}{\partial a_1} = f(r-1, s) - f(r, s), \quad \frac{\partial \log f(r, s)}{\partial a_2} = f(r-2, s) - f(r, s),$$

$$\frac{\partial \log f(r, s)}{\partial a_3} = f(r, s-1) - f(r, s), \quad \frac{\partial \log f(r, s)}{\partial a_4} = f(r, s-2) - f(r, s),$$

$$\frac{\partial \log f(r, s)}{\partial a_5} = f(r-1, s-1) - f(r, s).$$

Hence

$$\left.\begin{aligned}
\frac{\partial \log L}{\partial a_1} &= \sum_{r,s} n_{rs} \left[\frac{f(r-1, s)}{f(r, s)} - 1\right], \quad \frac{\partial \log L}{\partial a_2} = \sum_{r,s} n_{rs} \left[\frac{f(r-2, s)}{f(r, s)} - 1\right] \\
\frac{\partial \log L}{\partial a_3} &= \sum_{r,s} n_{rs} \left[\frac{f(r, s-1)}{f(r, s)} - 1\right], \quad \frac{\partial \log L}{\partial a_4} = \sum_{r,s} n_{rs} \left[\frac{f(r, s-2)}{f(r, s)} - 1\right] \\
\frac{\partial \log L}{\partial a_5} &= \sum_{r,s} n_{rs} \left[\frac{f(r-1, s-1)}{f(r, s)} - 1\right]
\end{aligned}\right\} \tag{8.4.19}$$

and a typical element of the information matrix is given by

$$E\left[-\frac{\partial^2 \log L}{\partial a_1^2}\right] = n \sum_{r,s} f(r, s) \left[\frac{f(r-1, s)}{f(r, s)} - 1\right]^2.$$

These equations are helpful for an iterative solution of the maximum likelihood equations. The equations for the parameter set in b's (or in μ, σ^2, λ_1, λ_2, λ_3) can be written down along the same lines.

Example 8.4.1

Kemp and Papageorgiou (1982) fit the five parameter Hermite distributions given in (8.4.5) and (8.4.6) to the accident data of Cresswell and Frogatt (1963) with X_1 as the accidents in the first period and X_2 as the accidents in the second period.

x_1 \ x_2	0	1	2	3	4	5	6	7	Total
0	117 (118.8)	96 (91.8)	55 (49.9)	19 (20.3)	2 (7.0)	2 (2.1)	0 (0.5)	0 (0.1)	291
1	61 (60.0)	69 (79.0)	47 (50.4)	27 (24.0)	8 (9.1)	5 (3.0)	1 (0.8)	0 (0.2)	218
2	34 (28.3)	42 (38.4)	31 (29.1)	13 (15.2)	7 (6.4)	2 (2.2)	3 (0.7)	0 (0.2)	132
3	7 (9.2)	15 (14.9)	16 (12.1)	7 (7.0)	3 (3.1)	1 (1.2)	0 (0.4)	0 (0.1)	49
4	3 (2.7)	3 (4.6)	1 (4.2)	1 (2.6)	2 (1.2)	1 (0.5)	1 (0.2)	1 (0.1)	13
5	2 (0.7)	1 (1.3)	0 (1.2)	0 (0.8)	0 (0.4)	0 (0.2)	0 (0.1)	0 (0.0)	3
6	0 (0.2)	0 (0.3)	0 (0.3)	0 (0.2)	1 (0.1)	0 (0.1)	0 (0.0)	0 (0.0)	1
7	0 (0.0)	0 (0.1)	0 (0.1)	1 (0.1)	0 (0.0)	0 (0.0)	0 (0.0)	0 (0.0)	1
Total	224	226	150	68	23	11	5	1	708

The summary statistics for the above table are:

$$\bar{x}_1 = 1.0014, \quad m_{2,0} = 1.1952,$$

$$\bar{x}_2 = 1.2910, \quad m_{0,2} = 1.5984, \quad m_{1,1} = 0.3261.$$

These yield the moment estimates:

$$\tilde{b}_1 = 1.0014, \quad \tilde{b}_2 = .09688, \quad \tilde{b}_3 = 1.2910,$$

$$\tilde{b}_4 = .15372, \quad \tilde{b}_5 = .3263$$

and

$$\tilde{a}_1 = .4813, \quad \tilde{a}_2 = \tilde{b}_2 = .09688, \quad \tilde{a}_3 = .6572,$$

$$\tilde{a}_4 = \tilde{b}_4 = .15372, \quad \tilde{a}_5 = \tilde{b}_5 = .3263.$$

Kemp and Papageorgiou (1982) also provide the maximum likelihood estimates for the two sets of parameters:

$$\hat{a}_1 = .5053, \ \hat{b}_1 = 1.0014, \ \hat{a}_2 = \hat{b}_2 = .1109, \ \hat{a}_3 = .7728, \ \hat{b}_3 = 1.2910,$$

$$\hat{a}_4 = \hat{b}_4 = .1219, \ \hat{a}_5 = \hat{b}_5 = .2744.$$

In the above table the expected numbers taken from Kemp and Papageorgiou (1982) and are based on the maximum likelihood estimates. Using these expected values for the χ^2 goodness-of-fit test, they obtain $\underline{X}^2 = 25.0$ on 28 degrees of freedom with a prob-value of 0.3078; hence, the fit is good.

8.4.4 Computer simulation

As in section 8.3.3, observations from (8.4.8) can be generated using the marginal distribution of X_2 and the conditional distribution of X_1 given x_2. Recall that X_2 has a univariate Hermite distribution with parameters $\beta_1 = (\lambda_2 + \lambda_3)(\mu - \sigma^2(\lambda_2 + \lambda_3)$ and $\beta_2 = \frac{1}{2}\sigma^2(\lambda_2 + \lambda_3)^2$. It can be shown [Kemp and Kemp (1965) and Jain (1983)] that a univariate Hermite random variable can be expressed as the convolution

$$X_2 = W_1 + 2\,W_2, \tag{8.4.20}$$

where

$$W_1 \sim P(\beta_1)$$
$$W_2 \sim P(\beta_2).$$

The algorithm for simulating pairs of observations from (8.4.8), can be summarized as:

(i) Generate a realization x_2 from X_2 using the convolution in (8.4.20).

(ii) For this given value x_2, generate a realization of $X_1 = Y_1 + Y_2 + Y_3$, where

$$Y_1 \sim B(x_2, p), \quad p = \lambda_3/(\lambda_2+\lambda_3);$$

$Y_2 \sim$ univariate Hermite obtained as $U_1 + 2\,U_2$ with

$$U_1 \sim P(\lambda_1\,[\mu + \sigma^2\gamma])$$
$$U_2 \sim P(\tfrac{1}{2}\sigma^2\lambda_1^2)$$

and

Y_3 which is a mixture of $B(x_2 - 2r, P)$, $r = 0, 1, 2, \ldots, [\frac{x_2}{2}]$ with the r-th component having the weight function

$$\omega_r(x) = \frac{\dfrac{x!}{(x-2r)!\,r!}\,\alpha^{-r}}{\displaystyle\sum_{s=0}^{[\frac{x}{2}]} \frac{x!}{(x-2s)!\,s!}\,\alpha^{-s}}, \qquad \alpha = 2\{\mu/\sigma - \sigma(\lambda_2+\lambda_3)\}^2.$$

and

$$P = \frac{\sigma^2\lambda_1}{\mu-\sigma^2(\lambda_2+\lambda_3)} .$$

The realized value of Y_3 is obtained as discussed in section 8.3.3.

Example 8.4.2

Using the technique described above, 1000 pairs of observations were generated from a bivariate Hermite distribution with parameters $\mu = 1, \sigma^2 = 1, \lambda_1 = 0.4, \lambda_2 = 0.3$ and $\lambda_3 = 0.1$.

x_1 \ x_2	0	1	2	3	4	Total
0	612 (618.78)	45 (37.13)	33 (28.96)	1	0 (2.42)	691
1	67 (49.50)	87 (89.6)	21 (26.08)	4 (5.30)	2 (1.34)	181
2	43 (51.48)	38 (34.77)	13 (14.95)	4 (4.86)	1 (1.32)	99
3	4 (4.01)	13 (9.43)	4 (6.48)	4	0 (3.34)	25
4	0 (2.14)	3 (2.75)	1 (1.91)	0	0 (3.45)	4
Total	726	186	72	13	3	1000

To assess the performance of the simulation technique, the distribution was fitted assuming the parameters to be known The fitted values are given in parantheses in the above table. The χ^2 goodness-of-fit statistic for this test was 21.68 on 21 degrees of freedom with a prob-value of 0.418.

8.5 Inverse Gaussian-bivariate Poisson distribution

Good (1953) is credited with having developed the univariate inverse Gaussian-Poisson distribution. A systematic study of the generalized distribution was later presented by Sichel (1975). Under the parameterization of Stein *et al.* (1987) the mixing distribution is taken to be

$$g(\lambda) = \frac{\xi^{-\gamma}\lambda^{\gamma-1}}{2K_\gamma[(\alpha^2+\xi^2)^{\frac{1}{2}} - \xi]} \exp[-\tfrac{1}{2}\{[(\alpha^2+\xi^2)^{\frac{1}{2}}-\xi]\{\tfrac{\lambda}{\xi} + \tfrac{\xi}{\lambda}\}], \tag{8.5.1}$$

where $\alpha > 0$, $\xi > 0$, $-\infty < \gamma < \infty$ and $K_\gamma(z)$ is the modified Bessel function of the third kind of order γ and argument z [Abramovitz and Stegun (1968)]. For $\gamma = -1/2$, the pdf $g(\lambda)$ becomes the inverse Gaussian (IG) distribution.

Using the pdf $g(\lambda)$ in the univariate Poisson case Stein and Juritz (1987) have given the pf of the random variable X as

$$f(x) = \frac{[\{(\alpha^2+\xi^2)^{\frac{1}{2}} - \xi\}/\alpha]^{\gamma}}{K_\gamma[(\alpha^2+\xi^2)^{\frac{1}{2}} - \xi]} \frac{[\xi\{(\alpha^2+\xi^2)^{\frac{1}{2}} - \xi\}/\alpha]^{x}}{x!} K_{\gamma+x}(\alpha). \qquad x = 0, 1, 2, \ldots \tag{8.5.2}$$

Several forms of the bivariate distribution can be developed by compounding the bivariate Poisson distribution with the generalized inverse Gaussian distribution of the form discussed by Jorgensen (1982). We refer to Rubinstein (1985) for more general results in this connection. We will consider here the particular case of the inverse Gaussian when $\gamma = -1/2$.

8.5.1 Some properties of the distribution for $\gamma = -\frac{1}{2}$

The pgf of the compound distribution is obtained by compounding the Poisson pgf

$$\Pi(t_1, t_2|\tau) = \exp[\tau\{\lambda_1 (t_1 - 1) + \lambda_2 (t_2 - 1) + \lambda_3 (t_1 t_2 - 1)\}]$$

with the random variable τ having the pdf

$$g(z) = [\frac{\lambda}{2\pi z^3}]^{\frac{1}{2}} \exp\{-\frac{\lambda}{2\mu^2 z}(z-\mu)^2\}, \quad z > 0. \tag{8.5.3}$$

An alternative representation for the pdf is obtained with the reparameterization $\xi = \mu$, $\omega = \lambda/\mu$

$$g(z) = [\frac{\omega\xi}{2\pi z^3}]^{\frac{1}{2}} \exp\{-\frac{\omega}{2\xi z}(z-\xi)^2, \qquad z > 0. \qquad (8.5.4)$$

In what follows we shall be using the two forms of the pdf for the inverse Gaussian distribution interchangeably. From the context it should be clear as to the form under reference. However, the distribution will be abbreviated as IG(μ, λ) from (8.5.3). For details on the univariate inverse Gaussian distribution we refer to Tweedie (1957).

From theorem 8.2.1 the unconditional pgf of (X_1, X_2) is

$$\Pi(t_1, t_2) = M[\lambda_1 (t_1 - 1) + \lambda_2 (t_2 - 1) + \lambda_3 (t_1 t_2 - 1)],$$

where

$$M(t) = \exp [\frac{\lambda}{\mu}\{1 - (1 - \frac{2\mu^2 t}{\lambda})^{\frac{1}{2}}\}], \ \lambda > 0, -\infty < \mu < \infty. \qquad (8.5.5)$$

Hence, the pgf of (X_1, X_2) is

$$\Pi(t_1, t_2) = \exp\Big[\frac{\lambda}{\mu}\{1 - (1 - \frac{2\mu^2}{\lambda}[\lambda_1 (t_1 - 1) + \lambda_2 (t_2 - 1) + \lambda_3 (t_1 t_2 - 1)])^{\frac{1}{2}}\}\Big]. \qquad (8.5.6)$$

Alternatively, the pgf can be written as

$$\Pi(t_1, t_2) = \exp\Big[\omega\{1 - (1 - \frac{2\xi}{\omega}[\lambda_1 (t_1 - 1) + \lambda_2 (t_2 - 1) + \lambda_3 (t_1 t_2 - 1)])^{\frac{1}{2}}\}\Big]. \qquad (8.5.7)$$

Marginal distributions

Setting $t_2 = 1$ and $t_1 = 1$, successively in the joint pgf, the marginal pgf's are found to be

$$\Pi_{X_i}(t) = \exp [\frac{\lambda}{\mu}\{1 - [1 - 2\frac{\mu^2}{\lambda}(\lambda_i + \lambda_3)(t-1)]^{\frac{1}{2}}\}], \quad i = 1, 2 \qquad (8.5.8)$$

or, alternatively,

$$\Pi_{X_i}(t) = \exp [\omega\{1 - [1 - 2\frac{\xi}{\omega}(\lambda_i + \lambda_3)(t-1)]^{\frac{1}{2}}\}], \quad i = 1, 2 \qquad (8.5.9)$$

each of which is the univariate inverse Gaussian-Poisson distribution. It is also seen that the random variables X_1 and X_2 cannot be independent under any circumstances.

Probability function

The following theorem will be of help in the use of the results in section 8.2.1 where we need to determine the r-th differential coefficient of the moment generating function M(t).

Theorem 8.5.1

The rth differential coefficient of the function

$$M(t) = \exp[\,\omega\{1 - (1 - \frac{2\xi t}{\omega})^{\frac{1}{2}}\}]$$

with respect to t is given by

$$M^{(r)}(t) = M(t)\frac{\xi^r}{\omega^{r-1}}\sum_{k=0}^{r-1} C_k^{(r)}\frac{\omega^k}{\alpha^{r-(k+1)/2}}, \tag{8.5.10}$$

where $\alpha = 1 - \frac{2\xi t}{\omega}$ and

$$C_k^{(r)} = \frac{(2r-2-k)!}{(r-1-k)!\,k!\,2^{r-1-k}}. \tag{8.5.11}$$

Proof:

The differential coefficients for r = 1, r = 2 are readily found to be

$$M^{(1)}(t) = M(t)\frac{\xi}{\alpha^{1/2}}$$

and

$$M^{(2)}(t) = M(t)\frac{\xi^2}{\omega}[\omega/\alpha + 1/\alpha^{3/2}].$$

For using induction to establish the general result, we assume that the equation (8.5.10) is valid for r. Differentiating it once with respect to t, we have

$$M^{(r+1)}(t) = M(t)\frac{\xi}{\alpha^{1/2}}\frac{\xi^r}{\omega^{r-1}}\sum_{k=0}^{r-1} C_k^{(r)}\frac{\omega^k}{\alpha^{r-(k+1)/2}}$$

$$+ M(t)\frac{\xi^r}{\omega^{r-1}}\sum_{k=0}^{r-1} C_k^{(r)}\frac{(2r-k-1)\omega^{k-1}\xi}{\alpha^{r-(k-1)/2}}. \qquad (8.5.11)$$

The right-hand side of (8.5.11) is $M(t)\,\xi^{r+1}/\omega^r$ times the series

$$\sum_{k=0}^{r-1} C_k^{(r)}\frac{\omega^{k+1}}{\alpha^{r-k/2}} + \sum_{k=0}^{r-1} C_k^{(r)}\frac{(2r-k-1)\omega^k}{\alpha^{r-(k-1)/2}}. \qquad (8.5.12)$$

For $k = 0$ the coefficient of ω^k is found from the second term of (8.5.12) to be $C_0^{(r+1)}/\alpha^{r+\frac{1}{2}}$. For $k = r$ the corresponding coefficient is found from the first term in (8.5.12) to be $C_r^{(r+1)}/\alpha^{(r+1)/2}$. For $1 \le k \le r - 1$, the coefficient is

$$\frac{C_{k-1}^{(r)}\,\alpha^{\frac{1}{2}}}{\alpha^{\frac{2r+2-k}{2}}} + \frac{C_k^{(r)}\,(2r-k-1)}{\alpha^{\frac{2r+1-k}{2}}} = \frac{C_k^{(r+1)}}{\alpha^{r+1-\frac{(k+1)}{2}}}.$$

Hence the result. □

From equation (8.2.12) the pf of the distribution is given by

$$f(r, s) = \frac{\lambda_1^r\lambda_2^s}{r!\ s!}\sum_{k=0}^{\min(r,s)}\binom{r}{k}\binom{s}{k}\,k!\,M^{(t-k)}(\gamma)\,\delta^k.$$

Substituting for the derivative in the summation, the pf becomes

$$f(r, s) = \frac{\lambda_1^r\lambda_2^s}{r!\ s!}M(\gamma)\sum_{k=0}^{\min(r,s)}\binom{r}{k}\binom{s}{k}\,k!\,\delta^k\frac{\xi^{n-k}}{\omega^{n-k-1}}\cdot$$

$$\sum_{i=0}^{n-k-1} C_i^{(n-k)}\,\omega^i\,\tau^{-\frac{2(n-k)-i-1}{2}}, \qquad (8.5.13)$$

where $n = r + s$ and $\tau = 1-2\xi\gamma/\omega$. Recall that $M(\gamma) = \exp[\omega(1-\tau^{\frac{1}{2}})]$. Then it is possible to represent the pf in terms of the modified (spherical) Bessel function of the third kind defined by the series

$$K_{n+\frac{1}{2}}(z) = K_{\frac{1}{2}}(z) \sum_{m=0}^{n} \frac{(n+m)!\,(2z)^{-m}}{m!\,(n-m)!}\,, \tag{8.5.14}$$

where $K_{\frac{1}{2}}(z) = \pi^{\frac{1}{2}} \exp(-z)/(2z)^{\frac{1}{2}}$. For such representations and related results see Erdelyi *et al.* (1953, Vol. 2, p. 10, equation 41) or Abramowitz and Stegun (1968, p. 444).

Thus substituting for the coefficient $C_i^{(n-k)}$ in equation (8.5.13) and changing the index of summation as m = n-k-i-1, we have

$$\sum_{m=0}^{k^*} \frac{(k^*+m)!}{m!\,(k^*-m)!\,2^m}\,\omega^{k^*-m}\,\tau^{-[\frac{k^*+m+1}{2}]}$$

$$= \omega^{k^*}\,\tau^{-[\frac{k^*+1}{2}]} \sum_{m=0}^{k^*} \frac{(k^*+m)!}{m!\,(k^*-m)!}\,(2\omega\tau^{\frac{1}{2}})^{-m}, \tag{8.5.15}$$

where $k^* = n-k-1$. Using (8.5.14) the right-hand side of (8.5.15) can be written as

$$\frac{\omega^{k^*}}{\tau^{[\frac{k^*+1}{2}]}}\;\frac{K_{k^*+\frac{1}{2}}(\omega\tau^{\frac{1}{2}})}{K_{\frac{1}{2}}(\omega\tau^{\frac{1}{2}})}\,.$$

Therefore the pf in equation (8.5.13) is

$$f(r, s) = \frac{\lambda_1^r\lambda_2^s}{r!\,s!}\exp[\omega(1-\tau^{\frac{1}{2}})]. \sum_{k=0}^{\min(r,s)} \binom{r}{k}\binom{s}{k}\,k!\,\delta^k\,[\frac{\xi}{\tau^{\frac{1}{2}}}]^{k^*+1}\;\frac{K_{k^*+\frac{1}{2}}(\omega\tau^{\frac{1}{2}})}{K_{\frac{1}{2}}(\omega\tau^{\frac{1}{2}})}\,.$$

Since $K_{\frac{1}{2}}(z) = \pi^{\frac{1}{2}} \exp(-z)/(2z)^{\frac{1}{2}}$, this pf can be simplified as

$$f(r, s) = \frac{\lambda_1^r\lambda_2^s}{r!\,s!}\,\frac{1}{K_{\frac{1}{2}}(\omega)} \sum_{k=0}^{\min(r,s)} \binom{r}{k}\binom{s}{k}\,k!\,\delta^k\,[\frac{\xi^{k^*+1}}{[\tau^{\frac{1}{2}}]^{k^*+\frac{1}{2}}}]\;K_{k^*+\frac{1}{2}}(\omega\tau^{\frac{1}{2}}).$$

(8.5.16)

When $\lambda_3 = 0$, the pf simplifies to

$$f(r, s) = \frac{\lambda_1^r \lambda_2^s}{r!\ s!} \frac{1}{K_{\frac{1}{2}}(\omega)} \left[\frac{\xi^{r+s}}{[\tau^{\frac{1}{2}}]^{r+s-\frac{1}{2}}}\right] K_{[r+s-1]+\frac{1}{2}}(\omega\tau^{\frac{1}{2}}),$$

which shows that the random variables X_1 and X_2 are not independent.

Recurrence relations

From the pgf $f(0, 0) = \exp[\omega\{1-\tau^{\frac{1}{2}}\}]$ while (8.5.16) yields

$$f(1, 0) = \frac{\lambda_1 \xi}{\tau^{\frac{1}{2}}} f(0, 0) \quad , \quad f(0, 1) = \frac{\lambda_2 \xi}{\tau^{\frac{1}{2}}} f(0, 0).$$

The other probabilities in the first row and column can be evaluated using the recurrence relation for the Bessel function [see Abramowitz and Stegun (1968), p. 444]:

$$K_{r+\frac{1}{2}}(z) = K_{(r-2)+\frac{1}{2}}(z) + \frac{2r-1}{z} K_{(r-1)+\frac{1}{2}}(z).$$

Hence, for $r \geq 2$,

$$f(r, 0) = \frac{(\lambda_1 \xi)^2}{r(r-1)\tau} f(r\text{-}2, 0) + \frac{(2r-3)\lambda_1 \xi}{r\omega\tau} f(r-1, 0),$$

and, for $s \geq 2$,

$$f(0, s) = \frac{(\lambda_2 \xi)^2}{s(s-1)\tau} f(0, s\text{-}2) + \frac{(2s-3)\lambda_2 \xi}{s\omega\tau} f(0, s\text{-}1).$$

Using (8.2.15) general recurrence formulas can be determined for $f(r, s)$ when $r \geq 1$, $s \geq 1$.

Factorial moments

The factorial moment generating function is seen to be

$$G(t_1, t_2) = \exp\left[\frac{\lambda}{\mu}\{1 - (1 - \frac{2\mu^2}{\lambda}[(\lambda_1 + \lambda_3)\, t_1 + (\lambda_2 + \lambda_3)\, t_2 + \lambda_3\, t_1 t_2])^{\frac{1}{2}}\}\right]$$

from which the factorial moments are found to be

$$\mu_{[r,s]} = (\lambda_1 + \lambda_3)^r(\lambda_2 + \lambda_3)^s \sum_{k=0}^{\min(r,s)} \frac{r!\ s!\ \tau^k}{(r-k)!\ (s-k)!\ k!}\ \mu'_{r+s-k}, \qquad (8.5.17)$$

where μ'_k is the k-th raw moment of the inverse Gaussian distribution (8.5.3) and $\tau = \lambda_3/[(\lambda_1 + \lambda_3)(\lambda_2 + \lambda_3)]$. In particular, the covariance is

$$\mu_{1,1} = \xi\ (\lambda_1 + \lambda_3)(\lambda_2 + \lambda_3)\ [\frac{\xi}{\omega} + \frac{\lambda_3}{\{(\lambda_1 + \lambda_3)(\lambda_2 + \lambda_3)\}}].$$

8.5.2 Conditional distributions and regression

From equation (8.2.17) the conditional distribution of X_1 given $X_2 = x_2$ is the convolution of the random variables

(i) $Y_1 \sim B(x_2, \lambda_3/(\lambda_2 + \lambda_3)$

(ii) Y_2 of which the pgf is $\dfrac{M^{(x_2)}[\lambda_1(t-1)-\lambda_2-\lambda_3]}{M^{(x_2)}[-\lambda_2-\lambda_3]}$.

To study the latter pgf, we recall the expansion

$$M^{(x_2)}[\lambda_1(t-1)-\lambda_2-\lambda_3] = M[\lambda_1(t-1)-\lambda_2-\lambda_3]\,\frac{\xi^{x_2}}{\omega^{x_2-1}} \sum_{k=0}^{x_2-1} C_k^{(x_2)}\ \frac{\omega^k}{\alpha^{x_2-(k+1)/2}},$$

where $\alpha = 1-2\,\frac{\xi}{\omega}\,\{\lambda_1(t-1)-\lambda_2-\lambda_3\}$. Similarly, writing $\beta = 1 + 2\,\frac{\xi}{\omega}\,\{\lambda_2+\lambda_3\}$, we have

$$M^{(x_2)}[-\lambda_2-\lambda_3] = M[-\lambda_2-\lambda_3]\,\frac{\xi^{x_2}}{\omega^{x_2-1}} \sum_{k=0}^{x_2-1} C_k^{(x_2)}\ \frac{\omega^k}{\beta^{x_2-(k+1)/2}}\ .$$

Taking the ratio of these two series, the pgf of Y_2 is obtained as

$$\frac{M[\lambda_1(t-1)-\lambda_2-\lambda_3]}{M[-\lambda_2-\lambda_3]} \frac{\sum_{k=0}^{x_2-1} C_k^{(x_2)} \omega^k [1-2\frac{\xi}{\omega}\{\lambda_1(t-1)-\lambda_2-\lambda_3\}]^{-(x_2-\{k+1\}/2)}}{\sum_{k=0}^{x_2-1} C_k^{(x_2)} \omega^k [1+2\frac{\xi}{\omega}\{\lambda_2+\lambda_3\}]^{-(x_2-\{k+1\}/2)}} .$$

The first term can be simplified and expressed in a standard form as

$$\exp \eta\{1-[1-2\frac{\zeta}{\eta}\lambda_1(t-1)]^{\frac{1}{2}}\},$$

where $\eta = \omega[1+2\frac{\xi}{\omega}(\lambda_2+\lambda_3)]^{\frac{1}{2}}$ and $\zeta = \xi/[1+2\frac{\xi}{\omega}(\lambda_2+\lambda_3)]^{\frac{1}{2}}$. This is the pgf of the compounding $P(\lambda_1\tau) \underset{\tau}{\wedge} IG(\xi/[1+2\frac{\xi}{\omega}(\lambda_2+\lambda_3)]^{\frac{1}{2}}, \xi\omega)$. The second term, on the other hand, becomes the ratio

$$\sum_{k=0}^{x_2-1} D_k^{(x_2)} [1-\frac{2\xi\lambda_1/\omega}{1+2\xi(\lambda_2+\lambda_3)/\omega}(t-1)]^{-(x_2-\{k+1\}/2)},$$

where

$$D_k^{(x_2)} = \frac{C_k^{(x_2)}\eta^k}{\sum_{m=0}^{x_2-1} C_m^{(x_2)}\eta^m} \qquad k = 0, 1, 2, ..., x_2-1. \qquad (8.5.18)$$

Thus the conditional distribution of X_1 given $X_2 = x_2$ is the convolution of :

(i) $Y_1 \sim B(x_2, \lambda_3/(\lambda_2+\lambda_3))$

(ii) $Z_1 \sim P(\lambda_1\tau) \underset{\tau}{\wedge} IG(\zeta, \eta\zeta)$

where

$$\eta = \omega[1+2\frac{\xi}{\omega}(\lambda_2+\lambda_3)]^{\frac{1}{2}}$$

and

$$\zeta = \xi/[1+2\frac{\xi}{\omega}(\lambda_2+\lambda_3)]^{\frac{1}{2}}.$$

The pdf of the latter distribution is given by (8.5.3) with the parameter μ replaced by ζ and λ by $\eta\zeta$.

(iii) Z_2 ~ the mixture of x_2 negative binomial random variables with the weight of the kth component being $D_k^{(x_2)}$ given by equation (8.5.18). The parameters of the kth component are $(x_2 - \{k+1\}/2)$ and

$$p = \frac{1 + \frac{2\xi}{\omega}(\lambda_2 + \lambda_3)}{1 + \frac{2\xi}{\omega}(\lambda_1 + \lambda_2 + \lambda_3)}$$

for $k = 0, 1, 2, \ldots, x_2-1$.

Regression

Taking the conditional expectations of the components of the convolution in the above the regression of X_1 on X_2 is

$$E[X_1 | x_2] = x_2[\lambda_3/(\lambda_2 + \lambda_3)] + \zeta\lambda_1 + p \sum_{k=0}^{x_2-1} D_k^{(x_2)} [x_2 - \{k+1\}/2],$$

which is not linear in the regressor x_2.

8.5.3 Estimation

In view of the complicated nature of the pf, we will examine only the method of moments in this case. There are five parameters in the distribution; namely μ, λ, λ_1, λ_2 and λ_3. Recall that the moments are

$$E[X_i] = \mu(\lambda_i + \lambda_3),$$

$$\text{Var}[X_i] = \mu(\lambda_i + \lambda_3)\,[1 + \frac{\mu^2}{\lambda}(\lambda_i + \lambda_3)] \qquad i = 1, 2$$

and

$$\text{Cov}[X_1, X_2] = \mu\,(\lambda_1 + \lambda_3)(\lambda_2 + \lambda_3)\,[\frac{\mu^2}{\lambda} + \frac{\lambda_3}{\{(\lambda_1 + \lambda_3)(\lambda_2 + \lambda_3)\}}].$$

For ease of computation it is preferable to introduce a reparameterization

$$\xi_i = \mu(\lambda_i + \lambda_3), \quad i = 1, 2, \qquad \xi_3 = \mu\lambda_3, \quad \xi_4 = \mu/\lambda.$$

Then we have four parameters to be estimated in this case. Replacing the population moments by their sample equivalents and solving for the parameters ξ_i, we have

$$\tilde{\xi}_1 = m_{1,0}, \quad \tilde{\xi}_2 = m_{0,1},$$

$$\tilde{\xi}_4 = \frac{1}{2}\left[\frac{m_{2,0} - m_{1,0}}{m_{1,0}^2} + \frac{m_{0,2} - m_{0,1}}{m_{0,1}^2}\right]$$

and

$$\tilde{\xi}_3 = m_{1,1} - \tilde{\xi}_1\tilde{\xi}_2\tilde{\xi}_4.$$

Example 8.5.1

Using the bus driver data in example 8.4.1, a bivariate inverse Gaussian Poisson was fitted using the method of moments estimates.

x_1 \ x_2	0	1	2	3	4	5	6	7	Total
0	117 (108.32)	96 (96.66)	55 (49.14)	19 (18.94)	2 (6.21)	2	0 (2.53)	0	291
1	61 (73.51)	69 (81.31)	47 (49.90)	27 (22.74)	8 (8.67)	5 (2.94)	1 (1.29)	0	218
2	34 (28.42	42 (37.92)	31 (27.64)	13 (14.73)	7 (6.47)	2 (2.50)	3 (1.29	0	132
3	7 (8.33)	15 (13.15)	16 (11.20)	7 (6.89)	3 (3.46)	1	0 (2.43)	0	49
4	3 (2.08)	3 (3.81)	1 (3.74)	1 (2.63)	2 (1.49)	1	1 (1.25)	1	13
5	2	1	0	0	0	0	0	0	3
6	0	0	0	0	1 (6.41)	0	0	0	1
7	0	0	0	1	0	0	0	0	1
Total	224	226	150	68	23	11	5	1	708

The summary statistics are given by

$$\bar{x}_1 = 1.0014, \quad \bar{x}_2 = 1.2910, \quad m_{2,0} = 1.1952, \quad m_{0,2} = .5984$$

and

$$m_{1,1} = 0.3263.$$

Using these statistics we obtain the moment estimates:

$$\hat{\xi}_1 = 1.0014 \quad \hat{\xi}_2 = 1.2910 \quad \hat{\xi}_3 = 0.0822 \quad \hat{\xi}_4 = 0.1889.$$

The calculated value of the goodness-of-fit statistic is $\overline{X}^2 = 25.19$ with d. f. = 33 - 4 - 1 = 28. Since the prob-value = 0.6175, the fit for this distribution is good.

8.5.4 Simulation

For simulating observations from (8.5.9), the marginal distribution of X_2 and the conditional distribution of X_1 given x_2 are used. The parameters were specified in terms of μ, λ, λ_1, λ_2 and λ_3 with $\xi = \mu$ and $\omega = \lambda/\mu$.

From (8.5.8) the marginal distribution of X_2 is univariate inverse Gaussian. An observation from this distribution was generated using an algorithm given by Michael *et al.* (1976). Using this realization x_2, a value of X_1 was generated using the convolution $X_1 = Y_1 + Z_1 + Z_2$. The actual algorithm for the generating the data is quite similar to that described previously. Here, however, Z_2 is a mixture of negative binomial random variables. The negative binomial random variables were generated using the compounding of Poisson and gamma random variables discussed in section 5.10.

Using

$$\lambda = 2.0, \mu = 1.0\ \lambda_1 = 0.4, \lambda_2 = 0.3, \lambda_3 = 0.1,$$

the following 1000 observations were generated from (8.5.9). Using the known values of the parameters, a goodness-of-fit test was performed to assess the performance of the simulation procedure. The calculated χ^2 was 13.135 on 20 degrees of freedom with a prob-value of 0.872, indicating that the simulation procedures are performing as expected.

x_1	x_2: 0	1	2	3	4	Total
0	515 (504.96)	112 (112.91)	15 (17.33)	1 (2.76)	1	644
1	149 (150.55)	81 (83.83)	23 (21.09)	3 (4.96)	0	256
2	31 (30.81)	23 (28.12)	8 (11.71)	5 (4.14)	0	67
3	8 (5.65)	10 (7.28)	2 (4.37)	2 (2.41)	1	23
4	1 (1.01)	2 (1.70)	3 (1.35)	0 (1.09)	0	6
5	0	0	1	1	0	2
6	0	0	0 (1.85)	1	0	1
7	0	0	1	0	0	1
Total	704	228	53	13	2	1000

9

SOME MISCELLANEOUS RESULTS

9.1 Introduction

This chapter deals with two distinct types of bivariate distributions. The first class have been recently developed to describe the joint distributions of accidents incurred by individuals in consecutive time intervals under assumptions not common to the standard distributions considered in the earlier chapters. The second class of distributions have been introduced with a view to unifying some of the bivariate discrete distribution theory. Unfortunately, the two classes share little other than that the literature is sparse and the applications limited. The common component to them is the appearance of special functions of applied mathematics. From the development of the bivariate distributions discussed in the preceding chapters, it is obvious that they share a common genesis arising sometimes as a limiting form of the standard sampling schemes with or without replacement. In other instances they are constructed by the process of compounding or generalizing. In this chapter we will introduce new models and a unifying knot to tie them together.

9.2 Bivariate Waring distribution

Irwin (1968) introduced two new concepts in the study of accidents, namely, proneness and liability. These ideas have been generalized to the bivariate situation by Xekalaki (1984b) in dealing with accidents incurred by an individual over two nonoverlapping periods of time. Under this approach the individuals in the population are assumed to be subject to two factors:

(i) Proneness to accidents which essentially characterizes the "idiosyncratic" predisposition to accidents. This is denoted by the parameter ν. The parameter will be considered to be intrinsic to an individual. So it is subject to random variation over the population. The type of distribution for ν considered here will be presented later.

(ii) Each is, in addition, exposed to an external risk of accident characterized by the parameters λ_1 and λ_2 over the periods under consideration.

The assumptions regarding the distributional properties are, generalizing those of Irwin,

(a) The joint pgf of (X, Y), given λ_1, λ_2 and ν is

$$\Pi(t_1, t_2|\lambda_1, \lambda_2, \nu) = \exp[\frac{\lambda_1}{\nu}(t_1-1) + \frac{\lambda_2}{\nu}(t_2-1)]. \qquad \lambda_1, \lambda_2, \nu > 0 \qquad (9.2.1)$$

(b) The liability parameters λ_1 and λ_2, given ν, are assumed to be independently distributed with the pdf's

$$g_1(\lambda_1|\nu) = \frac{e^{-\frac{\lambda_1}{\nu}} \lambda_1^{\tau-1}}{\Gamma(\tau)\, \nu^{\tau}} \quad \text{and} \quad g_2(\lambda_2|\nu) = \frac{e^{-\frac{\lambda_2}{\nu}} \lambda_2^{\kappa-1}}{\Gamma(\kappa)\, \nu^{\kappa}},$$

$$\lambda_1, \lambda_2, \tau, \kappa > 0 . \qquad (9.2.2)$$

Then the unconditional joint distribution of (X, Y), given ν, is seen to have the conditional pgf

$$\Pi(t_1, t_2|\nu) = [1-\nu(t_1-1)]^{-\tau}\,[1-\nu(t_2-1)]^{-\kappa}, \qquad (9.2.3)$$

the pgf of the double negative binomial distribution.

(c) The parameter characterizing the proneness of the individual is assumed to have the beta distribution

$$h(\nu) = \frac{\Gamma(\alpha+\beta)}{\Gamma(\alpha)\,\Gamma(\beta)} \frac{\nu^{\alpha-1}}{(1+\nu)^{\alpha+\beta}}, \qquad \nu > 0.$$

Then the pgf of the joint unconditional distribution of the random variables (X, Y) is seen to be

$$\Pi(t_1, t_2) = \int_0^\infty \frac{\Gamma(\alpha+\beta)}{\Gamma(\alpha)\,\Gamma(\beta)} \frac{v^{\alpha-1}}{(1+v)^{\alpha+\beta}} \frac{dv}{[1-v(t_1-1)]^{\tau}\,[1-v(t_2-1)]^{\kappa}}.$$

Changing the variable in this integral to $z = v/(1+v)$, it can be cast as

$$\int_0^1 \frac{\Gamma(\alpha+\beta)}{\Gamma(\alpha)\,\Gamma(\beta)} z^{\alpha-1}(1-z)^{\beta+\tau+\kappa-1}(1-zt_1)^{-\tau}(1-zt_2)^{-\kappa}dz.$$

From Erdelyi *et al.* (1953, Vol. 1, p. 231) this integral is seen to be related to the integral representation of the hypergeometric function of two arguments. Upon introducing the appropriate constants, we have

$$\Pi(t_1, t_2) = \frac{\beta_{(\tau+\kappa)}}{(\alpha+\beta)_{(\tau+\kappa)}} F[\alpha; \tau, \kappa; \alpha+\beta+\tau+\kappa; t_1, t_2], \tag{9.2.4}$$

where $x_{(r)} = \Gamma(x+r)/\Gamma(x)$.

The function on the right-hand side of (9.2.4) can be represented in a double infinite series as

$$F[\alpha; \tau, \kappa; \alpha+\beta+\tau+\kappa; t_1, t_2] = \sum_{r=0}^{\infty} \sum_{s=0}^{\infty} \frac{\tau_{(r)}}{r!} \frac{\kappa_{(s)}}{s!} \frac{\alpha_{(r+s)}}{\gamma_{(r+s)}} t_1^r t_2^s, \tag{9.2.5}$$

where $\gamma = \alpha+\beta+\tau+\kappa$. See Erdelyi *et al.* (1953, Vol. 1, p. 224).

The pgf in (9.2.4) is referred to as the bivariate generalized Waring distribution. In this section we shall abbreviate the distribution as BVWD. The term arises from the fact that the series is a bivariate extension of Waring's expansion. It is of interest to note the following special cases that may arise in future applications

(i) If $t_1 = t_2 = t$, then

$$F[\alpha; \tau, \kappa; \gamma; t, t] = {}_2F_1(\alpha, \tau+\kappa; \gamma; t)$$

$$= \sum_{r=0}^{\infty} \frac{\Gamma(\alpha+r)}{\Gamma(\alpha)} \frac{\Gamma(\tau+\kappa+r)}{\Gamma(\tau+\kappa)} \frac{\Gamma(\gamma)}{\Gamma(\gamma+r)} \frac{t^r}{r!},$$

(9.2.6)

which is Gauss' hypergeometric function.

(ii) If, in the above t = 1, then further simplification is afforded and

$$F[\alpha; \tau, \kappa; \gamma, 1, 1] = \frac{(\gamma-\tau-\kappa)_{(\tau+\kappa)}}{(\gamma-\tau-\kappa-\alpha)_{(\tau+\kappa)}},$$

as can be seen from the form assumed by Gauss' hypergeometric function in this case. See equation (46) in Erdelyi *et al.* (1953, Vol. 1, p. 104). This yields $\Pi(1, 1) = 1$.

9.2.1 Properties of the BVWD

Probability function

The coefficient of $t_1^r t_2^s$ is found from the expansions in (9.2.4) and (9.2.5). Thus the probability function of (X, Y) is

$$f(r, s) = \frac{\beta_{(\tau+\kappa)}}{(\alpha+\beta)_{(\tau+\kappa)}} \frac{\tau_{(r)}}{r!} \frac{\kappa_{(s)}}{s!} \frac{\alpha_{(r+s)}}{\gamma_{(r+s)}}, \quad r = 0, 1, \ldots; s = 0, 1, \ldots \tag{9.2.7}$$

where $\gamma = \alpha+\beta+\tau+\kappa$. This pf will be referred to as the bivariate Waring distribution (BVWD) with parameters $(\alpha; \tau, \kappa; \beta)$.

Marginal distributions

If (X, Y) have the BVWD$(\alpha; \tau, \kappa; \beta)$, then each of the marginal distributions of X and Y is a univariate Waring distribution (UVWD) with

$$X \sim \text{UVWD}(\alpha, \tau; \beta) \quad \text{and} \quad Y \sim \text{UVWD}(\alpha, \kappa; \beta),$$

where the pf of UVWD(a, b; c) is given by

$$f(r) = \frac{c_{(b)}}{(a+c)_{(b)}} \frac{a_{(r)}b_{(r)}}{(a+b+c)_{(r)}} \frac{1}{r!}, \quad r = 0, 1, \ldots . \tag{9.2.8}$$

The result follows from the joint pgf by setting the appropriate argument equal to 1. Thus the marginal of X has the pgf

$$\Pi_X(t) = \frac{\beta_{(\tau+\kappa)}}{(\alpha+\beta)_{(\tau+\kappa)}} \sum_{r=0}^{\infty} \sum_{s=0}^{\infty} \frac{\tau_{(r)}}{r!} \frac{\kappa_{(s)}}{s!} \frac{\alpha_{(r+s)}}{\gamma_{(r+s)}} t_1^r, \tag{9.2.9}$$

from which the pf is found as the coefficient of t_1^r to be

$$f_X(r) = \frac{\beta_{(\tau+\kappa)}}{(\alpha+\beta)_{(\tau+\kappa)}} \frac{\tau_{(r)}}{r!} \sum_{s=0}^{\infty} \frac{\kappa_{(s)}}{s!} \frac{\alpha_{(r+s)}}{\gamma_{(r+s)}}. \tag{9.2.10}$$

The series in (9.2.10) can be rearranged as

$$\sum_{s=0}^{\infty} \frac{\kappa_{(s)}}{s!} \frac{\alpha_{(r+s)}}{\gamma_{(r+s)}} = \frac{\alpha_{(r)}}{\gamma_{(r)}} \sum_{s=0}^{\infty} \frac{(\alpha+r)_{(s)}}{s!} \frac{\kappa_{(s)}}{(\gamma+r)_{(s)}}. \tag{9.2.11}$$

From equation (46) of Erdelyi *et al.* (1953, Vol. 1, p. 104) the series on the right hand side of (9.2.11) sums to

$$\frac{\Gamma(\gamma+r)\Gamma(\gamma-\alpha-\kappa)}{\Gamma(\gamma-\alpha)\Gamma(\gamma-\kappa+r)}.$$

Substituting these in the pf (9.2.10), we have

$$f_X(r) = \frac{\beta_{(\tau+\kappa)}}{(\alpha+\beta)_{(\tau+\kappa)}} \frac{\tau_{(r)}}{r!} \frac{\alpha_{(r)}}{\gamma_{(r)}} \frac{\Gamma(\gamma+r)\Gamma(\gamma-\alpha-\kappa)}{\Gamma(\gamma-\alpha)\Gamma(\gamma-\kappa+r)},$$

or

$$f_X(r) = \frac{\beta_{(\tau)}}{(\alpha+\beta)_{(\tau)}} \frac{\tau_{(r)}\alpha_{(r)}}{(\alpha+\beta+\tau)_{(r)}} \frac{1}{r!}, \qquad r = 0, 1, \dots, \tag{9.2.12}$$

which is the pf of a UVWD(α,τ; β). Similarly the marginal of Y can be seen to be UVWD(α, κ; β).

Distribution of the sum

The distribution of the sum $Z = X + Y$ of the random variables (X, Y) having the BVWD(α; τ, κ; β) is UVWD(α; $\tau+\kappa$; β). This follows from the fact that the pgf of the sum is $\Pi(t, t)$; that is,

$$\Pi_z(t) = \frac{\beta_{(\tau+\kappa)}}{(\alpha+\beta)_{(\tau+\kappa)}} \sum_{r=0}^{\infty} \sum_{s=0}^{\infty} \frac{\tau_{(r)}}{r!} \frac{\kappa_{(s)}}{s!} \frac{\alpha_{(r+s)}}{\gamma_{(r+s)}} t^{r+s} . \qquad (9.2.13)$$

Therefore, the pf is

$$h(z) = \frac{(\beta)_{(\tau+\kappa)}}{(\alpha+\beta)_{(\tau+\kappa)}} \frac{\alpha_{(z)}}{\gamma_{(z)}} \sum_{\substack{r=0 \\ r+s=z}}^{\infty} \sum_{s=0}^{\infty} \frac{\tau_{(r)}}{r!} \frac{\kappa_{(s)}}{s!} ,$$

which yields upon summing the series

$$h(z) = \frac{\beta_{(\tau+\kappa)}}{(\alpha+\beta)_{(\tau+\kappa)}} \frac{\alpha_{(z)} (\tau+\kappa)_{(z)}}{\gamma_{(z)}} \frac{1}{z!} , \qquad z = 0, 1, \dots . \qquad (9.2.14)$$

Conditional distributions

From Theorem 1.3.1 the pgf of the conditional distribution of X given Y = y is

$$\Pi_x(t|y) = \frac{\Pi^{(0,y)}(t,0)}{\Pi^{(0,y)}(1,0)} .$$

For the determination of the conditional distribution as well as the factorial moments it is useful to find the partial derivative of order (x, y) of $\Pi(t_1, t_2)$ with respect to its arguments. This can be seen to be

$$\Pi^{(x,y)}(t_1, t_2) = \frac{\beta_{(\tau+\kappa)}}{(\alpha+\beta)_{(\tau+\kappa)}} \sum_{r=x}^{\infty} \sum_{s=y}^{\infty} \frac{\tau_{(r)}}{r!} \frac{\kappa_{(s)}}{s!} \frac{\alpha_{(r+s)}}{\gamma_{(r+s)}} (r-x+1)_{(x)} (s-y+1)_{(y)} t_1^{r-x} t_2^{s-y} .$$

$$(9.2.15)$$

From (9.2.15) taking the appropriate ratio the pgf of X given Y = y is

$$\Pi_x(t|y) = \frac{\displaystyle\sum_{r=0}^{\infty} \frac{\tau_{(r)}}{r!} \frac{\kappa_{(y)}}{y!} \frac{\alpha_{(r+y)}}{\gamma_{(r+y)}} t^r}{\displaystyle\sum_{r=0}^{\infty} \frac{\tau_{(r)}}{r!} \frac{\kappa_{(y)}}{y!} \frac{\alpha_{(r+y)}}{\gamma_{(r+y)}}}$$

$$= \frac{(\gamma-\alpha-\tau)_{(\tau)}}{(\gamma-\tau+y)_{(\tau)}}\ {}_2F_1[\tau, \alpha+y; \gamma+y; t] \cdot$$

The probability function of X given Y = y is therefore

$$g(x|y) = \frac{(\gamma-\alpha-\tau)_{(\tau)}}{(\gamma-\tau+y)_{(\tau)}} \frac{(\alpha+y)_{(x)}\tau_{(x)}}{(\gamma+y)_{(x)}\ x!}, \qquad x = 0, 1, \ldots \tag{9.2.16}$$

which shows that the conditional distribution of X given Y = y is UVWD(α+y, τ; γ–α–τ).

Similarly, the conditional of Y given X = x is UVWD(α+x, κ; γ–α–κ).

The regressions of X on Y and of Y on X are

$$E[X|y] = \frac{\tau(\alpha+y)}{\gamma-\alpha-\tau-1} \quad \text{and} \quad E[Y|x] = \frac{\kappa(\alpha+x)}{\gamma-\alpha-\kappa-1}, \tag{9.2.17}$$

both of which are linear. Hence

$$\rho^2_{x,y} = \frac{\tau\kappa}{(\gamma-\alpha-\tau-1)(\gamma-\alpha-\kappa-1)} .$$

Factorial moments

By setting $t_1 = t_2 = 1$ in (9.2.15), the (r, s)th order factorial moment is seen to be,

$$\mu_{[r,s]} = \frac{\tau_{(r)}\,\kappa_{(s)}\,\alpha_{(r+s)}}{(\beta-r-s)_{(r+s)}}, \qquad r = 0, 1, \ldots ; s = 0, 1, \ldots \tag{9.2.18}$$

which shows that the condition for their existence is that $\beta > r+s$. In particular

$$E[X] = \frac{\alpha\tau}{\beta-1},\ E[Y] = \frac{\alpha\kappa}{\beta-1},$$

$$Var[X] = \frac{\alpha\tau(\alpha+\beta-1)(\tau+\beta-1)}{(\beta-1)^2(\beta-2)},\ Var[Y] = \frac{\alpha\kappa(\alpha+\beta-1)(\kappa+\beta-1)}{(\beta-1)^2(\beta-2)}$$

and

$$\text{Cov}(X, Y) = \frac{\alpha\kappa(\alpha+\beta-1)}{(\beta-1)^2(\beta-2)}.$$

Hence

$$\rho_{x,y} = \frac{(\tau\kappa)^{\frac{1}{2}}}{[(\beta+\tau-1)(\beta+\kappa-1)]^{\frac{1}{2}}} > 0 \cdot$$

9.2.2 Estimation

The method of moments using the factorial moments is suggested by Xekalaki (1984b) for estimation of the parameters. Writing n_{rs} for the observed frequency in the cell (r, s), the sample factorial moments can be determined as usual. Thus let $\bar{x}$, $\bar{y}$ represent the sample marginal moments. Let, in addition,

$$\bar{z} = \frac{1}{n}\sum_{r,s} n_{rs} r\,(r-1), \qquad \bar{w} = \frac{1}{n}\sum_{r,s} n_{rs} s\,(s-1), \qquad \bar{t} = \frac{1}{n}\sum_{r,s} n_{rs} r\, s.$$

Equating them to the population factorial moments in terms of the parameters, we have the estimating equations

$$\frac{\tilde{\alpha}\tilde{\tau}}{\tilde{\beta}-1} = \bar{x}, \qquad \frac{\tilde{\alpha}\tilde{\kappa}}{\tilde{\beta}-1} = \bar{y}, \qquad \frac{\tilde{\alpha}(\tilde{\alpha}+1)\tilde{\tau}\tilde{\kappa}}{(\tilde{\beta}-1)(\tilde{\beta}-2)} = \bar{t}$$

and

$$\frac{\tilde{\alpha}(\tilde{\alpha}+1)[\tilde{\tau}(\tilde{\tau}+1) + \tilde{\kappa}(\tilde{\kappa}+1)]}{(\tilde{\beta}-1)(\tilde{\beta}-2)} = \bar{z} + \bar{w}.$$

These equations can be solved to yield the estimates for the parameters. The covariance matrix of the estimates can be found as usual by applying the results in Chapter 1 on the moments of the sample factorial moments.

Example 9.2.1

The following data are quoted from Cresswell and Froggatt (1963). They refer to the accidents incurred by Belfast bus drivers over two nonoverlapping periods 1952-53 (X) and 1954-55 (Y).

	Y										
X	0	1	2	3	4	5	6	7	8	9	Total
0	8	9	4	3	5	0	0	0	0	0	29
1	13	15	14	7	4	1	0	0	0	0	54
2	6	16	10	8	2	3	2	0	1	0	48
3	0	8	7	3	5	1	0	0	1	0	25
4	2	4	1	3	3	1	0	0	0	0	14
5	1	1	0	1	1	1	0	1	0	0	6
6	0	1	1	0	1	1	0	0	0	1	5
7	0	0	0	0	0	0	0	0	0	0	0
8	0	1	0	0	0	0	0	0	0	0	1
9	0	0	1	0	0	0	0	0	0	0	1
Total	30	55	38	25	21	8	2	1	2	1	183

The use of the model is justified by considering the fact that, although the drivers included in the study were all driving in the same geographical region, they might have been subject to different hazards or they might react differently to the same hazards. In either case the application of the proneness-liability model seems a plausible one. In addition, the same individual may have been exposed to different driving conditions in the periods under study. These assertion are further strengthened by the fit of the data set to the distribution. The results of the computations are summarized below:

$$\bar{x} = 1.9563, \quad \bar{y} = 2.0437,$$

$$\bar{w} = 5.0492, \quad \bar{z} = 4.4372, \quad \bar{t} = 4.7159.$$

Hence the estimates of the parameters are

$$\tilde{\alpha} = 5.6338, \qquad \tilde{\beta} = 583.552, \qquad \tilde{\tau} = 202.286, \qquad \tilde{\kappa} = 211.327.$$

The χ^2 goodness-of-fit statistic is 9.4144 with 16 degrees of freedom having a prob-value of 0.894. This shows that the fit is good.

9.2.3 Models leading to the BVWD

Xekalaki (1984a) has examined several situations leading to the BVWD. While there are a variety of cases that seem to give rise to the distribution in practice, we restrict ourselves to the few that exemplify the models giving rise to the distribution.

Urn models

(i) Sampling from a trichotomous population: Let the elements be designated for definiteness as red (α), black (β) and white (γ) balls. A ball is drawn with replacement and in addition a ball of the same color is put into the urn. Let (X, Y) represent the number of the red and black balls drawn before the τ-th white ball is drawn. Then it can be shown that these random variables are jointly distributed with the probability function BVWD(τ; α, β; γ). This is readily seen to be a particular case of the Polya urn scheme.

(ii) Sampling from a dichotomous population: In this case, let the number of white balls in the urn be α while the number of black balls is β. The sampling scheme is the same as in the previous model. However, the random variables of interest are (X, Y) where

X = number of black balls drawn before the first κ white balls are drawn.

Y = number of black balls drawn before the next μ white balls are drawn *after* the first κ have been drawn.

Here the joint probability function of (X, Y) can be shown to be BVWD(β; κ, μ; α].

Mixing Models

Several schemes of compounding or mixing give rise to the BVWD. Central to these schemes is the negative binomial distribution with the univariate or bivariate beta distribution being the compounding variable. Compounding with the Poisson is also possible. In this section we will consider only the former situation.

The random variable with the pdf

$$g(x) = \frac{\Gamma(\alpha+\beta)}{\Gamma(\alpha)\Gamma(\beta)} \frac{x^{\alpha-1}}{(1+x)^{\alpha+\beta}}, \qquad x > 0 \qquad (9.2.19)$$

is referred to as the beta type II (α, β) random variable.

The bivariate analog of (9.2.19) has the pdf

$$g(x, y) = \frac{\Gamma(\alpha+\beta+\gamma)}{\Gamma(\alpha)\Gamma(\beta)\Gamma(\gamma)} \frac{x^{\alpha-1} y^{\beta-1}}{(1+x+y)^{\alpha+\beta+\gamma}}, \qquad x, y > 0. \qquad (9.2.20)$$

In this case (X, Y) is said to have the bivariate beta type II $[\alpha, \beta, \gamma]$ distribution.

Sometimes it may be necessary to consider the pair of random variables with the pdf

$$g(u, v) = \frac{\Gamma(\alpha+\beta+\gamma)}{\Gamma(\alpha)\Gamma(\beta)\Gamma(\gamma)} u^{\alpha-1} v^{\beta-1} (1-u-v)^{\gamma-1}, \qquad u > 0, v > 0, u+v < 1. \qquad (9.2.21)$$

In this case we say that the random variables (U, V) have the bivariate beta type I $[\alpha, \beta, \gamma]$ distribution.

With these definitions it is possible to introduce some of the models leading to the BVWD.

(i) *Negative binomial and bivariate beta type II* : Let (X, Y) have the bivariate negative binomial distribution with the pgf

$$\Pi(t_1, t_2|z, w) = [1 + z(1-t_1) + w(1-t_2)]^{-\alpha}, \qquad z, w > 0, \qquad (9.2.22)$$

while (Z, W) have the bivariate beta type II (κ, μ, β) distribution. The unconditional pgf of (X, Y) is

$$\Pi(t_1, t_2) = \int_0^\infty \int_0^\infty \Pi(t_1, t_2|z, w) g(z, w) \, dz \, dw, \qquad (9.2.23)$$

where the pdf g(z, w) is given in (9.2.20). The pgf in the integral (9.2.23) can be expanded as

$$(1+z+w)^{-\alpha}\sum_{x=0}^{\infty}\sum_{y=0}^{\infty}\frac{\Gamma(\alpha+x+y)}{x!y!\Gamma(\alpha)}\frac{z^x w^y}{(1+z+w)^{x+y}}t_1^x t_2^y.$$

Substituting in (9.2.23) and rearranging the terms, the double integral becomes

$$\int_0^{\infty}\int_0^{\infty}\frac{z^{\kappa+x-1}w^{\mu+y-1}}{(1+z+w)^{\kappa+\mu+\beta+\alpha+x+y}}\,dz\,dw = \frac{\Gamma(\kappa+x)\Gamma(\mu+y)\Gamma(\alpha+\beta)}{\Gamma(\alpha+\beta+\kappa+\mu+x+y)}.$$

Hence the joint pgf of (X, Y) is

$$\Pi(t_1, t_2) = \frac{\beta_{(\kappa+\mu)}}{(\alpha+\beta)_{(\kappa+\mu)}}F[\alpha; \kappa, \mu; \gamma; t_1, t_2], \tag{9.2.24}$$

from which the pf can be found to be

$$f(x, y) = \frac{\beta_{(\kappa+\mu)}}{(\alpha+\beta)_{(\kappa+\mu)}}\frac{\alpha_{(x+y)}}{\gamma_{(x+y)}}\frac{\kappa_{(x)}\mu_{(y)}}{x!\ y!}, \quad x = 0, 1, \ldots ; y = 0, 1, \ldots, \tag{9.2.25}$$

which is BVWD(α; κ, μ; β). Here $\gamma = \alpha+\kappa+\mu+\beta$.

(ii) *Negative binomial and bv beta type I* : If in the preceding we make the transformation

$$U = \frac{Z}{1+Z+W} \quad \text{and} \quad V = \frac{W}{1+Z+W},$$

then the distribution of (U, V) is that of the bivariate beta type I (κ, μ, β). Let the conditional pgf of (X, Y) be

$$\Pi(t_1, t_2|u, v) = (1-u-v)^{\alpha}[1-t_1u-t_2w]^{-\alpha}, \quad u > 0, v > 0, \ u+v < 1. \tag{9.2.26}$$

In this case it can be shown that the unconditional distribution of (X, Y) is the same as in the previous case.

(iii) *Double negative binomial and beta type II (α, β)* : Let the conditional pgf of (X, Y), given Z, be

$$\Pi(t_1, t_2|z) = [(1+z)-zt_1]^{-\alpha} [(1+z)-zt_2]^{-\alpha}, \quad z > 0, \tag{9.2.27}$$

while Z has the beta type II (α, β) distribution. Then the unconditional pgf of (X, Y)

$$\Pi(t_1, t_2) = \int_0^\infty \{(1+z)^{-(\kappa+\mu)} \sum_{x=0}^\infty \sum_{y=0}^\infty \frac{\Gamma(\kappa+x)\Gamma(\mu+y)}{\Gamma(\kappa)x!\,\Gamma(\mu)y!} [\frac{z}{1+z}]^{x+y} t_1^x t_2^y\} g(z)\, dz$$

$$= \sum_{x=0}^\infty \sum_{y=0}^\infty \frac{\Gamma(\alpha+\beta)\Gamma(\alpha+x+y)\Gamma(\beta+\kappa+\mu)\Gamma(\kappa+x)\Gamma(\mu+y)}{\Gamma(\alpha)\Gamma(\beta)\Gamma(\alpha+\beta+\kappa+\mu+x+y)\Gamma(\kappa)\Gamma(\mu)x!y!} t_1^x t_2^y, \tag{9.2.28}$$

which shows that the distribution is BVWD(α; κ, μ; β).

(iv) *Double negative binomial and beta typeI (α, β)* : As before, this distribution can be seen to reduce to that of the preceding model.

Exceedance Model

Let $X_{(r)}$ be the rth order statistic based on a sample of size n from a continuous population. Consider an independent *second* sample of size $k \le n$ from the same population. Define

U = Number of observations in the second sample $< X_{(r)}$,

V = Number of observations in the second sample that lie in the interval $(X_{(r)}, X_{(s)})$, $r < s$.

Let

p_1 = P{an observation in the second sample $\le X_{(r)}$},

p_2 = P{an observation in the second sample lies in the interval $(X_{(r)}, X_{(s)})$}.

Then, given $X_{(r)}, X_{(s)}$, (U, V) have the bivariate negative binomial distribution with the pf

$$P\{U = u, V = v|\, x_{(r)}, x_{(s)}\} = \frac{\Gamma(k+u+v)}{\Gamma(k)u!v!} p_1^u p_2^v (1-p_1-p_2)^k,$$

$$u = 0, 1, 2, \ldots; v = 0, 1, 2, \ldots. \tag{9.2.29}$$

However, p_1 and p_2 are themselves random variables with the joint pf

$$g(p_1, p_2) = \frac{\Gamma(n+1)}{\Gamma(r)\Gamma(s-r)\Gamma(n-s+1)} p_1^{r-1} p_2^{s-r-1} (1-p_1-p_2)^{n-s},$$

$$0 < p_1, p_2 < 1, p_1 + p_2 < 1.$$

[See Wilks (1962), p. 238, for details.] Hence the unconditional joint pf of (U, W) is BVWD(k; r, s; n–r–s+1).

9.2.4 Limiting distributions

Various limiting forms of the joint distribution can be obtained by considering the limiting forms of the ratios of the gamma functions appearing in the pgf (9.2.4) as the parameters tend to infinity. We recall that

$$\lim_{z\to\infty} \frac{\Gamma(z+\alpha)}{\Gamma(z+\beta)} = z^{\alpha-\beta}, \qquad \lim_{z\to\infty} \frac{\Gamma(z+\alpha)}{\Gamma(z)} = z^{\alpha}.$$

[Erdelyi *et al.* (1953, Vol. 1, p. 47, eqs. 1-5).]

(i) If in (9.2.4) we let α and β tend to infinity under the restriction $\alpha/(\alpha+\beta) = p$, then

$$\frac{\Gamma(\beta+\tau+\kappa)\Gamma(\alpha+\beta)\Gamma(\alpha+r+s)}{\Gamma(\beta)\Gamma(\alpha)\Gamma(\alpha+r+s+\beta+\tau+\kappa)} \approx p^{r+s}(1-p)^{\tau+\kappa}.$$

Substituting in (9.2.4) yields the limiting form for the pgf

$$\Pi(t_1, t_2) = \sum_{r=0}^{\infty} \sum_{s=0}^{\infty} \frac{\tau_{(r)}}{r!} \frac{\kappa_{(s)}}{s!} p^{r+s} (1-p)^{\tau+\kappa} t_1^r t_2^s$$

$$= (1-p)^{\tau}(1-pt_1)^{-\tau}(1-p)^{\kappa}(1-pt_2)^{-\kappa}. \qquad (9.2.30)$$

This is the pgf of a double negative binomial distribution.

(ii) If as before we let α and β tend to infinity and, in addition, τ and κ also are allowed to go to infinity under the restrictions

$$\frac{\alpha}{\alpha+\beta} \to 0, \quad \frac{\alpha\kappa}{\alpha+\beta} \to \kappa^* < \infty, \quad \frac{\alpha\tau}{\alpha+\beta} \to \tau^* < \infty,$$

then the limiting distribution is that of the double Poisson. This can be obtained from (9.2.30) by putting $p = \tau^*/\tau$ or κ^*/κ depending upon the index parameter and letting the corresponding index tend to infinity. We have

$$\Pi(t_1, t_2) = \exp[\tau^*(t_1-1) + \kappa^*(t_2-1)], \tag{9.2.31}$$

the pgf of the double Poisson distribution.

9.3 'Short' distributions

Cresswell and Froggatt (1963) consider a special model in the theory of accidents interpreted as the 'proneness' [Irwin (1964)] or as the 'spells' model. In this case the pgf of the number of accidents incurred by an individual is given by

$$\Pi(t) = \exp\{\lambda[e^{\theta(t-1)}-1] + \phi(t-1)\}, \tag{9.3.1}$$

which is the convolution of a Neyman type A random variable with the Poisson random variable. We shall use the abbreviation UVSD(ϕ, λ, θ) for this case. As pointed out by Papageorgiou (1986) the pf of the distribution can be written as

$$f(r) = \exp[\lambda(e^{-\theta}-1) - \phi]\, A_r(\phi, \theta, \lambda e^{-\theta}), \tag{9.3.2}$$

where

$$A_r(\phi, \theta, \lambda e^{-\theta}) = \sum_{i=0}^{r} \sum_{k=0}^{i} S(i,k)\, \frac{\phi^{r-i}\theta^i}{(r-i)!\, i!} (\lambda e^{-\theta})^k. \tag{9.3.3}$$

Bivariate extensions to this model has been presented by Papageorgiou (1986). In this section we generalize his models and put them in a more general setting. Basic to all these are the definitions

$$X = Z_1 + Z_3, \quad Y = Z_2 + Z_3. \tag{9.3.4}$$

where Z_1, Z_2 and Z_3 are distributed independently. More generally, it can be assumed that Z_1 and Z_2 are jointly distributed while Z_3 is distributed independently of them. Then the joint pgf of (X, Y) is seen to be

$$\Pi(t_1, t_2) = \Pi_{Z_1,Z_2}(t_1, t_2)\, \Pi_{Z_3}(t_1t_2). \tag{9.3.5}$$

If the random variables Z_1 and Z_2 are independent then the pgf reduces to

$$\Pi(t_1, t_2) = \Pi_{Z_1}(t_1)\Pi_{Z_2}(t_2)\Pi_{Z_3}(t_1t_2). \tag{9.3.5a}$$

By assigning to the random variables Z_1, Z_2 and Z_3 a variety of Poisson and Neyman Type A distributions, it is possible to construct a variety of bivariate 'short' distributions. Here we suggest four such bivariate 'short' distributions (BVSD).

Type I

Let the random variables $Z_i \sim NTA(\lambda_i, \theta_i)$ and independent, for i = 1, 2 while $Z_3 \sim P(\lambda)$. Then the joint pgf of (X, Y) is

$$\Pi(t_1, t_2) = \exp[\lambda_1\{e^{\theta_1(t_1-1)}-1\} + \lambda_2\{e^{\theta_2(t_2-1)}-1\} + \lambda(t_1t_2-1)]. \tag{9.3.6}$$

Type II

Let Z_i be independently distributed as $P(\lambda_i)$ for i = 1, 2 and let $Z_3 \sim NTA(\lambda, \theta)$. Then the joint distribution of the random variables (X, Y) has the pgf

$$\Pi(t_1, t_2) = \exp[\lambda_1(t_1-1) + \lambda_2(t_2-1) + \lambda\{e^{\theta(t_1t_2-1)}-1\}]. \tag{9.3.7}$$

Type III

The type I distribution can be generalized by taking (Z_1, Z_2) to be bivariate Neyman Type A random variables with the pgf

$$\exp[\lambda_1\{e^{[\theta_1(t_1-1)]}-1\} + \lambda_2\{e^{[\theta_2(t_2-1)]}-1\} + \lambda_3\{e^{[\theta_1(t_1-1)+\theta_2(t_2-1)]}-1\}].$$

This is the type II version of the bivariate Neyman type A distribution introduced by Holgate (1966) and studied extensively by Gillings (1974). On the other hand, let $Z_3 \sim P(\lambda)$. Then the joint pgf of (X, Y) is seen to be

$$\Pi(t_1, t_2) = \exp[\lambda_1\{e^{[\theta_1(t_1-1)]}-1\} + \lambda_2\{e^{[\theta_2(t_2-1)]}-1\} + \lambda_3\{e^{[\theta_1(t_1-1)+\theta_2(t_2-1)]}-1\} + \lambda(t_1t_2-1)]. \tag{9.3.8}$$

If $\lambda_3 = 0$ this simplifies to the Type I distribution.

Type IV

Finally, a generalization of the type II distribution can be obtained by taking the joint distribution of (Z_1, Z_2) to be bivariate Poisson with the parameters λ_1, λ_2 and λ_3 and the random variable $Z_3 \sim NTA(\lambda, \theta)$. From (9.3.2) the pgf of (X, Y) is

$$\Pi(t_1, t_2) = \exp[\lambda_1(t_1-1) + \lambda_2(t_2-1) + \lambda_3(t_1t_2-1) + \lambda\{e^{\theta(t_1t_2-1)}-1\}]. \tag{9.3.9}$$

While the first two types of the bivariate short distributions have been introduced in Papageorgiou (1986), the latter two forms are presented here to point out the possible other generalizations of these distributions.

9.3.1 Properties of the BVSD

Some basic properties of the distributions are discussed in this section. These include the marginal and conditonal distributions. It is shown that the types III and IV provide the generalizations of these properties; the results for the other two types of the distributions can be obtained from them.

Marginal distributions

We recall that the marginal pgf of the random variables can be obtained by setting the appropriate argument equal to one in the joint pgf. Thus for the joint distributions defined above the marginals are given by

	$\Pi_x(t)$	$\Pi_y(t)$
Type III	$e^{[\lambda_1^*\{e^{\theta_1(t-1)}-1\}+\lambda(t-1)]}$	$e^{[\lambda_2^*\{e^{\theta_2(t-1)}-1\}+\lambda(t-1)]}$
Type IV	$e^{[\lambda_1^*(t-1)+\lambda\{e^{\theta(t-1)}-1\}]}$	$e^{[\lambda_2^*(t-1)+\lambda\{e^{\theta(t-1)}-1\}]}$

where $\lambda_i^* = (\lambda_i+\lambda_3)$ for i = 1, 2. If we set the parameter $\lambda_3 = 0$, the distributions for types I and II are seen to be of the same form as above.

In each of these cases the marginals are readily seen to be univariate short distributions.

Factorial cumulants

The cumulants for the type III and type IV distributions are determined here as those for the other cases can be obtained from these by setting the parameter λ_3 equal to zero. Thus

Type III

$$H(t_1, t_2) = \lambda_1\{e^{\theta_1 t_1}-1\} + \lambda_2\{e^{\theta_2 t_2}-1\}+\lambda_3\{e^{\theta_1 t_1+\theta_2 t_2}-1\}$$
$$+\lambda[(t_1+1)(t_2+1)-1], \qquad (9.3.10)$$

which yields upon expansion

$$H(t_1, t_2) = \lambda_1 \sum_{r=0}^{\infty} \frac{\theta_1^r t_1^r}{r!} + \lambda_2 \sum_{s=0}^{\infty} \frac{\theta_2^s t_2^s}{s!} + \lambda_3 \sum_{u=0}^{\infty} \frac{(\theta_1 t_1+\theta_2 t_2)^u}{u!}$$
$$+ \lambda(t_1+t_2+t_1 t_2).$$

Collecting the appropriate coefficients of the powers of t_1 and t_2 we have

$$\kappa_{[1,0]} = \lambda_1^* \theta_1 + \lambda, \qquad \kappa_{[0,1]} = \lambda_2^* \theta_2 + \lambda, \qquad \kappa_{[1,1]} = \lambda_3 \theta_1 \theta_2 + \lambda,$$

$$\kappa_{[r,0]} = \lambda_1^* \theta_1^r \ \text{ for } r \geq 2, \qquad \kappa_{[0,s]} = \lambda_2^* \theta_2^s \ \text{ for } s \geq 2,$$

while for $r \geq 1$, $s \geq 1$ with $r+s \geq 3$

$$\kappa_{[r,s]} = \lambda_3 \theta_1^r \theta_2^s.$$

Type IV

$$H(t_1, t_2) = \lambda_1 t_1 + \lambda_2 t_2 + \lambda_3 (t_1 t_2 + t_1 + t_2) + \lambda \{ e^{\theta[t_1 t_2 + t_1 + t_2]} - 1 \}. \tag{9.3.11}$$

Upon collecting the appropriate coefficients in the expansion of this function, in powers of t_1 and t_2 we have

$$\kappa_{[1,0]} = \lambda_1^* + \lambda\theta, \qquad \kappa_{[0,1]} = \lambda_2^* + \lambda\theta, \qquad \kappa_{[1,1]} = \lambda_3 + \lambda\theta(1+\theta)$$

and for $r \geq 1$, $s \geq 1$, $r+s \geq 3$,

$$\kappa_{[r,s]} = \lambda \sum_{i=\max(r,s)}^{r+s} \frac{r!\, s!\, \theta^i}{(i-r)!\,(i-s)!\,(r+s-i)!}.$$

Setting $\lambda_3 = 0$ the corresponding expressions for these cumulants, given by Papageorgiou (1986) are obtained.

Conditional Distributions

The conditional distributions are found, as usual, with the help of the Theorem 1.3.1. We shall determine the conditional pgf's for the type III and type IV short distributions. The corresponding pgf's for type I and type II can be found from these by setting $\lambda_3 = 0$.

Type III

By a rearrangement of the terms, the pgf in this case can be written as

$$\Pi(t_1, t_2) = \exp[\overset{*}{\lambda}_2\{e^{\theta_2(t_2-1)}-1\}+\lambda(t_2-1)] \cdot$$

$$\exp[\{\lambda_1+\lambda_3 e^{\theta_2(t_2-1)}\}\{e^{\theta_1(t_1-1)}-1\}+\lambda t_2(t_1-1)]. \qquad (9.3.12)$$

Now, it is possible to write

$$\Pi_y(t|x) = \frac{\Pi^{(x,0)}(0, t)}{\Pi^{(x,0)}(0,1)} = \frac{\Pi^{(x,0)}(0, t)/x!}{\Pi^{(x,0)}(0,1)/x!}. \qquad (9.3.13)$$

We have for the numerator

$$\exp[\overset{*}{\lambda}_2\{e^{\theta_2(t-1)}-1\}+\lambda(t-1)]\exp[\{\lambda_1+\lambda_3 e^{\theta_2(t-1)}\}\{e^{-\theta_1}-1\}-\lambda t] \cdot$$

$$A_x[\lambda t; \theta_1; \{\lambda_1+\lambda_3 e^{\theta_2(t-1)}\}e^{-\theta_1}], \qquad (9.3.14)$$

from which the denominator is readily seen to be

$$\exp[\overset{*}{\lambda}_1\{e^{-\theta_1}-1\}-\lambda]A_x[\lambda; \theta_1; \overset{*}{\lambda}_1 e^{-\theta_1}]. \qquad (9.3.15)$$

Hence

$$\Pi_y(t|x) = \exp[(\lambda_2+\lambda_3 e^{-\theta_1})\{e^{\theta_2(t-1)}-1\}]\frac{A_x[\lambda t; \theta_1; \{\lambda_1+\lambda_3 e^{\theta_2(t-1)}\}e^{-\theta_1}]}{A_x[\lambda; \theta_1; \overset{*}{\lambda}_1 e^{-\theta_1}]}.$$

$$(9.3.16)$$

The pgf in (9.3.16) shows that the conditional distribution is the convolution of a Neyman type A distribution having the parameters $\lambda_2+\lambda_3 e^{-\theta_1}$ and θ_2 with the combinatorial distribution in the sense of Harper (1967). The special case of $\lambda_3 = 0$ considered by Papageorgiou can be readily obtained from this pgf. It can be seen that the regression is not linear in this case.

Type IV

As in the preceding case it is possible to rearrange the pgf of (X, Y) as

$$\Pi(t_1, t_2) = \exp[\lambda_2^*(t_2-1) + \lambda\{e^{\theta(t_2-1)}-1\}] \cdot$$
$$\exp[(\lambda_1 + \lambda_3 t_2)(t_1-1) + \lambda e^{\theta(t_2-1)}\{e^{\theta t_2(t_1-1)}-1\}]. \qquad (9.3.17)$$

The numerator of (9.3.13) can be shown to be

$$\exp[\lambda_2^*(t-1) + \lambda\{e^{\theta(t-1)}-1\}] \exp[-(\lambda_1 + \lambda_3 t) + \lambda e^{\theta(t-1)}\{e^{-\theta t}-1\}] \cdot$$
$$A_x[\lambda_1 + \lambda_3 t; \theta t; \lambda e^{-\theta}]. \qquad (9.3.18)$$

Therefore, the pgf in (9.3.13) is

$$\Pi_y(t|x) = \exp[\lambda_2(t-1)] \frac{A_x[\lambda_1+\lambda_3 t;\, \theta t;\, \lambda e^{-\theta}]}{A_x[\lambda_1+\lambda_3;\, \theta;\, \lambda e^{-\theta}]}, \qquad (9.3.19)$$

showing that the conditional distribution is the convolution of the Poisson distribution with the parameter λ_2 and the Harper type distribution. If we set $\lambda_3 = 0$ the result of Papageorgiou (1986) is obtained.

9.3.2 Estimation and fitting of the distribution

In view of the complex nature of the type III and IV distributions, we will take $\lambda_3 = 0$. This reduces the problem to the one considered by Papageorgiou; he has suggested the method of moments for estimating the parameters in this case.

Let the sample moments be represented as usual by $\bar{x}$, $\bar{y}$, $m_{2,0}$, $m_{0,2}$ and $m_{1,1}$. Then solving the moment equations yields the estimators for the parameters in the corresponding distributions as:

Type I

There are five parameters to be estimated in this case. So we use all of the above sample moments. These yield

$$\tilde{\lambda}_1 = \frac{(\bar{x} - m_{1,1})^2}{(m_{2,0} - \bar{x})}, \quad \tilde{\lambda}_2 = \frac{(\bar{y} - m_{1,1})^2}{(m_{0,2} - \bar{y})}$$

$$\tilde{\theta}_1 = \frac{m_{2,0} - \bar{x}}{(\bar{x} - m_{1,1})}, \quad \tilde{\theta}_2 = \frac{(m_{0,2} - \bar{y})}{(\bar{y} - m_{1,1})}$$

and $\tilde{\lambda} = m_{1,1}$. There are some restrictions that have to be satisfied by the statistics in order that the estimates be valid. These are $m_{2,0} > \bar{x} > m_{1,1}$ and $m_{0,2} > \bar{y} > m_{1,1}$. It should be pointed out that the parameters λ_i, θ_i, for $i = 1, 2$, have to be positive. However the parameter λ can be negative with restriction that

$$\lambda > -\min\left[\lambda_1\theta_1 e^{-\theta_1}, \lambda_2\theta_2 e^{-\theta_2}, \lambda_1\theta_1\lambda_2\theta_2 e^{-[\theta_1+\theta_2]}\right].$$

Type II

In this case there are only four parameters to be estimated. The parametric functions to be used will be $\mu'_{1,0} + \mu'_{0,1}$, $\mu_{2,0}$, $\mu_{0,2}$ and $\mu_{1,1}$. These yield

$$\tilde{\lambda}_1 = m_{2,0} - m_{1,1}, \qquad \tilde{\lambda}_2 = m_{0,2} - m_{1,1},$$

$$\tilde{\theta} = \frac{[m_{2,0} + m_{0,2} - \bar{x} - \bar{y}]}{[-(m_{2,0} + m_{0,2}) + \bar{x} + \bar{y} + 2m_{1,1}]}$$

and

$$\tilde{\lambda} = \frac{[m_{2,0} + m_{0,2} - (\bar{x} + \bar{y} + 2m_{1,1})]^2}{2[m_{2,0} + m_{0,2} - \bar{x} - \bar{y}]} .$$

The efficiency of the method of moments has been studied by Papageorgiou.

Recurrence relations

The following recurrence relations can be used for fitting the distributions when $\lambda_3 = 0$.

Type I

$$\left.\begin{aligned} (r+1)f(r+1, s) &= \lambda_1\theta_1 e^{-\theta_1}\sum_{k=0}^{r}\frac{\theta_1^k}{k!} f(r-k, s) + \lambda f(r, s-1) \\ (s+1)f(r, s+1) &= \lambda_2\theta_2 e^{-\theta_2}\sum_{k=0}^{s}\frac{\theta_2^k}{k!} f(r, s-k) + \lambda f(r-1, s) \end{aligned}\right\}. \tag{9.3.20}$$

Type II

$$\left.\begin{aligned} (r+1)f(r+1, s) &= \lambda\theta e^{-\theta}\sum_{k=0}^{\min(r,s-1)}\frac{\theta^k}{k!} f(r-k, s-k-1) + \lambda_1 f(r, s) \\ (s+1)f(r, s+1) &= \lambda\theta e^{-\theta}\sum_{k=0}^{\min(x-1,s)}\frac{\theta^k}{k!} f(r-k-1, s-k) + \lambda_2 f(r, s) \end{aligned}\right\}. \tag{9.3.21}$$

Example 9.3.1 Type I

The data quoted below are given in Table 6.18 of Cresswell and Froggatt (1963). Here X and Y refer to number of accidents sustained by bus drivers in two non-overlapping periods of time.

x \ y	0	1	2	3	4	5	6	7	Total
0	117 (124.1)	96 (87.3)	55 (44.5)	19 (18.4)	2 (6.6)	2 (2.1)	0 (0.6)	0 (0.2)	291
1	61 (63.1)	69 (84.8)	47 (51.1)	27 (23.8)	8 (9.3)	5 (3.2)	1 (1.0)	0 (0.3)	218
2	34 (25.0)	42 (38.2)	31 (30.0)	13 (15.7)	7 (6.7)	2 (2.5)	3 (0.8)	0 (0.3)	132
3	7 (8.1)	15 (13.9)	16 (12.0)	7 (7.2)	3 (3.3)	1 (1.3)	0 (0.5)	0 (0.1)	49
4	3 (2.3)	3 (4.3)	1 (4.0)	1 (2.6)	2 (1.3)	1 (0.5)	1 (0.2)	1 (0.1)	13
5	2 (0.6)	1 (1.2)	0 (1.2)	0 (0.8)	0 (0.4)	0 (0.2)	0 (0.1)	0 (0.0)	3
6	0 (0.1)	0 (0.3)	0 (0.3)	0 (0.2)	1 (0.1)	0 (0.1)	0 (0.0)	0 (0.0)	1
7	0 (0.0)	0 (0.1)	0 (0.1)	1 (0.1)	0 (0.0)	0 (0.0)	0 (0.0)	0 (0.0)	1
Total	224	226	150	68	23	11	5	1	708

$\tilde{\lambda}_1 = 2.376$, $\tilde{\lambda}_2 = 3.052$, $\tilde{\theta}_1 = 0.284$, $\tilde{\theta}_2 = 0.316$, $\tilde{\lambda} = 0.326$

χ^2 goodness-of-fit statistic on 29 degrees of freedom = 33.40 with prob-value = 0.2619.

Example 9.3.2 Type II

The following data are also taken from Cresswell and Froggatt (1963, Table 6.21). They pertain to the number of accidents incurred by bus drivers in non-overlapping periods of time in Ireland. Papageorgiou fits the Type II distribution to these data.

x \ y	0	1	2	3	4	5	6	7	Total
0	5 (4.5)	6 (6.2)	4 (4.2)	1 (1.9)	1 (0.7)	0 (0.2)	0 (0.0)	0 (0.0)	17
1	4 (4.5)	9 (8.0)	3 (6.7)	4 (3.6)	3 (1.4)	0 (0.4)	0 (0.1)	0 (0.0)	23
2	2 (2.3)	5 (4.9)	5 (5.3)	4 (3.6)	2 (1.8)	0 (0.7)	0 (0.2)	0 (0.1)	18
3	1 (0.8)	6 (1.9)	4 (2.7)	1 (2.4)	1 (1.5)	2 (0.7)	1 (0.3)	0 (0.1)	16
4	0 (0.2)	0 (0.6)	0 (1.0)	2 (1.0)	0 (0.9)	0 (0.5)	0 (0.2)	0 (0.1)	2
5	0 (0.0)	0 (0.1)	1 (0.3)	0 (0.4)	0 (0.4)	0 (0.3)	0 (0.2)	0 (0.1)	1
6	0 (0.0)	0 (0.0)	0 (0.1)	0 (0.1)	1 (0.1)	0 (0.1)	0 (0.1)	1 (0.1)	2
Total	12	26	17	12	8	2	1	1	79

$\tilde{\lambda}_1 = 1.009$, $\tilde{\lambda}_2 = 1.376$, $\tilde{\lambda} = 1.387$, $\tilde{\theta} = 0.432$

χ^2 goodness-of-fit statistic on 18 degrees of freedom = 20.57 with prob-value = 0.3016.

9.4 Power series type distributions

The univariate power series distribution is related to the expansion of the parametric function $u(\theta)$ in a power series as $\sum a(x)\theta^x$, with the summation extending over T, a subset of the set of nonnegative integers and $a(x) > 0$. Let $\{\theta > 0\} \in \Omega$, be the parameter space. We assume that $u(\theta)$ is finite and differentiable for such values of the parameter. Then Ω is the radius of convergence of the power series $u(\theta)$. The discrete random variable X with the probability function

$$f(x) = \frac{a(x)\,\theta^x}{u(\theta)}, \qquad x \in T,\ \{\theta > 0\} \in \Omega, \tag{9.4.1}$$

is said to have the univariate generalized power series distribution

(GPSD) with range T, series function $u(\theta)$, series parameter θ and the coefficient function $a(x)$. The pgf of the univariate power series distribution is readily seen to be $\Pi(t) = u(t\theta)/u(\theta)$.

The bivariate analog of the distribution is a natural extension of this definition to the 2-fold Cartesian product of the set of nonnegative integers.

Let T be a subset of the 2-fold Cartesian product of the set of nonnegative integers and $a(r, s) > 0$ over T. Also

$$u(\theta_1, \theta_2) = \sum \sum a(r, s)\theta_1^r \theta_2^s, \quad (r, s) \in T \tag{9.4.2}$$

with $\theta_1 \geq 0$, $\theta_2 \geq 0$ and $(\theta_1, \theta_2) \in \Omega$, a two dimensional space such that $u(\theta_1, \theta_2)$ is finite and differentiable for all such (θ_1, θ_2). Under these assumptions (9.4.2) is the power series expansion of u in terms of the parameters θ_1 and θ_2. It is clear that Ω is the region of convergence of the expansion. Then random variables (X, Y) with the probability function

$$f(r, s) = \frac{a(r, s)\theta_1^r \theta_2^s}{u(\theta_1, \theta_2)}, \quad (r, s) \in T, (\theta_1, \theta_2) \in \Omega, \tag{9.4.3}$$

are said to have the bivariate generalized power series distribution. Several of the classical distributions belong to this family of distributions. In addition, several bivariate discrete distributions have been suggested in the literature conforming with this definition. In this section we shall present some of the properties of the bivariate GPSD and other related distributions. The generalized power series distribution in the multivariate setting was developed by Khatri (1959) and later generalized by Patil (1965b). Reference may be made to Patil (1985b) for an illuminating introduction to the topic.

9.4.1 Bivariate GPSD

From the definition in (9.4.3) the probability generating function of the distribution is readily seen to be

$$\Pi(t_1, t_2) = \frac{u(t_1\theta_1, t_2\theta_2)}{u(\theta_1, \theta_2)}. \tag{9.4.4}$$

It is of interest to verify that some of the familiar bivariate discrete distributions belong to this class:

(i) *Trinomial*

Let the series function be of the form $u(\theta_1, \theta_2) = (1+\theta_1+\theta_2)^n$. Then the pgf of (X, Y) is

$$\Pi(t_1, t_2) = \left[\frac{1+t_1\theta_1+t_2\theta_2}{1+\theta_1+\theta_2}\right]^n,$$

which is the pgf of the trinomial distribution with the parameters n and $p_i = \frac{\theta_i}{1+\theta_1+\theta_2}$, i = 1, 2.

(ii) *Poisson*

Let the series function be of the form $u(\theta_1, \theta_2) = \exp(\theta_1+\theta_2)$. Then

$$\Pi(t_1, t_2) = \exp[(t_1-1)\theta_1+(t_2-1)\theta_2],$$

which is the pgf of the double Poisson distribution.

(iii) *Negative binomial*

If the series function is $u(\theta_1, \theta_2) = (1+\theta_1+\theta_2)^{-v}$. The resulting distribution is the bivariate negative binomial distribution.

Marginal Distributions

The marginal distributions can be obtained from (9.4.4) as

$$\Pi_x(t) = \frac{u(t\theta_1, \theta_2)}{u(\theta_1, \theta_2)} \quad \text{and} \quad \Pi_y(t) = \frac{u(\theta_1, t\theta_2)}{u(\theta_1, \theta_2)}. \tag{9.4.5}$$

These pgf's can be seen to be those of univariate GPSD's. To verify this, it is helpful to introduce the functions

$$u_1(r, \theta_2) = \sum_s a(r, s)\theta_2^s, \qquad u_2(s, \theta_1) = \sum_r a(r, s)\theta_1^r. \tag{9.4.6}$$

Now

$$u(\theta_1, \theta_2) = \sum_r \theta_1^r u_1(r, \theta_2) = v_1(\theta_1|\theta_2)$$

$$= \sum_s \theta_2^s u_2(s, \theta_1) = v_2(\theta_2|\theta_1).$$

In $v_1(\theta_1|\theta_2)$ the parameter θ_2 is treated as a constant, while in $v_2(\theta_2|\theta_1)$, the parameter θ_1 is taken to be a constant. Under this representation

$$\Pi_x(t) = \frac{v_1(t\theta_1|\theta_2)}{v_1(\theta_1|\theta_2)}, \qquad \Pi_y(t) = \frac{v_2(\theta_2 t|\theta_1)}{v_2(\theta_2|\theta_1)}, \tag{9.4.7}$$

both of which are the pgf's of univariate GPSD's.

Moments

The moments and the recurrence relations for the moments can be found from the defining forms. Repeated differentiation under the summation yields

$$\frac{\partial^{r_1+r_2} u(\theta_1,\theta_2)}{\partial\theta_1^{r_1}\partial\theta_2^{r_2}} = \sum_r \sum_s a(r, s) r^{[r_1]} s^{[r_2]} \theta_1^{r-r_1} \theta_2^{s-r_2},$$

from which we have

$$\mu_{(r_1,r_2)} = \frac{\theta_1^{r_1}\theta_2^{r_2}}{u(\theta_1,\theta_2)} \frac{\partial^{r_1+r_2} u(\theta_1,\theta_2)}{\partial\theta_1^{r_1}\partial\theta_2^{r_2}}. \tag{9.4.8}$$

Similarly, the ordinary moment around zero is seen to be

$$\mu'_{r_1,r_2} = \frac{1}{u(\theta_1,\theta_2)} \{\theta_2 \frac{\partial}{\partial\theta_2}\}^{r_2} \{\theta_1 \frac{\partial}{\partial\theta_1}\}^{r_1} u(\theta_1,\theta_2). \tag{9.4.9}$$

Upon differentiating with respect to θ_1

$$\frac{\partial}{\partial\theta_1}\mu'_{r_1,r_2} = \frac{1}{u(\theta_1,\theta_2)} \sum_r \sum_s a(r, s) r^{r_1+1} s^{r_2} \theta_1^{r-1} \theta_2^s$$

$$- \frac{1}{[u(\theta_1,\theta_2)]^2} \sum_r \sum_s a(r, s) r^{r_1} s^{r_2} \theta_1^r \theta_2^s. \qquad (9.4.10)$$

Recalling that

$$E[X] = \frac{\theta_1}{u(\theta_1,\theta_2)} \{\frac{\partial}{\partial\theta_1}\} u(\theta_1,\theta_2)$$

and rearranging the terms (9.4.10), we have

$$\mu'_{r_1+1,r_2} = \theta_1 \frac{\partial}{\partial\theta_1}\mu'_{r_1,r_2} + \mu'_{r_1,r_2} E[X]. \qquad (9.4.11)$$

Similarly,

$$\mu'_{r_1,r_2+1} = \theta_2 \frac{\partial}{\partial\theta_2}\mu'_{r_1,r_2} + \mu'_{r_1,r_2} E[Y]. \qquad (9.4.12)$$

Maximum Likelihood Estimation

Let (x_i, y_i), i = 1, 2, ... , n be a random sample of size n from the population. Then the likelihood equation for θ_1 is

$$\frac{\partial}{\partial\theta_1} \log L = \sum_{i=1}^{n} \frac{\partial}{\partial\theta_1} \log f(x_i, y_i) \qquad (9.4.13)$$

which becomes, upon substituting for the probability function,

$$\frac{\partial}{\partial\theta_1} \log L = \sum_{i=1}^{n} \frac{x_i - E[X]}{\theta_1};$$

or, $\bar{X}$ = E[X]. Similarly the equation with respect to the parameter θ_2 is readily seen to be $\bar{Y}$ = E[Y]. Thus the likelihood equations are precisely the same as the moment equations.

The asymptotic variance matrix of the estimators is obtained as usual by inverting the information matrix. The elements of this matrix reduce to

$$\left.\begin{aligned} E\left[\left\{\frac{\partial}{\partial\theta_1}\log f(r, s)\right\}^2\right] &= \frac{\text{Var }[X]}{\theta_1^2} \\ E\left[\left\{\frac{\partial}{\partial\theta_2}\log f(r, s)\right\}^2\right] &= \frac{\text{Var }[Y]}{\theta_2^2} \\ E\left[\left\{\frac{\partial}{\partial\theta_1}\log f(r, s)\right\}\left\{\frac{\partial}{\partial\theta_2}\log f(r, s)\right\}\right] &= \frac{\text{Cov }[X, Y]}{\theta_1\theta_2} \end{aligned}\right\}. \quad (9.4.14)$$

Independence of X and Y

The following theorems set up a necessary and sufficient condition for the random variables X and Y to be independent.

Theorem 9.4.1

Let X and Y have the bivariate GPSD with $u(\theta_1, \theta_2)$ as the series function. Also let X and Y each be marginally univariate GPSD with series functions $u_1(\theta_1)$ and $u_2(\theta_2)$, respectively. Then a necessary and sufficient condition for X and Y to be independent is that

$$u(\theta_1, \theta_2) = u_1(\theta_1)\, u_2(\theta_2).$$

Proof:

If X and Y are independent, then $f(x, y) = h(x)g(y)$; that is, for all $(x, y) \in T$,

$$\frac{a(x, y)\theta_1^x\theta_2^y}{u(\theta_1, \theta_2)} = \frac{a_1(x)\theta_1^x a_2(y)\theta_2^y}{u_1(\theta_1)u_2(\theta_2)}. \quad (9.4.15)$$

From (9.4.15), since a, a_1 and a_2 are independent of θ_1 and θ_2, it follows that

$$u(\theta_1, \theta_2) = u_1(\theta_1)u_2(\theta_2). \quad (9.4.16)$$

On the other hand, if this condition holds then

$$f(x, y) = \frac{a(x, y)\theta_1^x\theta_2^y}{u_1(\theta_1)u_2(\theta_2)}, \qquad (x, y) \in T, \tag{9.4.17}$$

and the marginal pf's are

$$h(x) = \frac{a_1(x)\theta_1^x}{u_1(\theta_1)}, \qquad g(y) = \frac{a_2(y)\theta_2^y}{u_2(\theta_2)}. \tag{9.4.18}$$

From (9.4.16)

$$\sum \sum a(x, y)\theta_1^x\theta_2^y = \sum a_1(x)\theta_1^x \sum a_2(y)\theta_2^y. \tag{9.4.19}$$

Equating the coefficients of $\theta_1^x\theta_2^y$ in (9.4.19) we have $a(x, y) = a_1(x)a_2(y)$ for all $(x, y) \in T$. Hence $f(x, y) = h(x)\, g(y)$, for all $(x, y) \in T$. □

Theorem 9.4.2

Let (X, Y) have the bivariate GPSD with the series function

$$u(\theta_1, \theta_2) = \sum_x \sum_y a(x, y)\theta_1^x\theta_2^y.$$

A necessary and sufficient condition for X and Y to be independent is that, for all $(x, y) \in T$, $a(x, y) = c_1(x)c_2(y)$ where c_1 and c_2 involve only the argument variables.

Proof:

Now

$$h(x) = \frac{\sum_y a(x, y)\theta_1^x\theta_2^y}{u(\theta_1, \theta_2)}, \qquad g(y) = \frac{\sum_x a(x, y)\theta_1^x\theta_2^y}{u(\theta_1, \theta_2)}.$$

Independence of X and Y implies that for all $(x, y) \in T$,

$$\frac{a(x, y)\theta_1^x\theta_2^y}{u(\theta_1, \theta_2)} = \frac{\sum_{y^*} a(x, y^*)\theta_1^x\theta_2^{y^*}}{u(\theta_1, \theta_2)} \frac{\sum_{x^*} a(x^*, y)\theta_1^{x^*}\theta_2^y}{u(\theta_1, \theta_2)}$$

or

$$a(x, y)u(\theta_1, \theta_2) = \sum_x a(x, y)\theta_1^x \sum_y a(x, y)\theta_2^y. \qquad (9.4.20)$$

Expanding the u function on the left hand side of (9.4.20) and equating the coefficients of $\theta_1^{x^*}$ and $\theta_2^{y^*}$ on the two sides of (9.4.20), we have

$$a(x, y)\, a(x^*, y^*) = a(x^*, y)\, a(x, y^*).$$

This yields

$$a(x, y) = \frac{a(x^*, y)}{\sqrt{a(x^*, y^*)}} \frac{a(x, y^*)}{\sqrt{a(x^*, y^*)}} = c_1(x)c_2(y)\,, \qquad (x, y) \in T$$

as was to be established.

Conversely, if a(x, y) can be decomposed as above, then

$$u(\theta_1, \theta_2) = \sum_x c_1(x)\theta_1^x \sum_y c_2(y)\theta_2^y = u_1(\theta_1)u_2(\theta_2).$$

But from the preceding theorem this is a necessary and sufficient condition for independence in this case. Hence the result. □

9.4.2 Mixtures of GPSD: Bates-Neyman model

Bates and Neyman (1952a) have studied the problem of mixtures or, compounding with independent Poisson distribution in the context of accident theory. Patil (1965a) has extended these results to the GPSD.

Let the bivariate GPSD be defined with the parameter space Ω being the region of convergence of the series function under the restriction $\theta_i \geq 0$, i = 1, 2. In the special case when the parameters are of the form $(\theta_1, \theta_2) = \lambda(a_1, a_2)$ with $\lambda \geq 0$ we say that Ω is radial with respect to the fixed vector (a_1, a_2). Thus the bivariate GPSD $(a_1\lambda, a_2\lambda)$ is a bivariate GPSD having the radial parameter space with respect to the vector (a_1, a_2). As usual, the distribution of (X, Y) is said to be a λ-mixture of a bivariate GPSD$(a_1\lambda, a_2\lambda)$ if the conditional distribution of (X, Y) given λ is of this form and λ itself is a random variable with the distribution function $G(\lambda)$. The latter is referred to as the mixing distribution. The unconditional distribution of (X, Y) is

$$f(x|\,z) = \int g(x, y|\lambda)dG(\lambda)$$

$$= a(x, y)\, a_1^x a_2^y \int \frac{\lambda^z\, dG(\lambda)}{u(a_1\lambda, a_2\lambda)}, \qquad (9.4.21)$$

where $z = x + y$.

In theorem 9.4.3 the conditional distribution of X given the sum $X + Y = z$ is developed.

Theorem 9.4.3

If the distribution of (X, Y) is a λ-mixture of a bivariate GPSD$(a_1\lambda, a_2\lambda)$, then the conditional distribution of X given $X + Y = z$ is given by the pf

$$f(x, z) = \frac{a(x, z\text{-}x)\, a_1^x\, a_2^{z-x}}{\sum_x a(x, z\text{-}x)\, a_1^x\, a_2^{z-x}}. \qquad (9.4.22)$$

Proof:

The marginal distribution of (X, Y) is given by (9.4.21). Using (9.4.21) the distribution of Z is given by

$$g(z) = \sum_x f(x, z\text{-}x)$$

$$= \sum_x a(x, y)\, a_1^x\, a_2^{z-x} \int \frac{\lambda^z\, dG(\lambda)}{u(a_1\lambda, a_2\lambda)}, \qquad (9.4.23)$$

which is independent of λ.

Hence,

$$h(x|z) = \frac{f(x, z\text{-}x)}{g(z)}$$

$$= \frac{a(x, z-x)\, a_1^x\, a_2^{z-x}}{\sum_x a(x, z-x)\, a_1^x\, a_2^{z-x}}, \tag{9.4.24}$$

as is given in (9.4.22). □

The following theorem relating the double Poisson distribution to the mixing in the bivariate GPSD is of interest.

Theorem 9.4.4

The mixed bivariate GPSD($a_1\lambda$, $a_2\lambda$) with support the nonnegative integers in R^2, having unit parameter space and independent components is double Poisson distribution if the λ-mixture is such that the conditional distribution of X given X+Y = z is binomial with parameters z and p which does not depend upon z.

Proof:

The pf of (X, Y) given λ is

$$f(x, y|\lambda) = \frac{a(x, y)\, (\lambda a_1)^x (\lambda a_2)^y}{\sum_{x,y} a(x, y)\, (\lambda a_1)^x (\lambda a_2)^y} \tag{9.4.25}$$

with the summation extending over the set of nonnegative integers in R^2. Since the random variables X and Y are independent, $a(x, y) = c_1(x)c_2(y)$. Let us assume that $c_1(0) = c_2(0) = 1$. Hence, from Theorem 9.4.3, the conditional pf of X given X+Y = z is

$$h(x|z) = \frac{c_1(x)c_2(z-x)a_1^x a_2^{z-x}}{\sum_x c_1(x)c_2(z-x)a_1^x a_2^{z-x}}. \tag{9.4.26}$$

By assumption of the theorem

$$h(x|z) = \binom{z}{x} p^x (1-p)^{z-x}. \tag{9.4.27}$$

Equating (9.4.26) and (9.4.27) we have

$$h(x|z) = \frac{c_1(x)c_2(z-x)a_1^x a_2^{z-x}}{\sum_x c_1(x)c_2(z-x)a_1^x a_2^{z-x}} = \frac{z!}{x!\,(z-x)!}\,p^x(1-p)^{z-x}. \qquad (9.4.28)$$

Writing D(z) for the denominator on the left and rearranging the terms in (9.4.28), we have

$$z!D(z) = b_1(x)\,b_2(z-x), \qquad (9.4.29)$$

where

$$b_1(x) = \frac{c_1(x)\,a_1^x\,x!}{p^x} \quad \text{and} \quad b_2(z-x) = \frac{c_2(z-x)\,a_2^{z-x}\,(z-x)!}{(1-p)^{z-x}}. \qquad (9.4.30)$$

Now, since $c_i(0) = 1$, from (9.4.30) we have $b_i(0) = 1$. To determine $b_1(x)$ and $b_2(x)$ for general x, in (9.4.29) we set

$z = 1$	$x = 0$	$b_2(1) = D(1) = d$
	$x = 1$	$b_1(1) = D(1) = d$
$z = 2$	$x = 0$	$b_2(2) = 2!\,D(2)$
	$x = 1$	$b_1(1)b_2(1) = 2!\,D(2)$
	$x = 2$	$b_1(2) = 2!\,D(2)$;

hence, $2!\,D(2) = d^2$. Therefore,

$$b_1(2) = b_2(2) = d^2.$$

Finally,

$$b_1(x) = d^x \quad \text{and} \quad b_2(z-x) = d^{z-x}.$$

Hence, from (9.4.30)

$$c_1(x) = \frac{1}{x!}\left\{\frac{dp}{a_1}\right\}^x \quad \text{and} \quad c_2(z-x) = \frac{1}{(z-x)!}\left\{\frac{d(1-p)}{a_2}\right\}^{z-x}.$$

Substituting in (9.4.25) and using the fact that the series in the denominator sums to $\exp\{dp\lambda + d(1-p)\lambda\}$, we have for the conditional pf of (X, Y)

$$f(x, y|\lambda) = \frac{\exp[-\theta_1\lambda]\,(\theta_1\lambda)^x}{x!}\;\frac{\exp[-\theta_2\lambda]\,(\theta_2\lambda)^y}{y!},$$

$$x = 0, 1, \ldots ; y = 0, 1, \ldots,$$

where $\theta_1 = dp$, $\theta_2 = d(1\text{-}p)$. This proves the result. □

A characteristic property of the Poisson distribution follows from Theorems 9.4.2 and 9.4.4.

Theorem 9.4.5

Conditional on λ, let (X, Y) be bivariate GPSD($a_1\lambda$, $a_2\lambda$) with independent components. Then, for each λ, the distribution of X and Y is Poisson if and only if the distribution of X given X+Y = z is binomial with parameters z and p, which does not depend upon z.

Proof:

Follows directly from Theorems 9.4.2 and 9.4.4. □

BIBILIOGRAPHY

Abdul - Razak, R. S. (1983). Power series distributions in mathematical statistics and applied probability. Ph.D. Thesis, Department of Statistics, Pennsylvania State University.

Aczel, J. (1975). Some recent applications of functional equations and inequalities to characterizations of probability distributions, combinatorics, information theory and mathematical economics. *Statistical Distributions in Scientific Work*, 3, 321-337.

functional equations; functional inequalities; combinatorics; information theory; mathematical economics; destructive process.

Ahmad, M. (1976). Estimation of Poisson probability function of dependent variables with applications. *Proceedings in Computational Statistics*, 2, 11-17.

hypergeometric function; Laguerre polynomials; Poisson probability function; truncation.

Ahmad, M. (1981). A bivariate hyper-Poisson distribution. *Statistical Distributions in Scientific Work*, 4, 225-230.

bivariate hyper-Poisson distribution; confluent hypergeometric function; method of moments.

Aitken, A. C. (1936). A further note on multivariate selection. *Proceeding of the Edinburgh Mathematical Society*, 5, 37-40.

bivariate Poisson as a limit of bivariate binomial.

Aitken, A. C. and Gonin, H. T. (1935). On fourfold sampling with and without replacement. *Proceedings of the Royal Society of Edinburgh*, 55, 114-125

fourfold sampling scheme; Type I bivariate binomial; Krawtchouk polynomials.

Aki, S. (1985). Discrete distributions of order k on a binary sequence. *Annals of the Institute of Statistical Mathematics*, 37, 205-224

geometric distribution; negative binomial distribution; Poisson distribution; binomial distribution; logarithmic series distribution; reliability theory.

Alzaid, A. A., Rao, C. R., and Shanbhag, D. N. (1986). Characterization of discrete probability distributions by partial independence. *Communications in Statistics - Theory and Methods*, 15, 643-656.

characterization by conditional distribution; generalized Polya-Eggenberger distribution; Moran's theorem, Poisson distribution; partial independence.

Arbous, A. G. and Kerrich, J. E. (1951). Accident statistics and the concept of accident proneness. *Biometrics*, 7, 340-432.

contingent probabilities; stochastic processes; compound Poisson distribution; contagious distributions; bivariate compound Poisson distribution; bivariate contagious distribution.

Arbous, A. G. and Sichel, H. S. (1954). New techniques for the analysis of absenteeism data. *Biometrika*, 41, 79-90.

univariate negative binomial; symmetrical bivariate negative binomial; model for absence-proneness.

Arnold, B. C. (1967). A note on multivariate distributions with specified marginals. *Journal of the American Statistical Association*, 62, 1460-1461.

multivariate distributions; marginals; multivariate geometric distribution; uniform distribution; normal distribution; positive correlations; second moments.

Arnold, B. C. (1975a). A characterization of the exponential distribution by multivariate geometric compounding. *Sankhya, A*, 37, 164-173.

multivariate geometric distribution; compound multivariate geometric distributions; characterization of the exponential distribution; Poisson processes.

Arnold, B. C. (1975b). Multivariate exponential distributions based on hierarchical successive damage. *Journal of Applied Probability*, 12, 142-147.

multivariate; exponential; geometric; shock model.

Balasubramanian, K. (1982). A bivariate distribution with certain properties. *Gujarat Statistical Review*, 9, 81-85.

Barrett, J. F. and Lampard, D. G. (1955). An expansion for some second order probability distributions and its application to noise problems. *IRE Trans. Information Theory*, IT-1, 10-15.

second-order probability distributions; orthogonal polynomials; time series; stationary Markov processes.

Bates, G. E. and Neyman, J. (1952a). Contributions to the theory of accident proneness, I: An optimistic model of correlation between light and severe accidents. *University of California Publications in Statistics*, 1, 215-254.

accident proneness; risks; accident; Poisson frequency distribution; bivariate negative binomial distribution.

Bates, G. E. and Neyman, J. (1952b). Contributions to the theory of accident proneness, II: True or false Contagion. *University of California Publications in Statistics*, 1, 256-275.

Berge, O. P. (1937). A note on a form of Tshebychev's theorem for two variables. *Biometrika*, 29, 405-406.

frequency function; moments of the first and second order; discontinuous frequency function; normal bivariate surface.

Beutler, F. J. (1983). A note on multivariate Poisson flows on stochastic processes. *Advances in Applied Probability*, 15, 219-220.

Markov step process; multivariate Poisson.

Bhattacharya, S. K. (1967). A result on accident proneness. *Biometrika*, 54, 324-325.

bivariate correlated Poisson; non-degenerate distribution.

Bhattacharya, S. K. and Holla, M. S. (1965). On a discrete distribution with special interest to the theory of accident proneness. *Journal of the American Statistical Association*, 60, 1060-1066.

Bayesian approach.

Bildikar, S. and Patil, G. P. (1968). Multivariate exponential-type distributions. *Annals of Mathematical Statistics*, 39, 1316-1326.

s-variate exponential-type distribution; multivariate exponential-type distributions; multivariate normal; multinomial; multivariate negative binomial; multivariate logarithmic series; statistical independence; characterization; canonical representation; characteristic function; moment generating function; cumulant generating function.

Birnbaum, Z. W., Raymond, J. and Zauckerman, H. S. (1947). A generalization of Tshebychev's inequality to two dimensions. *Annals of Mathematical Statistics*, 18, 70-79.

Tshebychev's inequality.

Block, H. W. (1977a). Multivariate reliability classes. *Applications of Statistics*, (P. R. Krishnaiah, ed.), North Holland, Amsterdam, 79-88.

multivariate exponential distributions and extensions; multivariate increasing failure rate; multivariate reliability classes and estimation procedures.

Block, H. W. (1977b). A family of bivariate life distributions. *Theory and Applications of Reliability : With Emphasis on Bayesian and*

Nonparametric Methods, (C. P. Tsokos and I. N. Shimi, eds.), Academic Press, 349-372.

characteristic function equation; equivalent compounding scheme; compartmental system; shock model; bivariate exponential distributions; bivariate geometric distributions.

Block, H. W., Savits, T. H. and Shaked, M. (1982). Some concepts of negative dependence. *Annals of Probability*, 10, 765-772.

multinomial, Dirichlet, multivariate hypergeometric; total positivity; negative and positive dependence; inequalities.

Block, H. W. and Paulson, A. S. (1984). A note on infinite divisibility of some bivariate exponential and geometric distributions arising from a compounding process. *Sankhya, A*, 46, 102-109.

infinite divisibility; bivariate exponential distribution; bivariate geometric distribution; Marshall-Olkin distribution.

Blum, M. L. and Mintz, A. (1951). Correlation versus curve fitting in research on accident proneness: Reply to Maritz. *Psychological Bulletin*, 48, 413-418.

accident proneness.

Boyles, R. A. and Samaniego, F. (1983). Maximum likelihood estimation for a discrete multivariate shock model. *Journal of the American Statistical Association*, 78, 445-448.

maximum likelihood estimation; shock model; multivariate Bernoulli; positive dependence.

Brown, L. D. and Farrell, R. H. (1985). Complete class theorems for estimation of multivariate Poisson means and related problems. *The Annals of Statistics*, 13, 706-726.

estimation; multivariate Poisson parameter; decision theory.

Cacoullos, T. and Papageorgiou, H. (1980). On some bivariate probability models applicable to traffic accidents and fatalities. *International Statistical Review*, 48, 345-356.

Poisson-binomial model; Bell polynomials; composite function; derivation of the conditional probability generating function; Poisson-Poisson model; bivariate probability generating function; conditional probability generating function; Poisson-Bernoulli model; bivariate Poisson distribution.

Cacoullos, T. and Papageorgiou, H. (1981). On bivariate discrete distributions generated by compounding. *Statistical Distributions in Scientific Work*, 4, 197-212.

bivariate distributions; compounding; conditioning; Bell polynomials; estimation.

Cacoullos, T. and Papageorgiou, H. (1982). Bivariate negative binomial-Poisson and negative binomial-Bernoulli models with an application to accident data. *Statistics and Probability: Essays in Honor of C. R. Rao*, (G. Kallianpur *et al.*, eds.), North Holland, Amsterdam, 155-168.

Poisson-Poisson, Poisson-binomial, negative binomial-Poisson, negative binomial-Bernoulli models; Bell polynomials; moments and cumulants; conditionality; regression; double-zero proportion method.

Cacoullos, T. and Papageorgiou, H. (1983). Characterizations of discrete distributions by a conditional distribution and a regression function. *Annals of the Institute of Statistical Mathematics*, 35, 95-104.

regression function; conditional distribution; binomial; Poisson; Pascal; right translation; characterizations; goodness-of-fit.

Campbell, J. T. (1934). The Poisson correlation function. *Proceedings of the Edinburgh Mathematical Society*, Series 2, 4, 18-26.

factorial moments; factorial moment generating functions; Poisson frequency function; Charlier's Type B function; double Poisson correlation.

Charalambides, Ch. A. (1981a). Bipartitional polynomials and their applications in combinatorics and statistics. *Discrete Mathematics*, 34, 81-84.

bipartitional polynomials; multivariate polynomials; bipartite number; combinatorics and statistics.

Charalambides, Ch. A. (1981b). Bivariate generalized discrete distributions and bipartitional polynomials. *Statistical Distributions in Scientific Work*, 4, 213-223.

bivariate generalized discrete distributions; bivariate Poisson; logarithmic series distribution; general binomial distribution; bipartitional polynomials; Bell partition polynomials.

Charalambides, Ch. A. (1984). Minimum variance unbiased estimation for zero class truncated bivariate Poisson and logarithmic series distribution. *Metrika*, 31, 115-123.

minimum variance unbiased estimation; truncated bivariate Poisson distribution; logarithmic series distribution; complete, sufficient statistics.

Charalambides, Ch. A. and Papageorgiou, H. (1981a). Bivariate Poisson binomial distributions. *Biometrical Journal*, 23, 437-450.

bivariate Poisson binomial distribution; bipartitional polynomials; method of moments; method of zero frequencies; ecological data.

Charalambides, Ch. A. and Papageorgiou, H. (1981b). On bivariate generalized binomial and negative binomial distributions. *Metrika*, 28, 83-92.

generalized general binomial (bivariate and univariate) distributions; generalized general negative binomial (bivariate and univariate) distributions; moments; conditional distributions; regression functions; bipartitional polynomials; recurrence relations; binomial-bivariate Poisson; bivariate binomial-Poisson.

Chatfield, C. (1975). A marketing application of a characterization theorem. *Statistical Distributions in Scientific Work*, 2, 175-185.

negative binomial distribution; beta distribution; Dirichlet distribution; characterization; purchasing model.

Chatfield, C., Ehrenberg, A. S. C., and Goodhardt, G. J. (1966). Progress on a simplified model of stationary purchasing behaviour. *Journal of the American Statistical Association*, 129, 317-367.

applications.

Churchill, K. E. and Jain, G. C. (1976). Further bivariate distributions associated with Lagrange expansion. *Biometrische Zeitschrift*, 18, 639-649.

Lagrange expansion; branching process; k-variate negative binomial distribution; multivariate Lagrange expansion; k-variate Borel-Tanner distribution; generalized bivariate negative binomial distribution; Poincare's bivariate generalization; Lagrange's power series expansion; Hermite type distributions; generalized bivariate logarithmic series distribution.

Consael, R. (1952). Sur les processus composes de Poisson a deux variables aleatoires. *Academie Royale de Belgigue, Classe des Sciences, Memoires*, 27, 4-43.

bivariate Poisson distribution.

Consul, P. C. (1975). Some new characterizations of discrete Lagrangian distributions. *Statistical Distributions in Scientific Work*, 3, 279-290.

Lagrangian Poisson distribution; quasi-binomial distribution; damage model; characterizations.

Consul, P. C. (1983a). Lagrange and related probability distributions. *Encyclopedia of Statistical Sciences*, 4, 448-454.

bivariate Lagrangian distribution, bivariate Borel-Tanner, bivariate Lagrangian Poisson, bivariate Lagrangian binomial.

Consul, P. C. (1983b). Lagrange expansions. *Encyclopedia of Statistical Sciences*, 4, 454-456.

Consul, P. C. and Shenton, L. R. (1973). On the multivariate generalization of the family of discrete Lagrange distributions. *Multivariate Statistical Inference,* (Kabe, D. G. and Gupta, R. P. eds.), North Holland, Amsterdam, 13-23.

Cressie, N., Davis, A. S., Folks, J. L. and Policello II, G. E. (1981). The moment generating function and negative integer moments. *The American Statistician*, 35, 148-150.

bias of ratio estimators; Laplace transforms; Mellin transform.

Crockett, N. G. (1979). A quick test of fit of a bivariate distribution. *Interactive Statistics*, (D. R. McNeil, ed.), North Holland, Amsterdam, 185-191.

bivariate correlated Poisson; chi-square test of fitness.

Dahiya, R. C. (1977). Estimation in a truncated bivariate Poisson distribution. *Communications in Statistics - Theory and Methods*, 6, 113-120.

maximum likelihood estimation; inefficiency of method of moments estimations.

Dahiya, R. C. (1979). Some bivariate models for traffic accidents. *Bulletin of the Greek Mathematical Society*, 20, 16-22.

Dahiya, R. C. (1986). Integer-parameter estimation in discrete distributions. *Communications in Statistics - Theory and Methods*, 15, 709-726.

integer-parameter estimation; maximum likelihood estimators; Poisson; binomial; bivariate model.

Dahiya, R. C. and Korwar, R. M. (1977). On characterizing some bivariate discrete distributions by linear regression. *Sankhya, A*, 39, 124-129.

trinomial and bivariate negative binomial distributions; linear regression; bivariate discrete random variables; double Poisson; bivariate discrete distributions.

Dwass, M. and Teicher, H. (1956). On infinitely divisible random vectors. *Annals of Mathematical Statistics*, 27, 461-470.

Doss, D. C. (1979). Definition and characterization of multivariate negative binomial distribution. *Journal of Multivariate Analysis*, 9, 460-464.

multivariate negative binomial; linear exponential; characterization.

Doss, D. C. and Graham, R. C. (1975a). A characterization of multivariate binomial distribution by univariate marginals. *Calcutta Statistical Association Bulletin*, 24, 93-100.

multivariate binomial distribution; univariate marginals; bivariate binomial distribution; multivariate linear exponential distribution.

Doss, D. C. and Graham, R. C. (1975b). Construction of multivariate linear exponential distributions from univariate marginals. *Sankhya, A*, 37, 257-268.

multivariate linear exponential distributions; univariate marginal distributions; moment generating function.

Downton, F. (1970). Bivariate exponential distributions in reliability theory. *Journal of the Royal Statistical Society, B*, 32, 408-417.

bivariate exponential distribution; reliability theory; "successive damage" model; "fatal shock" model; order statistics.

Eagleson, G. K. (1964). Polynomial expansions of bivariate distributions. *Annals of Mathematical Statistics*, 35, 1208-1215.

bivariate distributions; canonical variables; orthonormal polynomials; bivariate normal; bivariate gamma; Poisson; negative binomial; binomial; hypergeometric; identity in Hermite-Chebyshev polynomials; completeness; regressions; correlation coefficient; goodness-of-fit test.

Eagleson, G. K. (1969). A characterization theorem for positive definite sequences on the Krawtchouk Polynomials. *Australian Journal of Statistics*, 11, 29-38.

positive definite sequences; ultraspherical polynomials; Krawtchouk polynomials; canonical correlations; characterization theorem; symmetric, discrete bivariate distributions; dual Krawtchouk polynomials.

Edwards, C. B. and Gurland, J. (1961). A class of distributions applicable to accidents. *Journal of the American Statistical Association*, 56, 503-517.

accident proneness; correlated bivariate Poisson; compound correlated bivariate Poisson; bivariate negative binomial.

Fisher, R. A., Corbet, A. S. and Williams, C. B. (1943). The relations between the number of species and the number of individuals. *Journal of Animal Ecology*, 12, 42-58.

univariate logarithmic series distribution.

Gerber, H. U. (1980). A characterization of certain families of distributions via Esscher transforms and independence. *Journal of the American Statistical Association*, 75, 1015-1018.

exponential family; characterization of normal; Poisson.

Gerstenkorn, T. (1976). The multidimensional truncated Polya distribution. *Commentationes Mathematicae Universitatis Carolinae*,

Mathematical Institute of Charles University, Sokolovska 83, Praha, 19, 189-210.

multivariate power series distribution; equivalent families; sum of families of random vectors; theorem on addition; generating function; determining function; moments.

Gerstenkorn, T. (1981). On multivariate power series distributions. *Rev. Roum. Math. Pures et Appl.*, 26, 247-266.

multivariate power series distribution.

Ghosh, J. K., Kumar Sinha, B. and Kumar Sinha, B. (1977). Multivariate power series distributions and Neyman's properties for multinomials. *Journal of Multivariate analysis*, 7, 397-408.

positive and negative multinomial distributions; power series distributions; dispersion matrix; linear regression.

Gillings, D. B. (1974). Some further results for bivariate generalizations of the Neyman Type A distribution. *Biometrics*, 30, 619-628.

bivariate generalizations of the Neyman type A distribution; method of moments; maximum likelihood; minimum chi-square; recurrence relations; asymptotic variance matrix.

Gonin, H. T. (1966). Poisson and binomial frequency surfaces. *Biometrica*, 53, 617-619.

integrated tetrachoric series; normal bivariate surface for Poisson and binomial surfaces.

Good, I. J. (1953). The population frequencies of species and the estimation of population parameters. *Biometrika*, 40, 237-264.

Goodhardt, G. J., Ehrenberg, A. S. C. and Chatfield, C. (1984). The Dirichlet: a comprehensive model of buying behaviour (with discussion). *Journal of the Royal Statistical Society, A*, 147, 621-655.

buyer behavior; consumer purchasing; stochastic model; purchase incidence; brand choice; gamma distribution; Poisson distribution; multinomial distribution; Dirichlet distribution; multivariate beta distribution; beta binomial distribution; negative binomial distribution.

Gordon, F. S. and Gordon, S. P. (1975). Transcendental functions of a vector variable and a characterization of a multivariate Poisson distribution. *Statistical Distributions in Scientific Work*, 3, 163-172.

multivariate transcendental functions; multivariate Poisson distribution.

Gourieroux, C. and Monfort, A. (1979). On the characterization of a joint probability distribution by conditional distributions. *Journal of Econometrics*, 10, 115-118.

characterization; joint probability distribution; conditional distributions.

Greenwood, M. and Woods, H. M. (1919). On the incidence of industrial accidents upon individuals with special reference to multiple accidents. *Industrial Fatigue Research Board*, London, Report no. 4, 1-28.

accident proneness.

Greenwood, M. and Yule, G. U. (1920). An inquiry into the nature of frequency distribution representations of multiple happenings, with particular references to the occurrence of multiple attacks of disease of repeated accidents. *Journal of the Royal Statistical Society*, 83, 255-279.

pigeon-hole schema; Poisson series; generalized pigeon-hole schema; generalized Poisson series; infinitely compound Poisson distribution.

Griffiths, R. C. (1971). Orthonormal polynomials on the multinomial distribution. *Australian Journal of Statistics*, 13, 27-35.

orthogonal polynomials; multinomial distribution; Krawtchouk polynomials; contingency table.

Griffiths, R. C. (1972). Linear dependence in bivariate distributions. *Australian Journal of Statistics*, 14, 182-187.

Griffiths, R. C. (1974). A characterization of the multinomial distribution. *Australian Journal of Statistics*, 16, 53-56.

characterization; multinomial distribution; bivariate binomial distribution; orthonormal Krawtchouk polynomials.

Griffiths, R. C. (1975). Orthogonal polynomials on the negative multinomial distribution. *Journal of Multivariate Analysis*, 5, 271-277.

orthogonal polynomials; multivariate negative binomial distribution; Runge-type identity; Poisson, normal, gamma limit.

Griffiths, R. C. and Milne, R. K. (1978). A class of bivariate Poisson processes. *Journal of Multivariate Analysis*, 8, 380-395.

bivariate Poisson process; doubly stochastic Poisson process; Poisson-Charlier polynomials; probability generating functional.

Griffiths, R. C., Milne, R. K. and Wood, R. (1979). Aspects of correlation in bivariate Poisson distributions and processes. *Australian Journal of Statistics*, 21, 238-255.

correlations; bivariate Poisson distributions; processes; negative correlation; Frechet classes of bivariate distributions; extremal bivariate distributions.

Guldberg, A. (1934). On discontinuous frequency functions of two variables. *Skandinavisk Aktuarietidskrift*, 17, 89-117.

discontinuous frequency functions; Bernoulli frequency function; Pascal frequency function; hypergeometric frequency function.

Gupta, R. C. (1974). Mixtures of some discrete distributions. *South African Statistical Journal*, 8, 83-92.

mixtures; multivariate binomial; bivariate multinomial; generalized beta; factorial moments; conditional distributions.

Gupta, R. P. and Jain, G. C. (1976). A generalized bivariate logarithmic series distribution. *Biometrische Zeitschrift*, 18, 169-173.

bivariate negative binomial distribution; zero truncation; bivariate logarithmic series distribution; generalized bivariate negative binomial distribution; generalized bivariate logarithmic series distribution; Poincare's bivariate generalization of the Lagrange's expansion; goodness-of-fit; method of moments.

Gupta, R. P. and Jain, G. C. (1978a). Some applications of a bivariate Borel-Tanner distributions. *Biometrical Journal*, 20, 99-106.

probability distribution; Poisson input; constant service time; single server queue; traffic flow; semi-infinite discrete dam; branching processes; bivariate Borel-Tanner distribution; bivariate negative binomial distribution.

Gupta, R. P. and Jain, G. C. (1978b). A bivariate Borel-Tanner distribution and its approximation. *Journal of Statistical Computation and Simulation*, 6, 129-136.

bivariate Borel-Tanner distribution; approximation; bivariate negative binomial distribution.

Gurland, J. (1957). Some interrelations among compound and generalized distributions. *Biometrika*, 44, 265-268.

Pascal (negative binomial) distribution; compounding; gamma distribution; generalized Poisson distribution; logarithmic distribution.

Gurland, J. and Tripathi, R. (1975). Estimation of parameters on some extensions of the Katz family of discrete distributions involving hypergeometric functions. *Statistical Distributions in Scientific Work, 1, 59-82.*

discrete distributions; hypergeometric functions; estimation; efficiency.

Haight, F. A. (1965). On the effect of removing persons with N or more accidents from an accident prone population. *Biometrika*, 52, 298-300.

accident proneness; incomplete gamma, incomplete beta and hypergeometric functions; compound Poisson model.

Hamdan, M. A. (1972a). Canonical expansion of the bivariate binomial distribution with unequal marginal indices. *International Statistical Review,* 40, 277-280.

bivariate binomial distribution; different marginal indices; additive property of independent binomial random variables; bivariate factorial moment generating function; bivariate probability density function; canonical expansion.

Hamdan, M. A. (1972b). Estimation in the truncated bivariate Poisson distribution. *Technometrics,* 14, 37-45.

canonical form; Charlier's polynomials; bivariate Poisson; missing zeros; marginal parameters; correlation parameters; moment estimators.

Hamdan, M. A. (1973). A stochastic derivation of the bivariate Poisson distribution. *South African Statistical Journal,* 7, 69-71.

bivariate Poisson processes; bivariate Poisson distribution.

Hamdan, M. A. (1975). A note on the trinomial distribution. *International Statistical Review,* 43, 219-220.

bivariate distribution; trinomial distribution; bivariate binomial distribution; canonical expansion of Aitken and Gonin's bivariate binomial probability function; Krawtchouk's polynomials; independence.

Hamdan, M. A. and Al-Bayyati, H. A. (1969). A note on the bivariate Poisson distribution. *The American Statistician,* 23, 32-33.

bivariate Poisson probability function; limiting form of the bivariate binomial probability function.

Hamdan, M. A. and Al-Bayyati, H. A. (1971). Canonical expansion of the compound correlated bivariate Poisson distribution. *Journal of American Statistical Association,* 66, 390-393.

compound correlated bivariate Poisson; orthogonal polynomials; negative binomial distribution; canonical expansion; contingency table.

Hamdan, M. A. and Jensen, D. R. (1976). A bivariate binomial distribution and some applications. *Australian Journal of Statistics,* 18, 163-169.

bivariate binomial distributions; distribution function; conditional distributions; regression functions; bounds on joint probabilities; bivariate Poisson and Gaussian limits; applications of statistical methodology.

Hamdan, M. A. and Martinson, E. O. (1971). Maximum likelihood estimation in the bivariate binomial (0,1) distribution. *Australian Journal of Statistics,* 13, 154-158.

2x2 contingency table; bivariate binomial distribution; maximum likelihood estimates; asymptotic variance-covariance matrix.

Hamdan, M. A. and Nasro, M. O. (1986). Maximum likelihood estimation of the parameters of the bivariate binomial distribution. *Communications in Statistics - Theory and Methods*, 15, 747-754.

bivariate binomial; maximum likelihood estimation.

Hamdan, M. A. and Tsokos, C. P. (1971). A model for physical and biological problems: the bivariate-compounded Poisson distribution. *Review of the International Statistical Institute*, 39, 60-63.

bivariate-compound Poisson process; physical and biological problems; mathematical model.

Harper, L. H. (1967). Stirling behavior is asymptotically normal. *Annals of Mathematical Statistics*, 38, 410-414.

combinatorial distribution, Stirling's numbers.

Hartley, H. O. and Rao, J. N. K. (1968). A new estimation theory for sample surveys. *Biometrika*, 55, 547-557.

minimum variance unbiased estimator; maximum likelihood estimator; regression estimator; Bayes estimator.

Hawkes, A. G. (1972). A bivariate exponential distribution with applications to reliability. *Journal of the Royal Statistical Society, B*, 34, 129-131.

bivariate; exponential; reliability.

Hirschfeld, H. O. (Hartley, H. O.) (1935). A connection between correlation and contingency. *Proceedings of the Cambridge Philosophical Society*, 31, 520-524.

discontinuous bivariate distribution; correlation theory; regressions; finite sample; correlation coefficients; Pearson's mean square contingency.

Hoadley, B. (1969). The compound multivariate distribution and Bayesian analysis of categorical data from finite. *Journal of the American Statistical Association*, 64, 216-229.

Bayesian analysis; multivariate hypergeometric distribution; compound multinomial; analysis of variance; contingency table.

Hoadley, B. (1970). A method for finding joint moments which are applicable to a class of multivariate discrete distributions on a simplex. *Random Counts in Scientific Work* I, 47-56.

multivariate hypergeometric distribution; compound multinomial distribution; Bayesian analysis.

Holgate, P. (1964). Estimation for the bivariate Poisson distribution. *Biometrika*, 51, 241-245.

estimation; covariance parameter; bivariate Poisson distribution; method of moments; correlation; likelihood equation; maximum-likelihood method.

Holgate, P. (1966). Bivariate generalizations of Neyman's Type A distribution. *Biometrika*, 53, 241-244.

Neyman Type A distribution; bivariate case; ecological situations; method of moments estimators.

Irwin, J. O. (1953). On the 'transition probabilities' corresponding to any accident distribution. *Journal of the Royal Statistical Society, B*, 15, 87-89.

Irwin, J. O. (1963). The place of mathematics in medical and biological statistics. *Journal of the Royal Statistical Society, A*, 126, 1-44.

Irwin, J. O. (1968). The generalized Waring distribution applied to accident theory. *Journal of the Royal Statistical Society, A*, 131, 205-225.

Irwin, J. O. (1975). The generalized Waring distribution. *Journal of the Royal Statistical Society, A*, 138, 18-31 (Part I), 204-227 (Part II), 374-384 (Part III).

Ishii, G. and Hayakawa, R. (1960). On the compound binomial. *Annals of the Institute of Statistical Mathematics*, 12, 69-80.

bivariate beta binomial.

Jain, G. C. and Singh, N. (1975). On bivariate power series distributions associated with Lagrange expansion. *Journal of American Statistical Association*, 70, 951-954.

Lagrange expansion; bivariate power series distributions.

Jain, G. C. (1983). Hermite distributions. *Encyclopedia of Statistical Sciences*, 3, 607-611.

Janardan, K. G. (1973). Chance mechanisms for multivariate hypergeometric models. *Sankhya, A*, 35, 465-478.

chance mechanisms generating multivariate hypergeometric models; multivariate inverse hypergeometric; multivariate negative hypergeometric; multivariate negative inverse hypergeometric; multivariate Polya; multivariate inverse Polya.

Janardan, K. G. (1974). A characterization of multinomial and negative multinomial distributions. *Scandinavian Actuarial Journal*, 58-62.

multinomial, negative multinomial distributions; multivariate hypergeometric; multivariate inverse hypergeometric distribution; characterization theorems.

Janardan, K. G. (1975). Certain inference problems for multivariate hypergeometric models. *Communications in Statistics*, 4, 375-388.

multivariate (inverse; negative; negative inverse; unified) hypergeometric distributions; limiting forms of posterior distributions; sample surveys; Bayesian interpretation of estimators.

Janardan, K. G. (1976). Certain estimation problems for multivariate hypergeometric models. *Annals of the Institute of Statistical Mathematics*, 28, 429-444.

estimation; multivariate hypergeometric models; completeness; maximum likelihood estimates; multivariate negative hyper-geometric; multivariate negative inverse hypergeometric; Bayesian estimation; prior distribution; parametric super population; posterior expectations; variances; direct and inverse sampling procedures.

Janardan, K. G. (1978). On a generalized Markov-Polya distribution. *Gujarat Statistical Review*, 5, 16-32.

Janardan, K. G. and Patil, G. P. (1970a). Location of modes for certain univariate and multivariate discrete distributions. *Random Counts in Scientific Work*, 1, 57-76.

univariate discrete distributions; singular multivariate discrete distributions; non-singular multivariate discrete distributions.

Janardan, K. G. and Patil, G. P. (1970b). On the multivariate Polya distribution: a model of contagion for data with multiple counts. *Random Counts in Scientific Work*, 3, 143-162.

Polya-urn scheme; multivariate hypergeometric series distributions; pollen analysis; multivariate Polya distribution; structural properties and estimation problems.

Janardan, K. G. and Patil, G. P. (1971). The multivariate inverse Polya distribution: a model of contagion for data with multiple counts in inverse sampling. *Studi di Probabilita Statistica e Ricera Operative in Onore de G. Pompilj*, Toreno, 327-341.

Polya urn scheme; multivariate Polya distribution; multivariate inverse Polya distribution.

Janardan, K. G. and Patil, G. P. (1972). A unified approach for a class of multivariate hypergeometric models. *Sankhya, A*, 35, 363-376.

unified class of multivariate hypergeometric distributions; chance mechanisms; structural and statistical properties; estimation.

Jensen, D. R. (1971). A note on positive dependence and the structure of bivariate distributions. *SIAM Journal of Applied Mathematics*, 20, 749-752.

bivariate Gaussian; exponential; gamma; chi-square; binomial; Poisson; negative binomial; hypergeometric; probability inequality; series expansions in sets of orthonormal functions; positive dependence.

Jogdeo, K. (1968). Characterizations of independence in certain families of bivariate and multivariate distributions. *Annals of Mathematical Statistics*, 39, 433-441.

tests of independence; sample correlation coefficient; 2x2 contingency tables; bivariate distribution; uncorrelatedness; multivariate distributions; pairwise independence; mutual independence; characterization; positively quadrant dependent; negatively quadrant dependent; regression dependence; parametrization.

Jogdeo, K. (1975). Dependence concepts and probability inequalities. *Statistical Distributions in Scientific Work*, 1, 271-279.

positive dependence; probability inequalities; ordered families of bivariate distributions; associated random variables; multivariate unimodality; contaminated independence model.

Jogdeo, K. and Patil, G. P. (1975). Probability inequalities for certain multivariate discrete distribution. *Sankhya, B*, 37, 158-164.

multinomial distribution; probability inequalities; multivariate discrete distributions; Dirichlet distribution.

Johnson, N. L. (1957). Uniqueness of a result in the theory of accident proneness. *Biometrika*, 44, 530-531.

model for accident proneness; Pearson type III distribution; negative binomial distribution; Bayes' theorem; probability density function.

Johnson, N. L. and Kotz, S. (1976). On a multivariate generalized occupancy model. *Journal of Applied Probability*, 13, 392-399.

occupancy distribution; multivariate distribution.

Johnson, N. L. and Kotz, S. (1982). Developments in discrete distributions, 1969-1980. *International Statistical Review*, 50, 71-101.

bibliography.

Joshi, S. W. and Patil, G. P. (1974). Sum-symmetric power series distributions and minimum variance unbiased estimation. *Theory of Probability and its Applications*, 19, 587-594.

sum-symmetric power series distributions; minimum variance unbiased estimators; sums and products of parametric components; characterizations.

Kabe, D. G. (1976). Inverse moments of discrete distributions. *Canadian Journal of Statistics*, 4, 133-141.

inverse moments; discrete distributions; truncated binomial; truncated Poisson distribution; factorial moments; ordered discrete variates.

Kawamura, K. (1972). The diagonal distribution of the bivariate Poisson distribution. *Kodai Mathematical Seminar Reports*, 25, 379-384.

limiting distribution of a diagonal distribution of a bivariate binomial distribution.

Kawamura, K. (1973). The structure of bivariate Poisson distribution. *Kodai Mathematical Seminar Reports*, 25, 246-256.

two dimensional Poisson distribution; Poisson's theorem; two dimensional Bernoulli distribution; two dimensional binomial distribution; distribution of independent type.

Kemp, A. W. (1981a). Frugal methods of generating bivariate discrete random variables. *Statistical Distributions in Scientific Work*, 4, 321-329.

computer generation; minimal storage; chop-down search; optimal stacking; bivariate discrete distributions; bivariate logarithmic distribution.

Kemp, A. W. (1981b). Conditionality properties for the bivariate logarithmic distribution with an application to goodness of fit. *Statistical Distributions in Scientific Work*, 5, 57-73.

bivariate logarithmic series distribution; multivariate logarithmic series distribution; modes of genesis; chi-squared goodness-of-fit; objective grouping; homogeneous distributions.

Kemp, A. W. (1981c). Computer sampling from homogeneous bivariate discrete distributions. *ASA Proceedings of the Statistical Computing Section*, 173-175.

homogeneous bivariate discrete distributions; generation method; cumulative properties.

Kemp, A. W. and Kemp, C. D. (1966). An alternative derivation of the Hermite distribution. *Biometrika*, 53, 627-628.

Hermite distribution; compound Poisson distribution; compounding; normal.

Kemp, A. W. and Kemp, C. D. (1968). On a distribution associated with certain stochastic processes. *Journal of the Royal Statistical Society*, *B*, 30, 160-163.

Kemp, A. W. and Kemp, C. D. (1969). Branching and clustering models associated with the 'lost-games' distribution. *Journal of Applied Probability*, 6, 700-703.

Kemp, C. D. (1967). On a contagious distribution suggested for accident data. Biometrics, 23, 241-255.

'Short' distribution; Neyman Type A distribution; Poisson distribution; maximum likelihood equations; method of moments.

Kemp, C. D. and Kemp, A. W. (1965). Some properties of the 'Hermite' distribution. *Biometrika*, 52, 381-394.

Hermite distribution; cumulants and moments; maximum likelihood estimation; Poisson -binomial distribution.

Kemp, C. D. and Kemp, A. W. (1988). Rapid estimation for discrete distributions. *The Statistician*, 37, 243-255.

rapid estimation procedures; discrete distributions; empirical probability generating function; maximum likelihood estimation; simulation.

Kemp, C. D. and Loukas, S. (1978a). The computer generation of bivariate discrete random variables. *Journal of the Royal Statistical Society, A*, 141, 513-519.

bivariate discrete distributions; bivariate Poisson distribution; bivariate Hermite distribution; inverse interpolation; stochastic models; simulation.

Kemp, C. D. and Loukas, S. (1978b). Computer generation of bivariate discrete random variables using ordered probabilities. *ASA Proceedings of Statistical Computing Section*, 115-116.

generating bivariate discrete random variables; ordered probabilities; bivariate Poisson; Hermite distribution.

Kemp, C. D. and Loukas, S. (1981). Fast methods for generating bivariate discrete random variables. *Statistical Distributions in Scientific Work*, 4, 313-319.

computer generation; bivariate discrete distributions; alias generation method; non-sequential search procedures.

Kemp, C. D. and Papageorgiou, H. (1982). Bivariate Hermite distributions. *Sankhya, A*, 44, 269-280.

bivariate Hermite distribution.

Kerr, J. D. (1969). The estimation of the 'Short' distribution. *Biometrics*, 25, 417-420.

'Short' distribution; estimation using the zero-class frequency.

Khatri, C. G. (1959). On certain properties of power-series distributions. *Biometrika*, 46, 486-490.

power series.

Khatri, C. G. (1971). On multivariate contagious distributions. *Sankhya, B*, 33, 197-216.

multivariate contagious distributions; multivariate Poisson; multivariate Neyman type A; negative multinomial; Poisson-generalized multinomial; maximum likelihood method; multivariate negative binomial.

Khatri, C. G. (1978a). Characterization of some discrete distributions by linear regressions. *Journal of the Indian Statistical Association*, 16, 49-58.

non-degenerate discrete random variables; probability generating function; characterization.

Khatri, C. G. (1978b). Characterization of some multivariate distributions by conditional distributions and linear regression. *Journal of the Indian Statistical Association*, 16, 59-70.

characteristic function; multivariate normal; multivariate gamma; general multinomial distribution; linear regression and conditional distributions.

Khatri, C. G. (1982). Multivariate Lagrangian Poisson and multinomial distributions. *Sankhya, B*, 44, 259-269.

multivariate Lagrange's distribution; probability density functions of multivariate Lagrangian Poisson, multinomials and quasi-Polya.

Khatri, C. G. (1983a). Multivariate discrete exponential family of distributions and their properties. *Communications in Statistics - Theory and Methods*, 12, 877-894.

power series distribution; Lagrangian distributions; Lagrangian (negative) multinomial; logarithmic series distributions; maximum likelihood estimation; moments and cumulants.

Khatri, C. G. (1983b). Multivariate exponential discrete distributions and their characterization by the Rao-Rubin condition for the additive damage model. *South African Statistical Journal*, 17, 13-32.

generalized Markov-Polya distribution; hypergeometric distribution; Lagrangian multinomial; multivariate Lagrangian negative binomial; multivariate Lagrangian Poisson; Rao-Rubin condition; survival distribution.

Kocherlakota, K. and Kocherlakota, S. (1985). On some tests for independence in nonnormal situations: Neyman's C(α) test. *Communications in Statistics - Theory and Methods*, 14, 1453-1470.

construction of Neyman's C(α) test; bivariate Poisson; correlation.

Kocherlakota, K. and Kocherlakota, S. (1986). Statistical inference and related problems in discrete distributions. *Communications in Statistics - Theory and Methods*, 15 (editors).

Kocherlakota, S. (1988). On the compounded bivariate Poisson distribution: a unified approach. *Annals of the Institute of Statistical Mathematics*, 40, 61-76.

bivariate Poisson distribution; probability generating function; bivariate - Hermite; -negative binomial; -Poisson-inverse Gaussian; -Neyman type A; conditional distributions.

Kocherlakota, S. (1989). A note on the bivariate binomial distribution. *Statistics and Probability Letters*, 7, 21-24.

generating functions; properties; conditional distributions; regression.

Kocherlakota, S. (1991). Factorial cumulants: bivariate discrete distributions. *Gujarat Statistical Journal* (Khatri Memorial Volume)

bivariate factorial cumulants and moments; convolutions; trivariate reduction.

Kocherlakota, S. and Kocherlakota, K. (1986). Goodness of fit tests for discrete distributions. *Communications in Statistics - Theory and Methods*, 15, 815-829.

goodness-of-fit; empirical probability generating function; discrete distributions.

Kocherlakota, S. and Kocherlakota, K. (1990). "The bivariate logarithmic series distribution. *Communications in Statistics - Theory and Methods*, 19, 3387-3432.

bivariate negative binomial distribution; bivariate logarithmic series distribution; compounding; Fisher-limit; goodness-of-fit test; modified logarithmic series distribution; probability generating function; Neyman's C(α) test; Rao's efficient score test; simulation.

Kocherlakota, S. and Kocherlakota, K. (1991). Neyman's C(α) test and Rao's efficient score test for composite hypotheses. *Statistics and Probability Letters*, 11, 491-493.

Rao's efficient score test; Neyman's C(α) test, maximum likelihood estimates.

Kocherlakota, S. and Singh, M. (1982). On the behaviour of some transforms of the sample correlation coefficient : Discrete bivariate populations. *Communications in Statistics - Theory and Methods*, 11, 2017-2043.

bivariate Poisson; bivariate negative binomial; trinomial; $\tanh^{-1}$; $\sin^{-1}$; Samiuddin, Nair and Ruben transformation; Tarter-Kronmal density estimation; robustness.

Korn, E. L. (1986). Sample size tables for bounding small proportions. *Biometrics*, 42, 213-216.

binomial confidence interval; Clopper-Pearson confidence interval; phase I trial.

Kotz, S. (1975). Multivariate distributions at a cross road. *Statistical Distributions in Scientific Work*, 1, 247-270.

multivariate distributions; survey; expansions; numerical methods; Farlie-Morgenstern family.

Krishnamoorthy, A. S. (1951). Multivariate binomial and Poisson distributions. *Sankhya*, 11, 117-124.

multivariate binomial distribution; multivariate Poisson distribution; Charlier polynomials; G-polynomials; Hermite Polynomials.

Lai, C. D. and Vere-Jones, D. (1979). Odd Man Out - the Meixner hypergeometric distribution. *Australian Journal of Statistics*, 21, 256-265.

hyperbolic secant distribution; Meixner hypergeometric distribution; completeness; characterization;bivariate distributions.

Lal, D. N. (1955). A note on a form of Tshebychev's inequality for two or more variables. *Sankhya*, 15, 317-320.

Tshebychev's inequality; bivariate case.

Lampard, D. G. (1968). A stochastic process whose successive intervals between events form a first order Markov chain - I. *Journal of Applied Probability*, 5, 648-668.

stochastic point process; Markov chain; renewal processes; queueing theory; independent Poisson processes.

Lancaster, H. O. (1958). The structure of bivariate distributions. *Annals of Mathematical Statistics*, 29, 719-736.

problems of Hirschfeld (1935); description of contingency table by canonical variables and correlations; Pearson mean square contingency; eigen-function theory; Mehler's identity; extension of canonical theory to continuous marginal distributions; new test of goodness-of-fit; bivariate normal distribution.

Lancaster, H. O. (1963). Correlations and canonical forms of bivariate distributions. *Annals of Mathematical Statistics*, 34, 532-538.

expansion for the bivariate measure; marginal measures; canonical correlations and functions; independence.

Lancaster, H. O. (1965). Symmetry in multivariate distributions. *Australian Journal of Statistics*, 7, 115-126.

symmetry; multivariate distribution; permutations of random variables; orthonormal theory; coefficients of correlation; bivariate distribution.

Lancaster, H. O. (1974). Multivariate binomial distributions. *Studies in Probability and Statistics: Papers in honour of E. J. G. Pitman*, (E. J. Williams, ed.), North Holland, Amsterdam, 13-19.

stochastic dependence; joint binomial distributions; Krawtchouk polynomials; Mehler expansion.

Lancaster, H. O. (1975). Joint probability distributions in the Meixner classes. *Journal of the Royal Statistical Society, B*, 37, 434-443.

biorthogonal; bivariate; convolutions; correlation; dependence; distribution; Meixner; orthogonal; polynomials; regression.

Leckenby, J. D. and Kishi, S. (1984). The Dirichlet multinomial distribution as a magazine exposure model. *Journal of Marketing Research*, 21, 100-106.

Dirichlet multinomial distribution; beta binomial distribution; Kwerel-geometric distribution; Hofmans-geometric distribution.

Lee, P. A. (1975). Diagonal expansion in Hahn polynomials for the bivariate negative hypergeometric distribution. *Nanta Mathematica*, 8/1, 53-62.

bivariate negative hypergeometric distribution; Hahn orthogonal polynomials; Barrett-Lampard diagonal expansion.

Lee, P. A. (1979). Canonical expansion of a mixed bivariate distribution with negative binomial and gamma marginals. *Journal of the Franklin Institute*, 307, 331-339.

Mixed bivariate probability distribution; canonical expansion; Laguerre and Meixner orthogonal polynomials; chance mechanisms.

Lee, P. A. and Ong, S. H. (1986). The bivariate non-central negative binomial distributions. *Metrika*, 33, 1-28.

four bivariate generalizations (type I - IV) of the non-central negative binomial distribution; latent structure model; joint central and factorial moments; maximum likelihood estimates.

Leitner, R. E. and Hamdan, M. A. (1973). Some bivariate probability models applicable to traffic accidents and fatalities. *International Statistical Review*, 41, 87-100.

traffic accidents; fatalities; probability model; two bivariate probability models; Poisson distribution; bivariate distribution.

Loukas, S. and Kemp, C. D. (1986a). The index of dispersion test for the bivariate Poisson distribution. *Biometrics*, 42, 941-948.

bivariate Hermite distribution; bivariate Poisson distribution; chi-squared goodness-of-fit; Crockett's T-test; index of dispersion test; power; simulation.

Loukas, S. and Kemp, C. D. (1986b). The computer generation of bivariate binomial and negative binomial random variables. *Communications in Statistics - Simulation and Computation*, 15, 15-25 computer simulation; bivariate binomial distribution; bivariate negative binomial distribution.

Loukas, S. and Kemp, C. D. (1986c). On the chi-square goodness-of-fit statistic for bivariate discrete distributions. *The Statistician*, 35, 525-529.

grouping procedures; asymptotic power.

Loukas, S., Kemp, C. D. and Papageorgiou, H. (1986). Even point estimation for the bivariate Poisson distribution. *Biometrika*, 73, 222-223.

asymptotic efficiency; bivariate Poisson distribution; estimation; method of even points.

Loukas, S. and Papageorgiou, H. (1985). Estimation for the bivariate negative binomial distribution. *Journal of Statistical Computation and Simulation*, 22, 67-82.

bivariate negative binomial; estimation; method of even-points; asymptotic efficiency; recurrence relations.

Lukacs, E. (1979). Some Multivariate Statistical Characterization theorems. *Journal of Multivariate Analysis*, 9, 278-287.

characterizations of multivariate distributions; independence; regression; identity; multivariate stability theorems.

Lukacs, E. (1973). A characterization of the multivariate geometric distribution. *Multivariate Analysis III*, (P. R. Krishnaiah, ed.), Academic Press, New York, 199-208.

multivariate negative binomial distribution; characteristic function; multivariate geometric distribution; characterization theorem.

Lukacs, E. and Beer, S. (1977). Characterization of the multivariate Poisson distribution. *Journal of Multivariate Analysis*, 7, 1-12.

Poisson distribution; characterization.

Mahamunulu, D. M. (1967). A note on regression in the multivariate Poisson distribution. *Journal of the American Statistical Association*, 62, 251-258.

regression; p-variate Poisson distribution; linearity.

Mafoud, M. and Patil, G. P. (1982). On weighted distributions. *Statistics and Probability: Essays in Honor of C. R. Rao*, (G. Kallianpur *et al.* eds.), North Holland, Amsterdam, 479-492.

weighted distributions; characterizations; bivariate weighted sum-symmetric power series distributions.

Maritz, J. S. (1950). On the validity of inferences drawn from the fitting of Poisson and negative binomial distributions to observed accident data. *Psychological Bulletin*, 47, 434-443.

accident proneness; Poisson fit; negative binomial fit.

Maritz, J. S. (1952). Note on a certain family of discrete distributions. *Biometrika*, 39, 196-198.

Poisson distributions; accident proneness; factorial cumulant generating function.

Marshall, A. W. and Olkin, I. (1985). A family of bivariate distributions generated by the bivariate Bernoulli distribution. *Journal of the American Statistical Association*, 80, 332-338.

bivariate binomial distribution; bivariate hypergeometric distribution; bivariate Poisson distribution; bivariate geometric distribution; bivariate negative binomial distribution; bivariate exponential distribution; bivariate gamma distribution; associated random variables.

McKendrick, A. G. (1926). Applications of mathematics to medical problems. *Proceedings of the Edinburgh Mathematical Society*, 44, 98-130.

two-dimensional random variables; correlation.

Mellinger, C. D., Gaffey, W. R., Sylvester, D. L. and Manheimer, D. I. (1965). A mathematical model with applications to a study of accident repeatedness among children. *Journal of the American Statistical Association*, 60, 1046-1059.

Bates-Neyman model; childhood accidents; accident liability.

Michael, J. R., Schucany, W. R. and Hass, R. W. (1976). Generating random variables using transformations with multiple roots. *American Statistician*, 30, 88-89.

generation of inverse Gaussian random variables

Mintz, A. and Blum, M. L. (1949). A re-examination of the accident proneness concept. *Journal of Applied Psychology*, 33, 195-211.

accident proneness; accident liability.

Mista, K. (1967). Counterexamples to conjectures related to bivariate distributions. *Rep Stat. Appl. Res. JUSE*, 14, 34-35.

Poisson distribution; non-Poisson distribution; regression; bivariate normal distribution; non-normal distribution.

Mitchell, C. R. and Paulson, A. S. (1981). A new bivariate negative binomial distribution. *Naval Research Logistics Quarterly*, 28, 359-374.

new bivariate negative binomial distribution; probability function; positive or negative correlations; linear or nonlinear regressions; moments; maximum likelihood estimates.

Mohanty, S. G. (1972). On queues involving batches. *Journal of Applied Probability*, 9, 430-436.

combinatorial theory; queueing processes.

Moran, P. A. P. (1952). A characteristic property of the Poisson distribution. *Proceedings of the Cambridge Philosophical Society*, 48, 206-207.

Morgenstern, D. (1976). A two fold generalization of the Polya-distribution and its various interpretations. *Metrika*, 23, 117-122 (German)

Polya's distribution; generalization; generalized distribution of exceedances; waiting problem; law of succession; variances; covariances; generalized hypergeometric distribution.

Mosimann, J. E. (1962). On the compound multinomial distribution, the multivariate beta distribution, and correlation among proportions. *Biometrika*, 49, 65-82.

compound multinomial distribution; multivariate beta distribution; compound binomial distribution.

Mosimann, J. E. (1963). On the compound negative multinomial distribution and correlations among inversely sampled pollen counts. *Biometrika*, 50, 47-54.

inverse sampling; compound negative multinomial distribution; Pascal multinomial; covariation of pollen counts; beta-compound multinomial; beta-compound negative multinomial.

Nelsen, R. B. (1986). Properties of a uni-parameter family of bivariate distributions with specified marginals. *Communications in Statistics - Theory and Methods*, 15, 3277-3285.

bivariate distributions; copulas; Frechet bounds; Spearman's rho; Kendall's tau; simulation.

Nelsen, R. B. (1987). Discrete bivariate distributions with given marginals and correlation. *Communications in Statistics - Simulation and Computation*, 16, 199-208.

discrete Frechet bounds; dependent random variables; bivariate Poisson distribution; simulation.

Nelson, J. F. (1985). Multivariate gamma-Poisson models. *Journal of the American Statistical Association*, 80, 828-834.

Dirichlet distribution; gamma distribution; repeated events.

Nevill, A. M. and Kemp, C. D. (1975). On characterizing the hypergeometric and multivariate hypergeometric distributions. *Statistical Distributions in Scientific Work*, 3, 353-358.

characterization; completeness of families; hypergeometric; multivariate hypergeometric; binomial; multinomial distributions.

Neyman, J. (1939). On a new class of 'contagious' distributions applicable in entomology and bacteriology. *Annals of Mathematical Statistics*, 10, 35-57.

compounding, generalizing.

Neyman, J. (1959). Optimal asymptotic tests of composite statistical hypothesis. *Probability and Statistics: H. Cramer Volume*, (U. Grenander, ed.), New York, Wiley, 213-234.

C(α) test; nuisance parameters;

Newbold, E. M. (1926). A contribution to the study of the human factor in the causation of accidents. *Report of the Industrial Fatigue Research Board*, Number 34. London.

accident proneness, bivariate negative binomial, correlation.

Newbold, E. M. (1927). Practical applications of the statistics of repeated events, particularly to industrial accidents. *Journal of the Royal Statistical Society, A*, 90, 487-547.

accident proneness.

Nishida, T. and Ohi, F. (1979). Bivariate Erlang distribution functions. *Journal of the Japan Statistical Society*, 9, 103-108.

Ong, S. H. (1990). Mixture formulations of a bivariate negative binomial distribution. *Communications in Statistics - Theory and Methods*, 12, 1303-1322.

compound correlated bivariate Poisson; mixtures; latent structure models; non-central negative binomial; Wicksell-Kibble bivariate gamma; generalized negative binomial of Bhattacharya; diagonal expansions; grade correlation.

Ong, S. H. (1991). The computer generation of bivariate binomial variates with given marginals and correlation. To appear in *Communications and Statistics - Simulation and Computation.*

mixture; conditional distribution; common elements; simulation; varying dependence; bivariate negative binomial; bivariate gamma.

Ong, S. H. and Lee, P. A. (1985). On the bivariate negative binomial distribution of Mitchell and Paulson. *Naval Research Logisitics Quarterly*, 32, 457-465.

bivariate negative binomial distribution; accident proneness; Wicksell-Kibble bivariate gamma distribution; linear birth-and-death process.

Ong, S. H. and Lee, P. A. (1986a). Bivariate non-central negative binomial distribution: Another generalization. *Metrika*, 33, 29-46.

bivariate generalization (type V) of the non-central negative binomial distribution; latent structure model; accident proneness model; reversible stochastic counter model.

Ong, S. H. and Lee, P. A. (1986b). On a generalized non-central negative binomial distribution. *Communications in Statistics - Theory and Methods*, 15, 1065-1079.

series expansion; mixtures; negative binomial Markov chain; birth-and-death process; correlated compound bivariate Poisson; goodness-of- fit.

Ord, J. K. (1975). A characterization of a dependent bivariate Poisson distribution. *Statistical Distributions in Scientific Work*, 3, 291-298.

characterization; dependent bivariate Poisson model; survival patterns; damage.

Panaretos, J. (1980). A characterization of a general class of multivariate discrete distributions. *Analytic Function Methods in Probability Theory Colloquia Mathematica Societatis Janos Bolyai*, (B. Gyeres, ed.), North Holland, Amsterdam, 21, 243-252.

characterization; multivariate discrete distributions; multiple Poisson distribution; Shanbhag's extension of the Rao-Rubin condition; binomial and negative binomial distributions.

Panaretos, J. (1981a). A characterization of the negative multinomial distribution. *Statistical Distributions in Scientific Work* , 4, 331-339.

negative multinomial distribution; multivariate inverse hypergeometric distribution; truncated negative multinomial distribution; Rao-Rubin condition; Shanbhag's lemma.

Panaretos, J. (1981b). On the joint distribution of two discrete random variables. *Annals of the Institute of Statistical Mathematics*, 33, 191-198.

conditional distribution; power series distribution; binomial distribution; characterization.

Panaretos, J. (1982). On a structural property of finite distributions. *Journal of the Royal Statistical Society, B*, 44, 209-211.

Patil and Seshadri's theorem; finite distributions; independence.

Panaretos, J. (1983). An elementary characterization of the multinomial and the multivariate hypergeometric distributions. *Stability Prob-*

lems for Stochastic Models, (V. V. Kalashnikov and V. M. Zolotarev, eds.), Springer - Verlag, New York, 156-164.

characterization; multinomial; multivariate hypergeometric.

Panaretos, J. and Xekalaki, E. (1980). A characteristic property of certain discrete distributions. *Analytic Function Methods in Probability Theory Colloquia Mathematica Societatis Janos Bolyai*, (B. Gyeres, ed.), North Holland, Amsterdam, 21, 253.

characterization.

Panaretos, J. and Xekalaki, E. (1986). On generalized binomial and multinomial distributions and their relation to generalized Poisson distributions. *Annals of the Institute of Statistical Mathematics*, 38, 223-231.

cluster binomial distribution; cluster multinomial distribution; generalized Poisson distribution.

Papageorgiou, H. (1979). Zero frequency estimation for bivariate generalized Poisson distributions. *Bulletin of the International Statistical Institute*, 48, 397-400.

bivariate generalized Poisson distributions; maximum likelihood; minimum chi-square; method of moments; zero-frequency method; double-zero proportion; bivariate negative binomial distribution; Hermite distributions; bivariate Neyman A (types I, II and III) distributions.

Papageorgiou, H. (1983). On characterizing some bivariate discrete distributions. *Australian Journal of Statistics*, 25, 136-144.

bivariate (correlated) Poisson; binomial; negative binomial; logarithmic distribution; conditional distribution; regression function; characterization.

Papageorgiou, H. (1985a). Characterizations of some discrete distribution models arising in Pollen studies. *Biometrical Journal*, 27, 935-939.

characterization; multinomial mixtures; multiple binomial mixtures; regression; pollen analysis.

Papageorgiou, H. (1985b). On a bivariate Poisson-geometric distribution. *Zastosowania Matematyki Applicationes Mathematicae*, 18, 541-547.

bivariate Poisson-geometric model; Poisson distribution; Poisson-geometric distribution; estimation.

Papageorgiou, H. (1986). Bivariate 'short' distributions. *Communications in Statistics - Theory and Methods*, 15, 893-906.

bivariate 'short' distribution; conditional distribution; method of moments; accident data.

Papageorgiou, H. and Kemp, C. D. (1983). Conditionality in bivariate generalized Poisson distributions. *Biometrical Journal*, 25, 757-763.

bivariate generalized Poisson distributions; conditional distributions; Bell polynomials; bivariate Neyman A.

Papageorgiou, H. and Kemp, C. D. (1988a). Conditional even point estimation for bivariate discrete distributions. *Communications in Statistics - Theory and Methods*, 17, 3403-3412.

bivariate Poisson, bivariate negative binomial, estimation, asymptotic efficiency.

Papageorgiou, H. and Kemp, C. D. (1988b). A method of estimation for some bivariate discrete distributions. *Biometrical Journal*, 30, 993-1001.

estimation using conditional means; family data; correlation.

Papageorgiou, H., Kemp, C. D. and Loukas, S. (1983). Some methods of estimation for the bivariate Hermite distribution. *Biometrika*, 70, 479-484.

accident data; asymptotic efficiency; bivariate Hermite distribution; estimation; method of even points; method of moments; method of zero frequencies.

Papageorgiou, H. and Loukas, S. (1987). Estimation based on conditional distributions for the bivariate negative binomial - Poisson distribution. *ASA Proceedings of the Statistical Computing Section*, 284-286.

bivariate negative binomial-Poisson distribution; conditional even-points method; conditional moments method; asymptotic efficiencies small sample comparisons.

Papageorgiou, H. and Loukas, S. (1988a). Conditional even point estimation for bivariate discrete distributions. *Communications in Statistics: Theory and Methods*, 17, 3403-3412.

bivariate Poisson; bivariate negative binomial; estimation; conditional even points; asymptotic efficiency.

Papageorgiou, H. and Loukas, S. (1988b). On estimating the parameters of a bivariate probability model applicable to traffic accidents. *Biometrics*, 44, 495-504.

accident data; asymptotic efficiency; bivariate negative binomial-Poisson; estimation; simulation.

Patel, H. I. (1973). Quality control methods for multivariate binomial and Poisson distributions. *Technometrics*, 15, 103-112.

time-dependence; Markov chain; autocorrelation; singular covariance matrix; conditional independence; factor analysis; normal approximation.

Patel, H. I. and Trivedi, S. J. (1969). An application of a multivariate extended Poisson distribution in 2 x 2 contingency tables. *Journal of the Indian Statistical Association*, 7, 136-145.

2 x 2 contingency tables; Poisson distribution; likelihood ratio test.

Patel, S. R. (1978). Minimum variance unbiased estimation of multivariate modified power series distribution. *Metrika*, 25, 155-162.

minimum variance unbiased estimation; multivariate modified power series distribution; generalized negative multinomial distribution.

Patel, S. R. (1979a). MVUE of left truncated multivariate modified power series distribution. *Metrika*, 26, 87-94.

minimum variance unbiased estimation; left truncated multivariate modified power series distribution; truncation point; parameter.

Patel, S. R. (1979b). On multivariate generalized negative binomial distribution. *Metron*, 37/3, 59-69.

Patel, S. R. (1981). On multivariate generalized logarithmic series distribution. *Metron*, 39/3, 53-64.

multivariate generalized logarithmic series distribution; structural properties; estimation; univariate and multivariate discrete models; moment properties; regression and correlation analysis; method of maximum likelihood; unbiased minimum variance.

Patel, S. R. and Shah, S. M. (1978). Minimum variance unbiased estimation of left truncated multivariate power series distributions. *Metrika*, 25, 209-218.

minimum variance unbiased estimation; left truncated multivariate power series distribution; truncation point; parameter of the left truncated multivariate power series distribution.

Patil, G. P. (1964). On certain compound Poisson and compound binomial distributions. *Sankhya, A*, 26, 293-294.

compound Poisson, Poisson-normal.

Patil, G. P. (1965a). Certain characteristic properties of multivariate discrete probability distributions akin to the Bates-Neyman model in the theory of accident proneness. *Sankhya, A*, 27, 259-270.

accident proneness; multivariate negative binomial distribution; multivariate discrete probability distributions; characterizations.

Patil, G. P. (1965b). On multivariate generalized power series distribution and its application to the multinomial and negative multinomial. *Classical and Contagious Discrete Distributions*, 225-238.

generalized power series distribution; multivariate; characterization; maximum likelihood estimate; minimum variance unbiased estimate; range; multinomial; negative multinomial.

Patil, G. P. (1985a). Multivariate logarithmic series distribution. *Encyclopedia of Statistical Sciences*, 6, 83-85.

logarithmic series distribution; modified power series distributions; multivariate power series distributions; power series distributions.

Patil, G. P. (1985b). Multivariate power series distributions. *Encyclopedia of Statistical Sciences*, 6, 104-108.

factorial series distributions; modified power series distributions; multinomial distributions; multivariate distributions; multivariate logarithmic series distributions; Poisson distributions; power series distributions.

Patil, G. P. and Bildikar, S. (1966). On minimum variance unbiased estimation for the logarithmic series distribution. *Sankhya A*, 28, 239-250.

minimum variance unbiased estimation; univariate logarithmic series distribution.

Patil, G. P. and Bildikar, S. (1967). Multivariate logarithmic series distribution as a probability model in population and community ecology and some of its statistical properties, *Journal of American Statistical Association*, 62, 655-674.

multivariate logarithmic series distribution; marginal and conditional distributions; moment properties; regression and correlation analysis; modal property; parameter estimation; maximum likelihood; unbiased minimum variance; ecology.

Patil, G. P. and Ratnaparkhi, M. V. (1975). Problems of damaged random variables and related characterizations. *Statistical Distributions in Scientific Work*, 3, 255-270.

additive damage model; multiplicative damage model; Rao-Rubin condition; linear regression; conditional distributions.

Patil, G. P. and Ratnaparkhi, M. V. (1977). Certain characterizations with linearity of regression in additive damage models. *Journal of Multivariate Analysis*, 7, 598-601.

multivariate additive damage model; survival distributions; linearity of regression; double binomial distribution; double inverse hypergeometric distributions.

Patil, G. P. and Ratnaparkhi, M. V. (1986). Multivariate Polya distribution. *Encyclopedia of Statistical Sciences*, 1, 58-63.

Patil, G. P., Rao, C. R., and Ratnaparkhi, M. V. (1986). On discrete distributions and their use in model choice for observed data. *Communications in Statistics - Theory and Methods*, 15, 907-918.

discrete weighted distributions; bivariate weighted distributions; weight functions; size biased distributions; model specification; form invariance; mixtures; stochastic ordering; Bayesian inference; posterior and weighted posterior distributions; bivariate Poisson-negative hypergeometric distribution.

Patil, G. P. and Wani, J. K. (1965). Maximum likelihood estimation for the complete and truncated logarithmic series distributions. *Classic and Contagious Distributions*, 398-409.

univariate logarithmic series distribution.

Patil, S. A., Patel, D. I. and Kovner, J. L. (1977). On bivariate truncated Poisson distribution. *Journal of Statistical Computation and Simulation*, 6, 49-66.

canonical form; Charlier's polynomials; bivariate Poisson; factorial moments; missing zeroes and ones; marginal parameters; correlation parameter; moment estimators.

Paul, S. R. and Ho, N. I. (1989). Estimation in the bivariate Poisson distribution and hypothesis testing concerning independence. *Communications in Statistics - Theory and Methods*, 18, 1123-1133.

efficiency; symmetric and asymmetric cases; independence; likelihood ratio tests; Neyman's C(α) test; simulation.

Paulson, A. S. (1973). A characterization of the exponential distribution and a bivariate exponential distribution. *Sankhya, A*, 35, 69-78.

bivariate geometric; bivariate negative binomial.

Paulson, A. S. and Uppuluri, V. R. R. (1972). A characterization of the geometric and a bivariate geometric distribution. *Sankhya, A*, 34, 297-300.

characterization; exponential distribution; bivariate exponential distribution; v-fold convolution; bivariate gamma distribution; Kibble's bivariate gamma.

Philippou, N. and Roussas, G. G. (1974). A note on multivariate logarithmic series distribution. *Communications in Statistics*, 3, 469-472.

sum of N independent random vectors each with a common multivariate logarithmic series is distributed as multivariate negative binomial provided N is distributed as a Poisson.

Rao, B. R. and Janardan, K. G. (1982). On the moments of multivariate discrete distributions using finite difference operators. *American Statistician*, 36, 381-383.

finite difference operators; multivariate discrete distributions; moments of random vectors.

Rao, C. R. (1947). Large sample tests of statistical hypotheses concerning several parameters with applications to problems of estimation. *Proceedings of the Cambridge Philosophical Society*, 44, 50-57.

efficient score tests.

Rao, C. R. and Rubin, H. (1964). On a characterization of the Poisson distribution. *Sankhya, A*, 26, 295-298.

discrete distribution; damage; binomial probability law; characterization; Poisson distribution.

Rao, C. R., Srivastava, R. C., Talwalker, S. and Edgar, G. A. (1980). Characterization of probability distributions based on a generalized Rao-Rubin condition. *Sankhya, A*, 42, 161-169.

characterization; non-degenerate; non-negative distribution; damage model; binomial; generalized Rao-Rubin condition.

Rubinstein, G. Z. (1985). Models for Count Data using the Sichel Distribution. Ph. D. Thesis, University of Cape Town, South Africa.

Sanathanan, L. (1972). Estimating the size of a multinomial population. *The Annals of Mathematical Statistics*, 43, 142-152.

multinomial distribution; maximum and conditional maximum likelihood estimates; asymptotic distributions.

Sarndal, C. E. (1964). A unified derivation of some nonparametric distributions. *Journal of the American Statistical Association*, 59, 1042-1053.

parameter-free distributions; random sampling; independent samples.

Sarndal, C. E. (1965). Derivation of a class of frequency distributions via Bayes' theorem. *Journal of the Royal Statistical Society, B*, 27, 290-300.

Bayes' theorem; multinomial; independent Poisson; independent negative binomial; independent gamma; Dirichlet; multivariate normal.

Seshadri, V. and Patil, G. P. (1964). A characterization of a bivariate distribution by the marginal and the conditional distributions of the same component. *Annals of the Institute of Statistical Mathematics*, 15, 215-221.

bivariate distribution; marginal distribution; conditional distributions; mixtures; identifiability; Laplace transform; definitions of a general character; bivariate generalizations; univariate distribution.

Shanbhag, D. N. (1974). An elementary proof for the Rao-Rubin characterization of the Poisson distribution. *Journal of Applied Probability*, 11, 211-215.

characterization; Poisson distribution.

Shanbhag, D. N. (1977). An extension of the Rao-Rubin characterization of the Poisson distribution. *Journal of Applied Probability*, 14, 640-646.

characterization; Poisson distribution; renewal theorem.

Shanbhag, D. N. and Panaretos, J. (1979). Some results related to the Rao-Rubin characterization of the Poisson distribution. *Australian Journal of Statistics*, 21, 78-83.

Rao-Rubin characterization; Poisson distribution; Moran's theorem.

Shanbhag, D. N. and Rajamannar, G. (1974). Some characterization of the bivariate distribution of independent Poisson variables. *Australian Journal of Statistics*, 16, 119-125.

bivariate Poisson distribution; variant of Talwalker's result; characterization.

Shanmugam, R. and Singh, J. (1981). Some bivariate probability models applicable to traffic accidents and fatalities. *Statistical Distributions in Scientific Work*, 6, 95-103.

bivariate probability models; generalized Poisson-quasi binomial model; estimation; goodness-of-fit; traffic accidents; fatalities data.

Shenton, L. R. and Consul, P. C. (1973). On bivariate Lagrange and Borel-Tanner Distributions and their use in queueing theory. *Sankhya, A*, 35, 229-236.

Lagrange distribution; Borel-Tanner distribution; generalized negative binomial; generalized Poisson; queueing systems; bivariate case.

Shoukri, M. M. (1982a). Minimum variance unbiased estimation in a bivariate modified power series distribution. *Biometrical Journal*, 24, 97-101.

bivariate modified power series distribution; minimum variance unbiased estimator; generalized bivariate negative binomial distribution.

Shoukri, M. M. (1982b). On a generalization for the double Poisson distribution. *Communications in Statistics - Theory and Methods*, 11, 151-164.

Lagrange expansion; moments; convolution; maximum likelihood estimator; minimum variance unbiased estimator; goodness-of-fit.

Shoukri, M. M. and Consul, P. C. (1982). Bivariate modified power series distribution: Some properties, estimation and applications. *Biometrical Journal*, 24, 789-801.

bivariate modified power series distribution; moments; convolution; maximum likelihood estimation; bivariate generalized negative binomial and Poisson distribution.

Sibuya, M. (1979). Generalized hypergeometric, digamma and trigamma distributions. *Annals of the Institute of Statistical Mathematics*, 31, 373-390.

digamma; trigamma; zero-truncated generalized hyper-geometric distributions; compounding; logarithmic series distributions; zeta distributions; Poisson distribution; gamma product-ratio distributions.

Sibuya, M. (1980). Multivariate digamma distribution. *Annals of the Institute of Statistical Mathematics*, 32, 25-36.

multivariate digamma distributions; properties; multivariate logarithmic series distributions; limits; estimation.

Sibuya, M. (1983). Generalized hypergeometric distributions. *Encyclopedia of Statistical Sciences*, 3, 330-334.

hypergeometric distribution; Pearson system of curves; Polya's urn; urn models.

Sibuya, M. and Shimizu, R. (1981). The generalized hypergeometric family of distributions. *Annals of the Institute of Statistical Mathematics*, 33, 177-190.

generalized hypergeometric family of distributions; multivariate generalized hypergeometric distributions.

Sibuya, M. Yoshimura, I. and Shimizu, R. (1964). Negative multivariate distribution. *Annals of the Institute of Statistical Mathematics*, 16, 409-426.

negative multinomial distribution; multivariate negative binomial distribution; multivariate Fisher's logarithmic series; negative hypergeometric distribution.

Sichel, H. S. (1975). On a distribution law for word frequencies, *Journal of the American Statistical Association*, 70, 542-547.

Srivastava, R. C. and Singh, J. (1975). On some characterization of the binomial and Poisson distributions based on a damage model. *Statistical Distributions in Scientific Work*, 3, 271-277.

damage model; Poisson and binomial distribution; characterizations; damage process.

Srivastava, R. C. and Srivastava, A. B. L. (1970). On a characterization of the Poisson distribution. *Journal of Applied Probability*, 7, 495-501.

Poisson distribution; binomial destructive model; characterization; binomial distribution.

Steck, G. P. (1968). A note on contingency-type bivariate distributions. *Biometrika*, 55, 262-264.

Plackett; family of bivariate distributions; density function; bivariate density; bounds; joint distribution function.

Stein, G. Z. and Juritz, J. M. (1987). Bivariate compound Poisson distributions. *Communications in Statistics - Theory and Methods*, 16, 3591-3607.

compound Poisson distribution; bivariate discrete distribution; Sichel distribution; inverse Gaussian-Poisson distribution; negative binomial distribution.

Stein, G. Z., Zucchini, W. and Juritz, J. M. (1987). Parameter estimation for the Sichel distribution and its multivariate extension. *Journal of American Statistical Association*, 82, 938-944.

compound Poisson; inverse Gaussian-Poisson; multivariate discrete distributions.

Steyn, H. S. (1951). On discrete multivariate probability functions. *Koninklijke Nederlandse Akademie Wetenschappen Proceedings, Series A*, 54, 23-30.

multivariate hypergeometric series; multivariate probability functions; definite integral; regression; factorial moments.

Steyn, H. S. (1955). On discrete multivariate probability functions of hypergeometric type. *Koninklijke Nederlandse Akademie Wetenschappen Proceedings, Series A*, 58, 588-595.

hypergeometric probability function; Van der Monde's method; negative factorial multinomial function; convergence; multinomial sampling; negative multinomial sampling; factorial multinomial sampling; negative factorial multinomial sampling.

Steyn, H. S. (1957). On regression properties of discrete systems of probability functions. *Koninklijke Nederlandse Akademie Wetenscha-ppen Proceedings, Series A*, 60, 119-127.

discrete functions with linear regression; discrete bivariate functions of hypergeometric type with non-linear regression.

Steyn, H. S. (1976). On the multivariate Poisson normal distribution. *Journal of the American Statistical Association*, 71, 233-236.

factorial moment generating function; multivariate Poisson normal (Hermite) distribution; limiting form; series expansion; marginal and conditional distributions; method of moments or least squares; correlated variables.

Steyn, H. S. and Wiid, J. B. (1958). On eightfold probability functions. *Koninklijke Nederlandse Akademie Wetenschappen Proceedings, Series A*, 61, 129-138.

trivariate probability functions; hypergeometric series.

Subrahmaniam, K. (1966). A test for 'intrinsic correlation' in the theory of accident proneness. *Journal of the Royal Statistical Society, B*, 35, 131-146.

correlated bivariate Poisson distribution; Bates-Neyman model; conditional distribution; regression; convolution; estimation; Neyman C(α) test.

Subrahmaniam, K. (1967). On a property of the binomial distribution. *Trabajos de Estadistica*, 18, 89-103.

binomial distribution; bivariate discrete distribution; convolution.

Subrahmaniam, K. and Gajjar, A. V. (1980). Robustness to nonnormality of some transformations of the sample correlation coefficient. *Journal of Multivariate Analysis*, 10, 69-78.

transforms; moments under nonnormality; correlation.

Subrahmaniam, K. and Subrahmaniam, K. (1973). On the estimation of the parameters in the bivariate negative binomial distribution. *Journal of the Royal Statistical Society, B*, 35, 131-146.

bivariate negative binomial distribution; method of moments; maximum likelihood estimation; efficiency.

Sym, R. (1971). Estimation for the Hermite distribution with special reference to time series. Ph. D. Thesis, Imperial College, University of London.

multivariate Hermite distribution.

Taillie, C., Ord, J. K., Mosimann, J. E. and Patil, G. P. (1979). Chance mechanisms underlying multivariate distributions. *Statistical Distributions in Ecological Work*, 157-191.

multivariate distributions; inverse sampling; covariation; mixtures; discrimination; chance mechanisms.

Taillie, C. and Patil, G. P. (1986). The Fibonacci distribution revisited. *Communications in Statistics - Theory and Methods*, 15, 951-960.

Fibonacci numbers; Fibonacci distribution; mode; power series distributions.

Tallis, G. M. (1962). The use of a generalized multivariate distribution in the estimation of correlation in discrete data. *Journal of the Royal Statistical Society, B*, 24, 530-534.

generalized multinomial distribution; correlation; estimation.

Tallis, G. M. (1964). Further model for estimating correlation in discrete data. *Journal of the Royal Statistical Society, B*, 26, 82-85.

multinomial variates; correlation.

Talwalker, S. (1970). A characterization of the double Poisson distribution. *Sankhya, A*, 32, 265-270.

characterization; double Poisson distribution; multiple Poisson distribution; damage probability distributions; binomial distributions.

Talwalker, S. (1975). Models in medicine and toxicology. *Statistical Distributions in Scientific Work* 2, 263-274.

Poisson; binomial; negative binomial; identifiability of a mixture; damage model; Cauchy's functional equation; Rao-Rubin condition; bivariate generalization of Neyman's Type A model.

Tarter, M. E., Holcomb, R. L. and Kronmal, R. A. (1967). After the histogram what? A description of new computer methods for estimation the population density. *Proceedings of Association of Computing Machines*, 22, 511-519.

density estimation.

Teicher, H. (1954). On the multivariate Poisson distributions. *Skandinavisk Aktuarietidskrift*, 37, 1-9.

bivariate Poisson; multivariate Poisson; characterization theorem; asymptotic normality.

Teicher, H. (1960). On the mixture of distributions. *Annals of Mathematical Statistics*, 31, 55-73.

m-mixtures; mixtures of additively closed families; convolution; normal distributions; product-measure mixture; Poisson distribution.

Teicher, H. (1961). Identifiability of mixtures. *Annals of Mathematical Statistics*, 31, 244-248.

mixtures; identifiable; one-parameter additively closed family of distributions; scale parameter mixtures; type III and uniform distribution.

Tracy, D. S. and Doss, D. C. (1980). A characterization of multivector multinomial and negative multinomial distribution. *Multivariate Statistical Analysis* (ed. R. P. Gupta), North Holland, Amsterdam, 281-289.

multivector distribution; linear exponential; negative multinomial.

Tsao, C. K. (1965). A moment generating function of the hypergeometric distribution. *Classic and Contagious Discrete Distributions*, 75-78.

bivariate hypergeometric distribution.

Tsui, K.-W. (1986). Multiparameter estimation for some multivariate discrete distributions with possibly dependent components. *Annals of the Institute of Statistical Mathematics*, 38, 45-56.

dependent; difference inequality; loss function; negative multinomial.

Tyan, S. G., Derin, H. and Thomas, J. B. (1976). Two necessary conditions on the representation of bivariate distributions by polynomials. *Annals of Statistics*, 4, 216-222.

bivariate distribution function; orthonormal polynomials.

Tweedie, M. C. K. (1957). Statistical properties of inverse Gaussian distributions, I and II, *Annals of Mathematical Statistics*, 28, 362-377, 696-705.

Wani, J. K. and Patil, G. P. (1975). Characterizations of linear exponential families. *Statistical distribution in Scientific Work*, 3, 423-431.

characterizations; linear exponential family.

Webb, W. B. and Jones, E. R. (1953). Some relations between two statistical approaches to accident proneness. *Psychological Bulletin*, 50, 133-136.

accident proneness; Poisson fit.

Wicksell, S. D. (1916). Some theorems in the theory of probability, with special reference to their importance in the theory of homograde correlations. *Svenska Aktuarieforeningens Tidskrift*, 165-213.

bivariate Bernoulli trials; bivariate binomial distribution; correlation.

Wicksell, S. D. (1923). Contributions to the analytical theory of sampling. *Arkiv for Matematik, Astronomi Och Fysik*, 17 , 1-46.

bivariate Bernoulli trials; bivariate binomial distribution.

Wiid, A. J. B. (1957-58). On the moments and regression equations of the fourfold negative and fourfold negative factorial binomial distributions. *Proceedings of the Royal Society of Edinburgh, Series A*, 65, 29-34.

fourfold - binomial (bivariate Bernoulli distribution); - factorial binomial distributions; - negative binomial; - negative factorial binomial.

Wishart, J. (1949). Cumulants of multivariate multinomial distributions. *Biometrika*, 36, 47-58.

(Bernoulli) multinomial distribution; cumulants; multivariate multinomial distribution.

Xekalaki, E. (1980). On characterizing the bivariate Poisson, binomial and negative binomial distributions. *Analytic Function Methods in Probability Theory, Colloquia Mathematica Societatis Janos Bolyai*, (B. Gyeres, ed.), 21, North Holland Publishing Co., Amsterdam, 369-379.

characterization; double Poisson; binomial; negative binomial; bivariate binomial.

Xekalaki, E. (1983a). The univariate generalized waring distribution in relation to accident theory: Proneness, Spells or Contagion? *Biometrics*, 39, 887-895.

univariate generalized Waring distribution; accident theory; accident proneness; accident risk exposure; contagion; Spells; discrimination.

Xekalaki, E. (1983b). Expressions for the probabilities of the bivariate Hermite distribution and related properties. *Zastosowania Matematyki Applicationes Mathematical*, 18, 35-41.

Xekalaki, E. (1984a). Models leading to the bivariate generalized Waring distribution. *Utilitas Math.*, 25, 263-290.

bivariate generalized Waring distribution; bivariate continuous Pearson distributions; urn, mixing, conditionality and exceedance models.

Xekalaki, E. (1984b). The bivariate generalized Waring distribution and its application to accident theory. *Journal of the Royal Statistical Society, A*, 147, 488-498.

bivariate generalized Waring distribution; accident theory; proneness; liability; Waring's expansion.

Xekalaki, E. (1985). Factorial moment estimation for the bivariate generalized Waring distribution. *Statistische Hefte*, 26, 115-129.

generalized Waring distribution; factorial moment method of estimation; asymptotic standard errors; variance components.

Xekalaki, E. (1986a). The multivariate generalized Waring distribution. *Communications in Statistics - Theory and Methods*, 15, 1047-1064.

generalized Waring distribution; Waring's expansion; accident proneness; accident liability.

Xekalaki, E. (1986b). The bivariate Yule distribution and some of its properties. *Statistics*, 17, 311-317.
bivariate Yule distribution; bivariate STER model; tail probabilities.

Yoshimura, I. (1964a). Unified system of cumulant recurrence relations. *Rep. Stat. Appl. Res., JUSE*, 11, 1-8.
multivariate cumulant recurrence relation; cumulant generating function; characterization; exponential type distribution.

Yoshimura, I. (1964b). A Complementary note on the multivariate moment recurrence relation. *Rep. Stat. Appl. Res., JUSE*, 11, 9-12.
multivariate moment recurrence relation; multivariate cumulant recurrence relation.

BOOKS

Abramowitz, M. and Stegun, I. A. (1968). *Handbook of Mathematical Functions*. National Bureau of Standards Applied Mathematics Series 55.

Cresswell, W. L. and Froggatt, P. (1963). *The Causation of Bus Driver Accidents. An Epidemiological Study.* Oxford University Press, London.

Douglas, J. B. (1981). *Analysis with Standard Contagious Distributions*, International Co-operative Publishing House, Fairland, Maryland.

Douglas, J. B. (1980). Analysis with standard contagious distributions. *Statistical Distributions*, 4, Fairland, Maryland: International Co-operative Publishing House.

Elandt-Johnson, R. C. (1971). *Probability Models and Statistical Methods in Genetics*, John Wiley and Sons, New York.

Erdelyi, a., Magnus, W., Oberhettinger, F. and Tricomi, F. G. (1953). *Higher Transcendental Functions, Volumes 1 and 2.* Bateman Manuscript Project of California Institute of Technology. McGraw-Hill Book Co. Inc., New York.

Feller, W. (1957). *An Introduction to Probability Theory and Its Applications* (Second Edition). John Wiley and Sons, New York.

Gradshteyn, I. S. and Ryzhik, I. M. (1980). *Table of Integrals, Series and Products*. Academic Press, New York.

Haight, F. A. (1966). *Handbook of the Poisson Distribution*, John Wiley and Sons, New York.

Hogg, R. V. and Craig, A. T. (1978). Introduction to Mathematical Statistics (Fourth edition), Macmillan Publishing Co., Inc., New York.

IMSL (1980). International Mathematical and Statistical Libraries, Inc. Houston, Texas.

Johnson, N. L. and Kotz, S. (1969). *Discrete Distributions*, Houghton Mifflin, Boston.

Johnson, N. L. and Kotz, S. (1977). *Urn Models and Their Applications,* John Wiley and Sons, New York.

Jorgensen, B. (1982). *Statistical Properties of the Generalized Inverse Gaussian Distribution.* Springer-Verlag, New York.

Kendall, M., Stuart, A. and Ord, J. K. (1983). *The Advanced Theory of Statistics*, Volume 3 (Fourth Edition), Macmillan Publishing Co., Inc., New York.

Kendall, M. and Stuart, A. (1979). *The Advanced Theory of Statistics,* Volume 2 (Fourth Edition), Macmillan Publishing Co., Inc., New York.

Kendall, M. and Stuart, A. (1977). *The Advanced Theory of Statistics,* Volume 1 (Fourth Edition), Macmillan Publishing Co., Inc., New York.

Kotz, S. and Johnson, N. L. (1982-89). *Encyclopedia of Statistical Sciences*, Volumes 1-9 and Supplement, John Wiley and Sons, New York.

Mardia, K. V. (1970). *Families of Bivariate Distributions*, Charles Griffin and Sons, London.

Patil, G. P. (1965). *Classical and Contagious Discrete Distributions,* Pergamon Press, Oxford, New York, Toronto, Paris, Frankfurt; and Statistical Publishing Society, Calcutta, India.

Patil, G. P. (1970). *Random Counts in Scientific Work, Vol. 1: Random Counts in Models and Structures*, The Pennsylvania State University Press, University Park, Pennsylvania.

Patil, G. P. (1970). *Random Counts in Scientific Work, Vol. 3: Random Counts in Physical Science, Geosciences, and Business*, The

Pennsylvania State University Press, University Park, Pennsylvania.

Patil, G. P., Kotz, S. and Ord, J. K. (1975). *Statistical Distributions in Scientific Work, Vol. 1: Models and Structures*, Reidel Publishing Company, Dordrecht, Boston, and London.

Patil, G. P., Kotz, S. and Ord, J. K. (1975). *Statistical Distributions in Scientific Work, Vol. 2: Model Building and Model Selection*, Reidel Publishing Company, Dordrecht, Boston, and London.

Patil, G. P., Kotz, S. and Ord, J. K. (1975). *Statistical Distributions in Scientific Work, Vol. 3: Characterizations and Applications*, Reidel Publishing Company, Dordrecht, Boston, and London.

Patil, G. P., Boswell, M. T., Joshi, S. W. and Ratnaparkhi, M. V. (1984). *Dictionary and Classified Bibliography of Statistical Distributions in Scientific Work, Vol. 1: Discrete Models.* International Co-operative Publishing House, Burtonsville, Maryland.

Patil, G. P., Boswell, M. T., Ratnaparkhi, M. V., and Roux, J. J. J. (1984). *Dictionary and Classified Bibliography of Statistical Distributions in Scientific Work, Vol. 3: Multivariate Models.* International Co-operative Publishing House, Burtonsville, Maryland.

Rao, C. R. (1973). Linear Statistical Inference and its Applications (Second edition), John Wiley and Sons, New York.

Stuart, A. and Ord, J. K. (1987). *Kendall's Advanced Theory of Statistics*, Volume 1 (Fifth Edition), Oxford University Press, Oxford.

Szego, G. (1959). *Orthogonal Polynomials.* Colloquium Publications, 23, American Mathematical Society, Providence.

Taillie, C., Patil, G. P., and Baldessari, B. A. (1981). *Statistical Distributions in Scientific Work, Vol. 4: Models, Structures, and Characterizations*, Reidel Publishing Company, Dordrecht, Boston, and London.

Taillie, C., Patil, G. P., and Baldessari, B. A. (1981). *Statistical Distributions in Scientific Work, Vol. 5: Inferential Problems and Properties*, Reidel Publishing Company, Dordrecht, Boston, and London.

Taillie, C., Patil, G. P., and Baldessari, B. A. (1981). *Statistical Distributions in Scientific Work, Vol. 6: Applications in Physical, Social, and Life Sciences*, Reidel Publishing Company, Dordrecht, Boston, and London.

Titterington, D. M., Smith, A. F. M. and Makov, U. E. (1985). Statistical Analysis of Finite Mixture Distributions. John Wiley and Sons, New York.

Wilks, S. S. (1962). *Mathematical Statistics.* John Wiley and Sons, New York.

KEY WORD INDEX

SUBJECT INDEX